T0180781

AMERICAN VACUUM SOCIETY CLASSICS

QUADRUPOLE MASS SPECTROMETRY AND ITS APPLICATIONS

AMERICAN VACUUM SOCIETY CLASSICS

H. F. Dylla, Series Editor-in-Chief

Basic Data of Plasma Physics: The Fundamental Data on Electrical Discharges in Gases
Sanborn C. Brown

Field Emission and Field Ionization
Robert Gomer

Vacuum Technology and Space Simulation
David H. Holkeboer, Donald W. Jones, Frank Pagano, and Donald J. Santeler

The Physical Basis of Ultrahigh Vacuum
P. A. Redhead, J. P. Hobson, and E. V. Kornelsen

Handbook of Electron Tube and Vacuum Techniques
Fred Rosebury

Vacuum Sealing Techniques
A. Roth

Ionized Gases
A. von Engel

Handbook of Materials and Techniques for Vacuum Devices
Walter H. Kohl

Plasma Diagnostics
W. Lochte-Holtgreven

Quadrupole Mass Spectrometry and Its Applications
Peter H. Dawson, Editor

AMERICAN VACUUM SOCIETY CLASSICS

QUADRUPOLE MASS SPECTROMETRY AND ITS APPLICATIONS

Editor
Peter H. Dawson

Library of Congress Cataloging-in-Publication Data
Quadrupole mass spectrometry and its applications / edited by
Peter H. Dawson
p. cm. -- (American Vacuum Society classics)
Originally published: Amsterdam : Elsevier Scientific Pub. Co.,
1976
Includes bibliographical references and index
ISBN 978-1-56396-455-8
1. Mass Spectrometry. I. Dawson, Peter H. II. Series.
QD96. M3Q3 1995 95-21857
543'.0873--dc20 CIP

Series Preface

The science of producing and measuring high and ultrahigh vacuum environments has fundamental interest for basic research in addition to a wide variety of important practical applications. Basic research involving particle physics, atomic and molecular physics, plasma physics, physical chemistry, and surface science often involves careful production, control, and measurement of a vacuum environment in order to perform experiments. Practical applications of vacuum science and technology are found in many materials processing techniques used for microelectronic, photonic, and magnetic materials and in the simulation of space and rarefied gas environments.

As in most of modern science, the rapid development of the field has been accompanied by a parallel growth in the related technical literature including specialized journals, monographs, and textbooks. There exist early publications in vacuum science and technology which have attained the status of indispensable references among practitioners, lecturers, and students of the field. Many of these "classic" publications have gone out of print and are currently unavailable to newcomers to the field. The present series, commissioned by the American Vacuum Society, and published by AIP Press to celebrate the 40th anniversary of the Society in 1993, is entitled "**American Vacuum Society Classics.**"

The American Vacuum Society Classics will reprint important books from the last four decades that continue to have significant impact on the modern development of the field. It is the goal of the American Vacuum Society Classics to reprint these books in a high quality and affordable format to ensure wide availability to the technical community, individual researchers, and students.

H. F. Dylla
Continuous Electron Beam
Accelerator Facility
Newport News, VA
and
College of William and Mary
Williamsburg, VA

CONTENTS

Preface xv

Principal symbols xix

Chapter I. Introduction, by P.H. Dawson 1

References 6

Chapter II. Principles of operation, by P.H. Dawson 9

A. The quadrupole field 9
(1) The geometry 9
(2) The applied potential 12
(3) The equations of motion 13
(4) The ion trajectories 15

B. The mass filter 19
(1) The stability diagram 19
(2) The resolution 22
(3) The mass filter aperture 24
(4) The mass filter acceptance 28
(5) Computed performance 31
(6) Ion exit 34
(7) The rectangularly driven mass filter 35
(8) A high-resolution device 36

C. The monopole 37
(1) Conditions for ion transmission 37
(2) Resolution 38
(3) Mass scanning 40
(4) Computed performance 42
(5) Ion energy 44
(6) Geometrical variations 45
(7) Apparent advantages and disadvantages 45

D. The quadrupole ion trap 46
(1) Conditions for ion trapping 46
(2) Mass-selective ion detection 49
(3) Mass-selective ion storage 52
(4) Computer simulations 54
(5) Novel features 56

E. Exact focussing devices 57
F. The oscillatory field time-of-flight spectrometer 61
G. Static quadrupole devices 62
References 63

Chapter III. Analytical theory, by P.H. Dawson 65
A. The Hill equation 66
B. The Mathieu equation 69
C. The complete solution 72
D. Auxiliary fields 74
E. Special cases 76
References 77

Chapter IV. Numerical calculations, by P.H. Dawson 79
A. Numerical integration 80
B. Matrix methods 83
(1) Matrix representation 83
(2) Phase-space dynamics 86
References 92

Chapter V. Fringing fields and other imperfections, by P.H. Dawson . 95
A. Fringing fields 96
(1) The mass filter 97
(2) The monopole 107
B. Systematic field faults 109
(1) The analytical approach 110
(2) Computer simulation 113
(3) Experimental evidence 116
(4) Round rods 117
(5) Field distortions and the monopole 118
References 119

Chapter VI. The mass filter: design and performance, by W.E. Austin, A.E. Holme and J.H. Leck 121
A. Basic considerations of mass range and resolution 121
B. Field imperfections and their effect on performance . . . 125

(1) Mechanical misalignments of the rods 125
(2) Contamination of the rods 128
(3) The use of circular rods to approximate hyperbolic fields . 129

C. The ion source 131
(1) Design and construction 131
(2) Alignment of the source 134

D. Ion detection 136
(1) The Faraday cup 136
(2) The electron multiplier 137
(3) High pressure limit of operation 140

E. Instrument sensitivity 141
(1) Dependence on resolution 141
(2) Dependence on mass 143
(3) The delayed d.c. ramp 144

F. Operation with non-sinusoidal fields 144

G. Electrical power supplies 145

References 151

Chapter VII. The monopole: design and performance, by R.F. Herzog 153

A. Introduction 153

B. Ion oscillations in a monopole 153

C. Ion entrance and exit 158

D. Mechanical construction of the monopole 159

E. The power supply 161
(1) Sweep unit 161
(2) Rf unit 164

F. Comparison of the experimental observations with theoretical expectations 164
(1) Mass scale and dispersion 164
(2) Scanning methods and extended mass range 165
(3) Peak width and peak shape 169
(4) Resolution and peak separation 171
(5) Sensitivity 171
(a) Entrance and exit apertures 172
(b) Voltage ratio 172
(c) Ion energy 172
(d) Focusing effects 173
(e) Sensitivity gain by a magnetic field 174

(f) Geometrical effects 176
G. Effect of a high pressure in the monopole section . . . 176
H. Negative ions 176
I. Power consumption 178
J. Conclusions. 179
References 180

Chapter VIII. Quadrupole ion traps, by J.F.J. Todd, G. Lawson and R.F. Bonner 181
A. Introduction 181
B. The three-electrode ion trap. Construction and instrumentation. 182
(1) The electrodes 182
(a) Geometry 182
(b) Materials and methods of mounting 183
(2) Ion creation 183
(3) The rf power supply 184
(4) Ion detection. 185
C. Other forms of ion trap 190
(1) The six-electrode trap 190
(2) The storage-ring trap 193
(3) The static ion trap 193
D. Survey of applications of ion traps 195
(1) Mass spectrometric applications 195
(2) The storage of microparticles 199
(3) The quadrupole ion storage source (quistor) 203
E. Experimental and theoretical aspects of ion containment . . 204
(1) Ion loss processes 204
(a) Mechanisms of ion loss 204
(b) The kinetics of ion loss 207
(2) A theoretical model of ion trapping 210
(3) Space charge and ion kinetic energies 214
(a) Space charge and the Dehmelt model 214
(b) Space charge effects along the axis of a quadrupole mass filter 216
(c) Fischer's treatment of space charge 216
(d) Oscillatory phenomena 217
(e) Ion kinetic energies 219
References 222

Chapter IX. Time-of-flight spectrometers, by J.P. Carrico 225
A. Introduction 225
B. Principle of operation 226
C. Ion displacement. 227
D. Arrival time. 230
E. Resolution 231
F. Experimental results 233
G. Proposed new design 234
H. New modes of operation 236
I. Conclusions. 239
References 239

Chapter X. Applications in atomic and molecular physics, by J.F.J. Todd 241
A. General applications 241
(1) Vacuum technology 241
(2) Surface studies 243
(3) Reaction studies involving solids or surfaces 247
(4) Gase phase studies 249
(a) The monitoring of ions and neutral species formed in swarms and plasma 249
(b) The sampling of ions from flames 252
(c) Beam studies 253
(d) Photoionization and photodetachment studies . . 255
(e) Electron impact induced ionization and excitation . . 256
B. Special applications 257
(1) The use of rf fields for "beam guides" 257
(a) The study of metastable ion lifetimes 257
(b) Determination of the integral cross-sections for ion–molecule reactions 258
(2) Applications of charged-particle traps 261
(a) Experiments with microparticles and droplets. . . 261
(b) Radiofrequency spectroscopy of trapped ions . . 262
(c) Electron–ion recombination studies 264
(d) Heavy ion plasma confinement 264
(e) The quadrupole ion storage source (quistor) . . . 265
References 267

Chapter XI. Applications to upper atmosphere research, by G.R. Carignan 273

A. Early history of mass spectrometric measurements in the upper atmosphere 273

B. The upper atmosphere of the earth 277

C. Techniques of in situ measurements 277
(1) Neutral gas analysis 277
(2) Ion analysis 280

D. Recent quadrupole applications and developments . . . 281

References 284

Chapter XII. Applications to gas chromatography, by M.S. Story . 287

A. Introduction 287

B. Gas chromatography–quadrupole mass spectrometry instrumental developments 287
(1) Resolution, mass range and peak shapes of the quadrupole . 287
(2) Gas chromatograph interfacing 288
(3) Sensitivity and dynamic range 290
(4) Electronic control requirements and capabilities . . . 294

C. Automation 296

D. New techniques 297
(1) Stable isotope mass fragmentography 297
(2) Chemical ionization 298
(3) Capillary columns 303

E. Conclusion 304

References 306

Chapter XIII. Medical and environmental applications, by G. Lawson . 307

A. Medical applications 307
(1) Respiratory gas analysis. 308
(2) Blood gas analysis 312
(3) Drug detection and analysis 313
(4) “Fingerprinting” of bacteria 318

B. Environmental monitoring 319
(1) Detection of pollutants in air. 320
(2) Water analysis 323
(3) Technological applications 326

C. Conclusions 328

Notes added in proof 329

References 331

Appendices 335

Appendix A. Parameters characterizing the acceptance ellipses for the mass filter with no fringing fields. 335

Appendix B. Approximate acceptance ellipse parameters for the mass filter for transmission at 50% of the initial phases in the presence of fringing fields of various lengths . . . 336

Appendix C. Paired values of a and q satisfying the conditions $\beta(a, q) = 1/d; \beta(-a, -q) = p/d$ for $2 < d < 13, p \leqslant d$ . . . 337

Appendix D. Coefficients from C_0 to C_6 used in the analytical solution to the equation of motion in the mass filter . . 338

Appendix E. An example of a simple computer program for calculating ion trajectories by numerical (Runge—Kutta) integration of the Mathieu equation. The program is written in Fortran IV G 339

Appendix F. List of quadrupole patents 340

Index 345

PREFACE

There has been a rapid proliferation of quadrupole mass spectrometers in the last decade. Quadrupoles are dynamic mass analysers and, while very simple in their physical geometry, are very complex in their behaviour. Recent theoretical and experimental advances have now provided a basic understanding of real (that is to say, imperfect) quadrupole devices.

This book proceeds from a general explanation of the action of radio-frequency quadrupole fields to the description of their utilization in mass analysers such as the quadrupole mass filter, the monopole, the three-dimensional quadrupole ion trap and various time-of-flight spectrometers and finally to the characteristic applications of quadrupoles.

A multi-author format was adopted, even though the intention was to produce a systematic text rather than a series of reviews. This provides a broader-than-usual viewpoint in the book and while there is, of necessity, a certain amount of recapitulation in various chapters, repetition has been avoided by the circulation of drafts between authors and many cross-references have been added.

Chapters I–V provide a unified approach to explaining the principles of operation of quadrupole devices. Chapter II is largely a qualitative account, illustrated by ion trajectories and computer simulations of performance. The mathematical background is dealt with in Chapter III. Chapter IV furnishes a detailed description of numerical methods of calculation of performance, including the recently developed application of phase–space dynamics. The very important and sometimes controversial subjects of fringing fields and other field imperfections are discussed in Chapter V.

Chapters VI–IX provide design and performance evaluations of the mass filter, the monopole, ion traps and time-of-flight instruments. These chapters have been contributed by authors with extensive practical experience with the respective instruments. The approach each takes is quite different, but this reflects the very different state of development in each field. For the mass filter, there is a considerable body of design data available. For the monopole, little data of any kind has been published and Chapter VII is largely a description of a particular design and its performance. For ion traps the present emphasis is on understanding the properties of the stored ions. Chapter IX is more speculative since quadrupole time-of-flight devices are in their infancy.

Chapters X–XIII are descriptions of four areas of application where quadrupole devices have made the greatest impact because of their particular advantages and disadvantages.

A project of this kind has involved the co-operation and assistance of many people. We are indebted to our colleagues who have provided data and illustrations used in the book. I would like to thank Rey Whetten for reading much of the manuscript and making many helpful suggestions and, especially, for the pleasure and inspiration that I derived from the years that we worked together on quadrupoles.

There follows a list of further acknowledgements and the detailed addresses of the authors.

Chapters I–V

P.H. Dawson,
Division of Physics,
National Research Council of Canada,
Ottawa, Canada.

I gratefully acknowledge the help of Alain Laverdiere of Laval University and Ryoichi Matsumura of NRC in preparing the illustrations and of Colette Verrette of Laval University in typing early drafts. I am deeply indebted to Jill Baker for the preparation of the final manuscript.

Chapter VI

W.E. Austin, A.E. Holme and J.H. Leck,
Department of Electrical Engineering and Electronics,
University of Liverpool,
Brownlow Hill,
Liverpool, U.K.

Chapter VII

R.F. Herzog,
Department of Physics and Astronomy,
University of Southern Mississippi,
Southern Station Box 5202,
Hattiesburg,
Mississippi 39401,
U.S.A.

Chapter VIII

J.F.J. Todd and G. Lawson,
Chemical Laboratory,
University of Kent,
Canterbury, Kent, U.K.

R.F. Bonner,
Department of Chemistry,
Trent University,
Peterborough,
Ontario, Canada.

We are indebted to our colleagues in the "dynamic" mass spectrometry group at the University of Kent for their helpful comments and criticisms, and particularly to Roger Mather who obtained some of the data quoted. The assistance of Maree Pollett and Joy Bower-Smith in typing and preparing the manuscript is gratefully acknowledged.

Chapter IX

J.P. Carrico,
Bendix Research Laboratories,
Bendix Center,
Southfield,
Michigan 40875,
U.S.A.

Discussions with Dr. R.K. Mueller are gratefully acknowledged. The assistance of S. Miller in the preparation of the manuscript is deeply appreciated.

Chapter X

J.F.J. Todd,
Chemical Laboratory,
University of Kent,
Canterbury, Kent, U.K.

It is a pleasure to acknowledge the help of my colleague Graham Lawson in preparing some of the material for the chapter, and of Maree Pollett, Sheila Cousins, and Linda Lawson in typing sections of the manuscript.

Chapter XI

G.R. Carignan,
Space Physics Research Laboratory,
Department of Electrical and Computer Engineering,
University of Michigan,
Ann Arbor,
Michigan 48105,
U.S.A.

The author is indebted to Wilson Brubaker, Rocco Narcissi, Nelson Spencer, and Ulf von Zahn for their thoughtful assistance in preparing the historical section of the chapter. I am especially indebted to Marti Moon for transforming my handwriting into a final manuscript. Copyright for Figs. 3–5 belongs to the American Geophysical Union.

Chapter XII

M.S. Story,
Finnigan Corporation,
845 West Maude Avenue,
Sunnyvale,
California 94086,
U.S.A.

I would especially like to acknowledge the tremendous creative and electronic contributions of William J. Fies of Finnigan Corporation without which much of the work discussed here would not have been possible.

Chapter XIII

Graham Lawson,
Chemical Laboratory,
University of Kent,
Canterbury, Kent, U.K.

I am indebted to my colleagues in the mass spectrometry group at the University of Kent for their many contributions to and criticisms of the earlier drafts of the chapter. I am particularly grateful to my wife for typing the manuscript.

PRINCIPAL SYMBOLS

a	parameter in the Mathieu equation of motion which depends upon U
a_m	stability limit in the Mathieu diagram (even solutions)
a_{2n}	parameter in the calculation of C_{2n}
b	parameter in the equation of motion which depends on an auxiliary field
b_m	stability limit in the Mathieu diagram (odd solutions)
b_{2n}	parameter in the calculation of C_{2n}
d	integer used to specify exact focusing conditions as in $\beta = p/d$
d_1	height of the monopole entrance aperture
d_2	height of the monopole exit aperture
e	electronic charge
f	applied frequency (in hertz)
f_0	lens focal length
g	parameter in the equation of motion which depends on an auxiliary magnetic field
h	constant relating the resolution to n^2
h_1, h_2	parameters in the Meissner equation of motion
h_∞	ion trap signal at saturation
k	parameter representing viscous drag
l'	minor axis of the acceptance ellipse
m	ionic mass (kg)
m_1	unit atomic mass (kg)
$m_{11}, m_{12}, m_{21}, m_{22}$	elements of the matrix M
n	number of rf cycles an ion spends in the field
n_0	number of rf cycles in a quarter wavelength of the fundamental ion motion
p_g	gas pressure
p	integer used to specify exact focusing conditions as in $\beta = p/d$
q	parameter in the Mathieu equation of motion which depends upon V
r	radial coordinate
r_0	field radius
s	an integer
t	time in seconds
t_j	arrival time
u	coordinate parameter representing x, y, z, or r

$\dot{u}$	$\mathrm{d}u/\mathrm{d}\xi$
v_z	velocity in the axial direction
x	transverse coordinate direction (conventionally that towards the positively biased electrodes)
x_m	maximum value of x during a trajectory
y	transverse coordinate direction (conventionally that towards the negatively biased electrodes)
y_m	maximum value of y during a trajectory
z	axial coordinate direction
z_0	field size parameter for the ion trap $= r_0/2^{0.5}$
z_f	distance of the focusing point in the focusing monopole
A	parameter representing gravity in the equation of motion of macro-particles suspended in a field
A_N	weighting factor in the general expression for potential
B	magnetic field strength
C, C'	components of a matrix representing a fringing field
C_{2n}	constants in the solution to the equation of motion depending on a and q
D	rod diameter
$\overline{D}$	pseudo-potential well depth
E	electric field
E'	auxiliary electric field
eE_z	ion energy in the axial direction
F	force
F_1, F_2	particular solutions to the equation of ion motion
G_1, G_2	parameters in the general solution to the equation of motion depending on u and $\dot{u}$, respectively
I	ion current in amperes
L	length of the analyser field
L'	major axis of the acceptance ellipse
M	specific ionic mass in amu
ΔM	peak width
$\mathbf{M}$	matrix representing one cycle of ion motion
N	ion density
N_∞	ion density at saturation
P	power loss
Q	quality factor of a coil
R	resolution ($M/\Delta M$)
S, S'	components of a matrix representing the fringing field
T	a small time interval ($\ll$ 1 rf cycle)
T_i	ion temperature
U	d.c. voltage applied between opposite sets of electrodes
U^*	system constant for the monopole

V	zero-to-peak rf voltage applied between opposite sets of electrodes
W	the Wronksian determinant
Z	ion displacement averaged over one rf cycle
α	ratio of the frequency of an auxiliary field to ω_0
α', α''	integration constants in the solution to the Mathieu equation
β	parameter characterizing the nature of ion motion in the stable region
γ	weighting factor for the z direction of the field
δ	displacement due to the ion micromotion
δ_0	duty cycle of a rectangular waveform
ϵ	emittance or acceptance of the mass filter
ϵ_1	$2u/v$
ϵ_0	permittivity
ϵ', ϵ''	construction tolerances
η	viscosity coefficient
θ	angular coordinate
θ_r	parameters in the Hill equation of motion
λ	weighting factor for the x direction of the field
λ_a	angular momentum about the z axis
μ	parameter characterizing the nature of the ion motion ($= i\beta$ in the stable region)
ξ	time expressed in terms of the applied field ($= \omega t/2$)
ξ_0	initial phase of the rf field
ξ_j	phase when G_2 becomes zero
ρ	the ratio between electron beam width and $2z_0$
ρ_n	terms in the recursion relationships between C_{2n} values
ρ'	parameter representing the relative magnitude of an auxiliary field
ρ_{MAX}	maximum space charge
σ	weighting factor for the y direction of the field
τ	characteristic period in the ion oscillation
τ_1	characteristic period for ion loss
ψ	pseudo-potential
ω	angular frequency of the applied field
ω_0	fundamental frequency of ion motion
ω_1, ω_2	higher frequency components of the ion motion
ω'	angular frequency of an auxiliary field
ω_p	plasma resonance frequency
A	parameter of the acceptance ellipse
B	parameter of the acceptance ellipse
Γ	parameter of the acceptance ellipse
Φ	electric potential
Φ_0	potential applied between opposite sets of electrodes
Φ_3, Φ_4	third and fourth order terms in the potential

CHAPTER I

INTRODUCTION

P.H. Dawson

The preparation of this book coincides with the twenty-first anniversary of the first publication concerning quadrupole mass spectrometry. This chapter attempts to delineate subsequent developments in the context of mass spectrometry as a whole, to put into perspective the current state of the art and to direct attention towards problems and challenges that remain.

The design and construction of quadrupole mass spectrometers remains something of an art. In recent years, however, the detailed theoretical investigation of "real" quadrupoles complete with field imperfections and fringing fields, has enabled us to at least enumerate the factors involved in obtaining high performance, even if it is difficult to specify ab initio all the design factors or even to comprehend observed abnormalities of behaviour. None the less, a basic technological proficiency has developed, often inadequately described in or difficult to retrieve from the scientific literature. It seems an opportune moment to try to assemble the information that is available into a basic text on the theory, design, construction, performance and applications of the various types of quadrupole field mass spectrometers.

Although there have been countless routine applications of quadrupole mass spectrometers, particularly in residual gas analysis, there are an unusual number of individualized applications in both science and technology which require the construction of specially designed or modified instruments with particular combinations of properties. Thus, for many users, not primarily interested in quadrupole mass spectrometers as such, the instrument cannot remain merely a "black box". It is the characteristic qualities leading to these specialized applications which justify the consideration (in the later part of this book) of quadrupole use separately from general mass spectrometric applications [1], although there is obviously a considerable overlap.

The very first mass spectrographs were Thomson's parabola instrument of 1910 and Aston's velocity-focussing device of 1919 which provided important information concerning isotopes and atomic structure. Dempster introduced the 180° deflection magnetic spectrometer and subsequent work was concentrated on improving design and achieving higher resolutions. Double-focussing mass spectrometers using radial electric field sectors before the magnetic field, in order to provide an ion beam of precise energy, were a

References pp. 6—7

major advance in this respect. Mass spectrometers with 90° and 60° magnetic deflection became more widespread after the work of Nier, beginning in 1940. However, before 1950, mass spectrometry remained a very specialized art, apparently fraught with difficulty and beyond the scope of many potential users in chemistry and engineering. Then, spurred by industrial interest in analytical applications, particularly in the petroleum industry, a rapid commercialization took place and a few standard types of magnetic spectrometers evolved. Emphasis changed from the physics of the devices to applications. New areas of analysis opened up; analysis of solids, application to the study of reaction kinetics, ultra-high resolution spectrometry to identify the molecular composition of complex molecules by accurate mass measurement, applications to organic chemistry and extensions to molecules of biological interest. Thus, to greatly simplify, forty years of sporadic development were followed by twenty years of widespread application to analytical problems.

The burgeoning applications in the 1950's also produced novel demands for instrument design and performance. Up to this point, mass spectrometers had been almost entirely magnetic deflection instruments; that is, "static" analysers where the magnetic or electric fields remain essentially constant during the passage of an ion. The new demands for particular combinations of qualities turned attention towards a new class of instruments, "dynamic" analysers [2, 3]. In such analysers, the ion separation is based on the time dependence of one of the system parameters. Blauth [2] has suggested three main sub-groups, energy-balance spectrometers, such as the omegatron; time-of-flight spectrometers; and path-stability spectrometers. Quadrupole spectrometers generally fit into the last category, although sometimes dynamic focussing is also involved (as in the monopole).

There are three principle types of mass analyser which utilize electrodynamic quadrupole fields; the mass filter, (Chapters II and VI) the monopole (Chapters II and VII) and the three-dimensional quadrupole ion trap (Chapters II and VIII). Some other types of analysers have also been proposed, as will also be described in Chapter II, but none has yet found general application. Special time-of-flight devices using quadrupole fields are described in Chapter IX.

The mass filter, as its name implies, allows an ion of a chosen mass-to-charge ratio (m/e) to pass through it while rejecting all others. The monopole partially utilizes the filtering action, but also depends upon a dynamic focussing to achieve the final mass separation. The quadrupole ion trap uses the filtering action to maintain ions of the chosen m/e range within a closed volume, rejecting all others. The chosen ions are accumulated within the volume and then their number measured after a suitable storage period.

The possibility of using an electrodynamic quadrupole field was first recognized by Paul and his colleagues at the University of Bonn and the first publication appeared in 1953 [4]. Proposals were put forward independently by Post of the University of California Radiation Laboratory and an instrument was constructed there but the work was not formally published [5].

The origin of the use of alternating gradient quadrupole fields lay in advances in quite a different field of physics, the design of high-energy accelerators. In 1952, Courant et al. [6] had announced the strong-focussing alternating gradient technique. Quadrupole magnetic fields were used to "squeeze" a proton beam first in one direction and then alternatively in a direction at right angles, and a strong focussing was achieved. The principle had been independently postulated by Christofilos two years earlier but his work had been overlooked. Electric quadrupole field focussing was shown to be feasible and analogous to the magnetic case by Blewett [7].

The first years of quadrupole mass filter development belonged to Paul's group and they hold basic patents on the device [8]. The work culminated in the classic paper on the mass filter by Paul et al. [9] and in the construction of a 5.82 m long mass filter intended for very high resolution ($\simeq$ 16,000) studies [10].

The promising characteristics of the quadrupole as a mass analyser which spurred development were sensitivity and moderate resolution in compact devices, the apparent mechanical simplicity, the light weight and absence of a cumbersome magnet, high-speed electronic scanning, the linear mass scale, and the possibility of trading off sensitivity against resolution by a simple adjustment of the circuitry. The availability of these features coincided with a leap forward in the routine application of high and ultra-high vacuum techniques and the (still) growing realization that when working with vacuum, it is not sufficient to measure the total pressure but essential to identify its components. There was, therefore, a growing demand for compact, simple, partial pressure analysers. The early 1960's also produced the great expansion in upper atmosphere and space research, and this financed a great deal of development work. Consequently, a very rapid commercialization became possible. The history of the monopole illustrates this very dramatically. The first publication on the monopole (yet another contribution by von Zahn [11]) was in 1963, but a commercial version was available by 1966 [12]. This was a period of feverish development for quadrupole devices. Much of the impetus has continued to come from the need to develop compact flyable instruments for ionospheric measurements, or for the analyses of space-cabin gases, astronaut's breath, and planetary atmospheres. This includes a remarkable series of studies and design innovations by Brubaker, the dominant figure in mass-filter studies in the past ten years. In the commercial field, in the late 1960's, the improvements tended to be in price rather than performance. A review written in 1968 [13] suggested a continuing but slower evolution of quadrupole field devices. This has been true in commercial terms but the establishment of a basic technological know-how seems to have inspired a number of imaginative concepts for more complex quadrupole structures, as is illustrated by the sketches of Fig. 1.1. Some of these

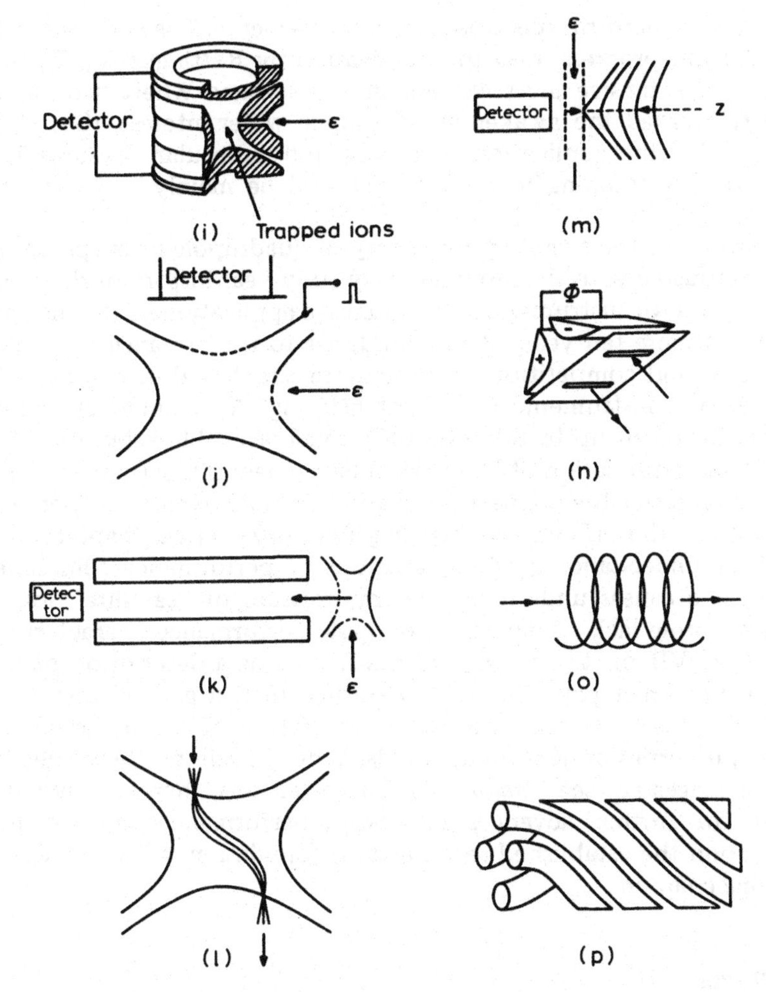

Fig. 1.1. Some actual and proposed quadrupole field devices. (a) The mass filter; (b) The mass filter with a delayed d.c. ramp (p. 105); (c) The mass filter driven rectangularly (p. 35); (d) The mass filter with a high-resolution attachment (p. 36); (e) The monopole; (f) The focussing monopole (p. 57); (g) The quadruple monopole (p. 45), end view; (h) A four-fold monopole (p. 45), end view; (i) The three-dimensional quadrupole ion trap with resonant ion detection; (j) The ion trap with external ion detection; (k) The ion trap as a mass filter source (p. 57); (l) A static three-dimensional quadrupole field used as an energy analyser (p. 62); (m) A time-of-flight spectrometer (p. 61); (n) A focussing "dipole" (p. 60); (o) A solenoid mass spectrometer (p. 63); (p) A static twisted quadrupole beam transport system (p. 62).

and the feasibility of the principle of ion storage was demonstrated by Berkling [14] and Fischer [15]. However, not much attention was paid to the development of this unusual device. It was very soon applied by Wuerker et al. [16] to trap macroscopic particles and by Dehmelt [17] to confine ions in order to perform spectroscopic measurements. The application to gas analysis did not progress until the publication by Rettinghaus [18] in 1967 and then the extensive investigations of Dawson and Whetten beginning in 1968 [19] marked the awakening of a much wider interest. (See Chapters VIII and X for the application to atomic and molecular physics.) The importance of the trapping technique may well lie mainly in its specialized applications.

This outline of the twenty-year history of quadrupole mass spectrometers is to be contrasted with the more leisurely sixty-year history of development of magnetic sector instruments. The intensive application of the instruments dates back a mere ten years. The boundaries to performance are still being pushed back, and competition with the more established technology of magnetic deflection instruments is still intensifying. A commercial instrument with a resolution of up to 8000 recently made available is but one example [20]. The miniaturization of low-performance mass filters (with 2 in. length) is another. And yet the technology is still not fully explored. There is some controversy on the influence of fringing field effects (see Chapter V). There is a lack of understanding of degradations of performance sometimes observed, of the causes underlying the transmission of "satellite" ion peaks. There is a simple lack of data on monopole performance characteristics, so that Chapter VII on the monopole has had to be a description of a set of measurements on a particular device rather than a general review of the technology. There has been almost no utilization of the remarkable exact focussing properties of quadrupole fields, although suitable device geometries have been suggested (see Chapter II). There remains, therefore, a great deal of scope for further advances in design, performance and applications, building upon the established technological foundation which we describe in the ensuing chapters.

REFERENCES

1 See, for example, F.A. White, Mass Spectrometry in Science and Technology, Wiley, New York, 1968.
2 E.W. Blauth, Dynamic Mass Spectrometers, Elsevier, Amsterdam, 1966.
3 D. Price (Ed.), Dynamic Mass Spectrometry, Vols. I and III, Heyden, London, 1970.
4 W. Paul and H. Steinwedel, Z. Naturforsch, A, 8 (1953) 448.
5 See R.F. Post, Univ. Calif. Radiat. Lab. Rep. U.C.R.L. 2209, 1953.
6 E.D. Courant, M.S. Livingston and H.S. Snyder, Phys. Rev., 88 (1952) 1190.
7 J.P. Blewett, Phys. Rev., 88 (1952) 1197.
8 W. Paul and H. Steinwedel, Ger. Pat. 944,900 (1956); U.S. Pat. 2,939,952 (1960).
9 W. Paul, H.P. Reinhard and U. von Zahn, Z. Phys., 152 (1958) 143.
10 U. von Zahn, Z. Phys., 168 (1962) 129; U. von Zahn, S. Gebauer and W. Paul, 10th Annu. Conf. Mass Spectrom., New Orleans, 1962 (not generally available).
11 U. von Zahn, Rev. Sci. Instrum., 34 (1963) 1.
12 J.B. Hudson and R.L. Watters, IEEE Trans. Instrum. Meas., 15 (1966) 94.

13 P.H. Dawson and N.R. Whetten, Advan. Electron. Electron Phys., 27 (1969) 59.
14 K. Berkling, Diplomarbeit, Physik. Institut, University of Bonn, 1956.
15 E. Fischer, Z. Phys., 156 (1959) 26.
16 R.F. Wuerker, H. Shelton and R.V. Langmuir, J. Appl. Phys., 30 (1959) 342.
17 H.G. Dehmelt, Phys. Rev., 103 (1956) 1125; H.G. Dehmelt, Advan. At. Mol. Phys., 3 (1968) 53.
18 G. Rettinghaus, Z. Angew. Phys., 22 (1967, 321.
19 P.H. Dawson and N.R. Whetten, J. Vac. Sci. Technol., 5 (1968) 1.
20 W. Fite, J. Vac. Sci. Technol., 11 (1974) 351.

CHAPTER II

PRINCIPLES OF OPERATION

P.H. Dawson

In Chapter III, details of the mathematical theory of quadrupole fields are presented and in Chapter IV we discuss the numerical calculation of quadrupole properties and methods of computer simulation. This chapter includes only that part of these two subjects which seems necessary for the comprehension of the principles of operation. Only instruments with perfect fields will be considered, but including instruments of finite diameter and length ("mathematical perfection" would require infinitely large devices). Fringing fields and field imperfections are considered in Chapter V. Details of ion sources, analyser construction and ion detection are described in subsequent chapters, as are the experimental details of performance.

An understanding of the general properties of quadrupole fields is the common starting point for a discussion of how the various instruments work. This is followed by separate sections on the mass filter, the monopole, the ion trap, exact focussing devices, the time-of-flight analyser, and finally some static quadrupole devices.

A. THE QUADRUPOLE FIELD

(1) *The geometry*

A quadrupole field is expressed by its linear dependence on the co-ordinate position. In the Cartesian co-ordinates x, y, and z

$$\mathrm{E} = E_0(\lambda x + \sigma y + \gamma z) \tag{2.1}$$

where λ, σ, and γ are weighting constants and E_0 is a position-independent factor which may be a function of time. Note that this field is uncoupled in the three directions which provides an immense simplification in analysis of ion motion (in a perfect field) since ion motion can be considered independently in each direction. For an ion in a quadrupole field, the force acting upon it, eE, increases according to the displacement of the ion from zero.

The field is subject to the restraints imposed by Laplace's equation

References pp. 63–64

(assuming no space charge within the electrode structure).

$$\nabla \cdot \mathbf{E} = 0 \tag{2.2}$$

so that

$$\lambda + \sigma + \gamma = 0 \tag{2.3}$$

The simplest ways of satisfying eqn. (2.3) are evidently

$$\lambda = -\sigma; \; \gamma = 0 \tag{2.4}$$

and

$$\lambda = \sigma; \; \gamma = -2\sigma \tag{2.5}$$

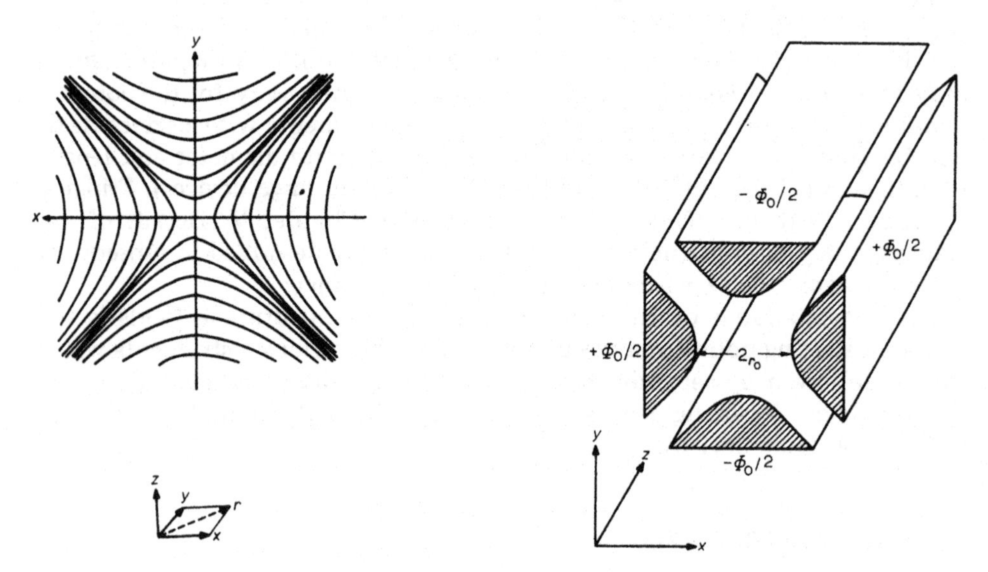

Fig. 2.1. Equipotential lines for a quadrupole field where $\phi = -1/2E_0\lambda(x^2 - y^2)$.
Fig. 2.2. The electrode structure required to generate the potential shown in Fig. 2.1. These are the ideal quadrupole mass filter electrodes having hyperbolic cross-sections.

To express these conclusions in more concrete terms, one must determine the form of the potentials to be applied. This is done by integration, since $-(\partial\Phi/\partial x) = E_x$ etc. The potential Φ can suitably be expressed as

$$\Phi = -\tfrac{1}{2} E_0 (\lambda x^2 + \sigma y^2 + \gamma z^2) \tag{2.6}$$

Consider first, the relation given in eqn. (2.4). Then

$$\Phi = -\tfrac{1}{2} E_0 \lambda (x^2 - y^2) \tag{2.7}$$

and the equipotential lines are as shown in Fig. 2.1. They are sets of rectangular hyperbolae in the xy plane with, geometrically, a four-fold symmetry about the z axis. Such potentials are generated by a set of four hyperbolic cylinders with adjacent electrodes oppositely charged as shown in Fig. 2.2. If the minimum distance between opposite electrodes is $2r_0$ and the potential between opposite electrodes is Φ_0, then eqn. (2.7) becomes

$$\Phi = \frac{\Phi_0(x^2 - y^2)}{2r_0^2} \tag{2.8}$$

since

$$\lambda = -\frac{1}{r_0^2}$$

In practice, round rods closely approximating the correct hyperbolic rods are generally used, but these introduce field faults and impose certain limitations (see Chapter V). Figure 2.2. represents the electrode structure of the mass filter.

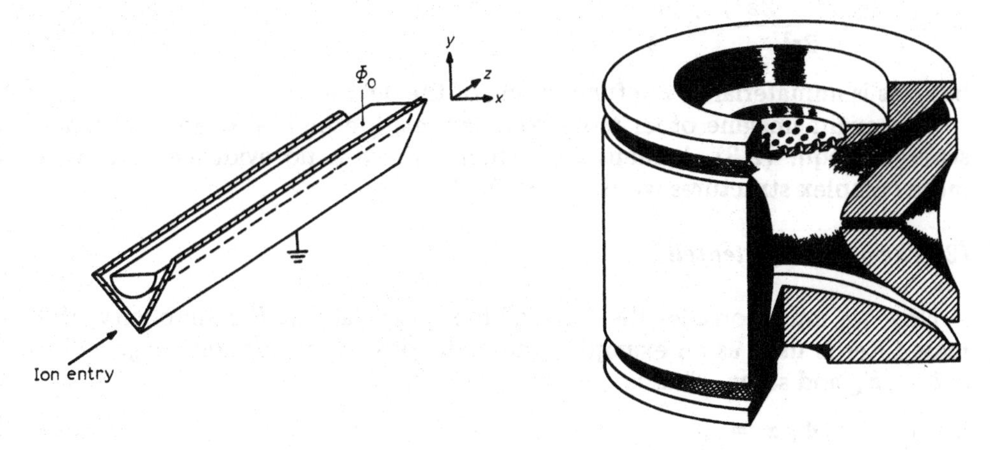

Fig. 2.3. The monopole mass spectrometer electrodes.
Fig. 2.4. The electrode structure required to produce the three-dimensional rotationally symmetric quadrupole fields used in the ion trap.

The monopole mass spectrometer has the same field as one quadrant of the mass filter as shown in Fig. 2.3. The instrument consists of a single rod and a right-angled V-shaped electrode in the ground plane. Retaining the convention that Φ_0 is the potential applied between the two electrodes of the instrument, the potential at any point would, however, be expressed as

$$\Phi = \Phi_0(x^2 - y^2)\,r_0^2 \tag{2.9}$$

Considering the second choice of constants, as given in eqn. (2.5), one obtains

$$\Phi = -\tfrac{1}{2}\,E_0\lambda(x^2 + y^2 - 2z^2) \tag{2.10}$$

or, using cylindrical coordinates

$$\Phi = -\tfrac{1}{2} E_0 \lambda (r^2 - 2z^2) \tag{2.11}$$

The equipotential lines are as illustrated in Fig. 2.29 for the xy plane and the rz plane. The appropriate electrode structure is illustrated in Fig. 2.4, the three-dimensional rotationally symmetric quadrupole ion trap. The two "end-caps" and the ring electrode have cross-sections in the rz plane which are complementary hyperbolae with a ratio of $2^{1/2}$ in the semi-axes (r_0/z_0).

Again, if Φ_0 is the potential applied between the ring and the end caps

$$\Phi = \frac{\Phi_0 (r^2 - 2z^2)}{2r_0^2} \tag{2.12}$$

Frequently, it is convenient to operate the ion trap with the end-caps grounded, so that only a single potential need be applied to the ring. Equation (2.12) is then more properly represented as

$$\Phi = \frac{\Phi_0 (r^2 - 2z^2)}{2r_0^2} + \frac{\Phi_0}{2} \tag{2.13}$$

but this is immaterial to the functioning of the device.

One might imagine other more complex relationships between λ, σ, and γ satisfying eqn. (2.3). Perhaps fortunately, there is no evidence that such more complex structures would be useful.

(2) *The applied potential*

Now we must consider the form of the potential Φ_0. For simplicity, the mass filter is used as an example. The equations of ion motion are given by $m\ddot{x} = eE_x$ and so on. That is

$$\ddot{x} + (e/mr_0^2)\Phi_0 x = 0 \tag{2.14}$$

$$\ddot{y} - (e/mr_0^2)\Phi_0 y = 0 \tag{2.15}$$

$$m\ddot{z} = 0 \tag{2.16}$$

On injecting ions into the mass filter with a certain velocity in the z direction, eqns. (2.14) and (2.15) give the ion motion in the xz and yz planes. If Φ_0 were merely constant, it is evident that there would be a simple harmonic motion in the xz plane for any ion and all ion trajectories would be "stable", i.e. remain finite in amplitude. However, in the yz plane, the ions would diverge from the z axis (called a defocussing) and eventually be lost. If, on the other hand, Φ_0 is a periodic function of time, the trajectories in both planes will alternately be deflected towards and away from the zero and one can envisage a stability in both planes providing that the periodicity is short enough and the ion is heavy enough that it cannot respond sufficiently during the defocussing part of the cycle to escape the device. Later, this condition for ion stability will be expressed explicitly in a stability diagram.

Finally, consider the combination of a direct component in Φ_0 and a periodic alternating component. Light ions are able to follow the alternating component. For the x direction, they tend to have unstable trajectories whenever the alternating component is larger than the direct component, and exhibit oscillations of ever increasing amplitudes. The x direction is therefore the equivalent of a high-pass mass filter. Only high masses would be transmitted to the other end of the quadrupole without striking the x electrodes. At the same time, in the y direction, heavy ions will be unstable because of the defocussing effect of the direct component, but some lighter ions will be stabilized by the alternating component if its magnitude and frequency are such as to "correct" the trajectory whenever its amplitude is tending to increase. Thus, one can describe the y direction as a low-pass mass filter. The two directions together give a mass filter with a certain pass-band. The combination of a direct and periodic alternating component for Φ_0 is therefore suitable for mass analysis.

Another visualization of the nature of stability in a mass filter or in a three-dimensional quadrupole ion trap is that of a ball on a saddle. As the ball begins to roll down the lower slopes of the saddle, the latter is inverted. If the frequency of inversion is well chosen, the ball can be trapped in the saddle.

The form chosen for Φ_0 has usually been $(U - V \cos \omega t)$; that is, a direct voltage U and a sinusoidal voltage of zero to peak amplitude V and angular frequency ω ($= 2\pi f$ where f is in Hertz). However, in principle any choice of Φ_0 is possible providing it is periodic and there has been an example of the use of a periodic *rectangular* waveform (see pp. 35 and 36 for a discussion of non-sinusoidal wave forms).

(3) *The equations of motion*

Returning to the equations of motion for the mass filter, they can now be expressed

$$\ddot{x} + (e/mr_0^2)(U - V \cos \omega t)x = 0 \tag{2.17}$$

$$\ddot{y} - (e/mr_0^2)(U - V \cos \omega t)y = 0 \tag{2.18}$$

Defining

$$a_u = a_x = -a_y = \frac{4eU}{m\omega^2 r_0^2} \tag{2.19}$$

$$q_u = q_x = -q_y = \frac{2eV}{m\omega^2 r_0^2} \tag{2.20}$$

and expressing time in terms of the parameter ξ where $\xi = \omega t/2$, both equations have the form

$$\frac{d^2 u}{d\xi^2} + (a_u - 2q_u \cos 2\xi)u = 0 \tag{2.21}$$

where u represents either x or y.

Note that the definition of a and q given in eqns. (2.19) and (2.20) differ by a factor of two from those used by some authors who define U and V as half the voltages applied between opposite pairs of rods.

Equations (2.17) and (2.18) are of the Mathieu type and eqn. (2.21) is the Mathieu equation in its canonical form. The Mathieu equation is a special case of the Hill equation (see Chapter III), viz.

$$\frac{d^2 u}{d\xi^2} + (a - 2q\psi(\xi))u = 0, \; \psi(\xi + \pi) = \psi(\xi) \tag{2.22}$$

The properties of the Mathieu equation are well-established and are considered in more detail in Chapter III. Only the nature of the solutions is discussed here.

To generalize the discussion which follows, consider also the monopole and the three-dimensional quadrupole ion trap. For the monopole, the equations of motion are

$$\ddot{z} = 0$$

$$\ddot{x} + (2e/mr_0^2)(U - V\cos \omega t)x = 0 \tag{2.23}$$

$$\ddot{y} - (2e/mr_0^2)(U - V\cos \omega t)y = 0 \tag{2.24}$$

with the additional geometrical constraint $y \geqslant |x|$. The two expressions again reduce to eqn. (2.21) if we define for the monopole

$$a_x = -a_y = \frac{8eU}{m\omega^2 r_0^2}$$

and

$$q_x = -q_y = \frac{4eV}{m\omega^2 r_0^2} \tag{2.25}$$

For the ion trap, the equations of motion are

$$\ddot{z} - (2e/mr_0^2)(U - V\cos \omega t)z = 0 \tag{2.26}$$

$$\ddot{r} + (e/mr_0^2)(U - V\cos \omega t)r = 0 \tag{2.27}$$

and $\ddot{\theta} = 0$ where θ is the angular coordinate (provided that the ions are formed with no initial velocity in the tangential direction, see p. 48).

Defining now

$$a_z = -2a_r = -\frac{8eU}{mr_0^2\omega^2} = -\frac{4eU}{mz_0^2\omega^2} \tag{2.28}$$

and

$$q_z = -2q_r = -\frac{4eV}{mr_0^2\omega^2} = -\frac{2eV}{mz_0^2\omega^2} \tag{2.29}$$

both the equations of motion again reduce to the general form

$$\frac{d^2u}{d\xi^2} + (a_u - 2q_u \cos 2\xi)u = 0 \tag{2.30}$$

or, more exactly

$$\frac{d^2u}{d\xi^2} + [a_u - 2q_u \cos 2(\xi - \xi_0)]u = 0 \tag{2.31}$$

The parameter ξ_0 takes account of the phase of the alternating field when the ion first experiences its influence. It is generally called the "initial phase". Nomenclature can be confusing since $2\xi_0$ is the initial rf phase. (i.e. in terms of ωt_0). Furthermore when $\xi' = [\pm(\pi/2) \pm \xi]$ is substituted for ξ in eqn. (2.30), the only change is that the $2q \cos 2\xi$ term becomes positive. Some spectrometrists have prefered to use this alternative form. The only difference then is in the definition of the initial phase, i.e. $\xi_0' = [\xi_0 \pm (\pi/2)]$.

In subsequent discussions ion velocities are usually defined as $\dot{u} = du/d\xi$.

The three instruments under discussion have in common that the single equation, (2.31), describes the ion motion in both coordinate directions of importance and that motion in the two directions is independent except for the constraints that for the mass filter and monopole $a_x = -a_y$ and $q_x = -q_y$ and for the ion trap $a_z = -2a_r$ and $q_z = -2q_r$. Note that these constraints derive directly from the Laplace equation.

(4) *The ion trajectories*

An understanding of quadrupole instruments demands some knowledge of the properties of the Mathieu equation. Solutions to this equation (see Chapter III for a detailed account) can be expressed by

$$u = \alpha' e^{\mu\xi} \sum_{n=-\infty}^{\infty} C_{2n} e^{2in\xi} + \alpha'' e^{-\mu\xi} \sum_{n=-\infty}^{\infty} C_{2n} e^{-2in\xi} \tag{2.32}$$

where α' and α'' are integration constants depending on the initial conditions; that is, u_0, $\dot{u}_0$, and ξ_0. The constants C_{2n} and μ depend on the values of a and q and *not* on the initial conditions. Thus we have the first important property of the Mathieu equation, i.e. the nature of the ion motion depends upon a and q but not on the initial conditions. All ions with the same (a, q) value (for a particular coordinate direction) have the same periodicity of motion. Furthermore, solutions of the form (2.32) are of two types depending upon the nature of μ. Stable solutions are those where μ remains finite as $\xi \rightarrow \infty$, and such solutions may be useful in our instruments providing that the value of u_{max} does not exceed r_0, the physical limit of our field. The second group of solutions, unstable ones, where μ increases without limit as $\xi \rightarrow \infty$, are not useful in these instruments.

There are four possibilities for μ.

(1) μ is real and non-zero. Instability arises from the $e^{\mu\xi}$ or $e^{-\mu\xi}$ factor.

(2) $\mu = i\beta$ is purely imaginery and β is not a whole number. These solutions are the periodic stable ones.

(3) μ is a complex number. The solutions are unstable (except for the trivial case $u_0 = \dot{u}_0 = 0$).

(4) $\mu = im$ is purely imaginery and m is an integer. The solutions are periodic but unstable. For $m = 2n$ the periodicity is π in ξ and for $m = (2n + 1)$ the periodicity is 2π. These solutions, called Mathieu functions of integral order, form the boundaries in (a, q) space between stable and unstable regions.

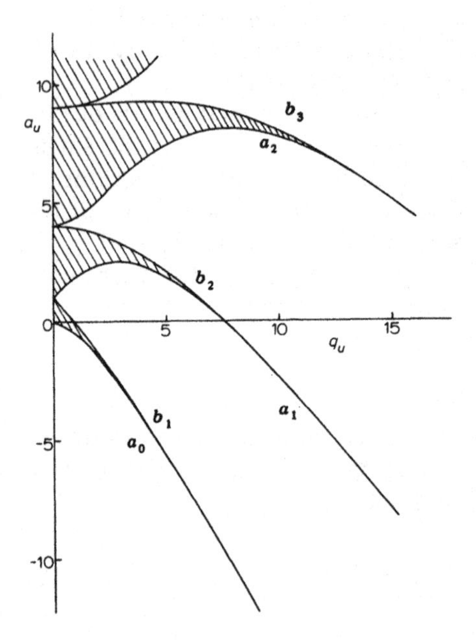

Fig. 2.5. The a–q stability diagram for the Mathieu equation considering a single coordinate direction. The shaded areas result in "stable" ion trajectories where the ion displacement always remains finite. The parameters a and q are defined in eqns. (2.28) and (2.29).

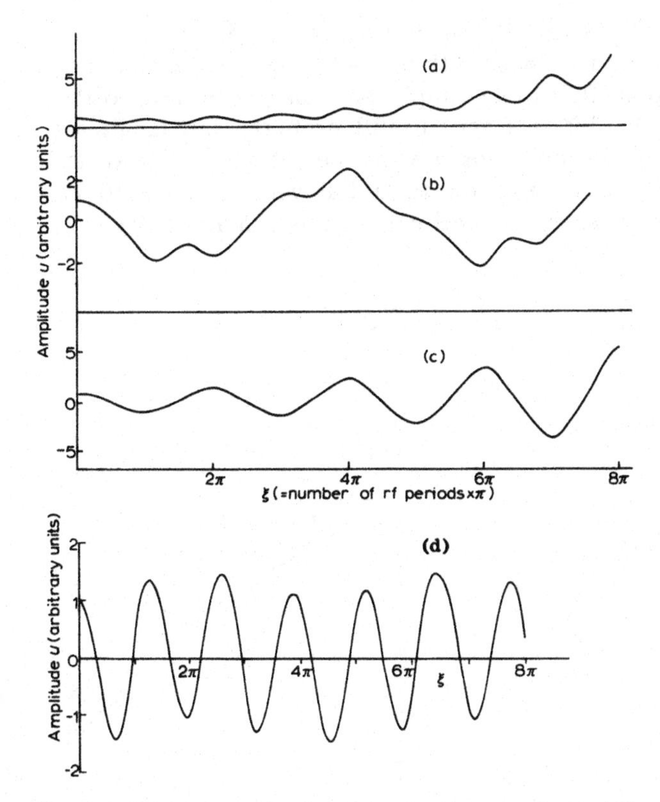

Fig. 2.6. Ion trajectories for several points in the stability diagram. (a) q = 0.55, a = −0.1625, near the lower stability limit of the first stable region; (b) q = 0.55, a = 0, within the first region of stability; (c) q = 0.55, a = 0.425, near the upper stability limit of the first stable region; (d) q = 0.55, a = 2.5, within the second region of stability.

Since μ depends only upon a and q, the conditions for stability can be represented on an a–q diagram or stability diagram as indicated in Fig. 2.5. The regions of stable or "bounded" trajectories are shown shaded. The stability limits are labelled a_m for the even solutions and b_m for odd solutions. As is evident from examining the original equation of motion, the stability regions are symmetric about the a axis.

In Fig. 2.6, the nature of the ion oscillation is illustrated for several points in the stability diagram, assuming the initial conditions $u_0 = 1$, $\dot{u}_0 = 0$, and $\xi_0 = 0$.

(a) the point $q = 0.55$, $a = -0.1625$ is near the lower limit of the first stability area; the periodicity is π (in terms of ξ).

(b) The point $a = 0$, $q = 0.55$ is within the first region of stability. The trajectory is complex but stable. The fundamental periodicity is $> 2\pi$.

(c) $a = 0.425$, $q = 0.55$. This is near the upper stability limit of the first stable region. The periodicity is 2π.

(d) $a = 2.5$, $q = 0.55$. This is within the second region of stability.

Since the equation of motion is linear with respect to the displacement u, the trajectories in quadrupole fields can simply be scaled according to the initial position (or velocity). This is a very useful property and facilitates computer simulation of quadrupole instrument performance. However, particular combinations of initial position to initial velocity have to be separately calculated (unless phase-space methods are used, Chapter IV).

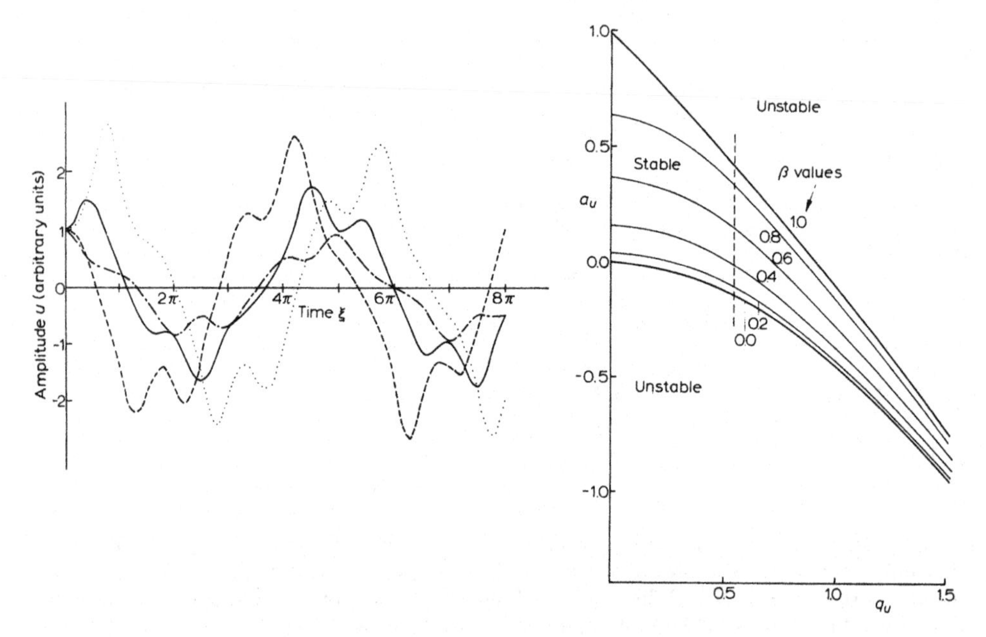

Fig. 2.7. Ion trajectories for four different initial field phases when $q = 0.55$ and $a = 0$. The fundamental nature of the ion motion remains unchanged.

Fig. 2.8. The lowest zone of stability for the Mathieu equation applied in a single coordinate direction. The parameter β is related to the characteristic frequencies making up an ion trajectory.

The independence of the nature of ion motion from the initial rf phase is illustrated in Fig. 2.7. The amplitude of the oscillation depends upon the phase but the nature of the motion does not. Note that the form of the stability diagram and the nature of the trajectories are both consistent with the qualitative discussion given earlier.

Because of the limited physical bounds of the usual quadrupole spectrometers, only the first stability region is of importance, although use of the second has also been proposed. The lowest zone is illustrated in more detail in Fig. 2.8, which includes some iso-β lines or contours. The calculation of the relationships between a, q, and β is directly related to the characteristic frequencies making up the ion trajectory. The ion motion has a fundamental frequency given by

$$\omega_0 = \beta\omega/2 \tag{2.33}$$

and higher frequencies

$$\omega_1 = (1-\beta/2)\omega; \ \omega_2 = (1+\beta/2)\omega; \ \text{etc.} \tag{2.34}$$

Thus the lower limit to the stability diagram ($\beta = 0$) represents trajectories composed of frequencies of oscillation of 0 and ω. The upper limit ($\beta = 1$), as we have seen, represents trajectories with fundamental frequencies of $\omega/2$. The trajectory shown in Fig. 2.6 (b) has a β of approximately 0.414 so that the frequency spectrum is given by

$$\omega_0 = 0.207; \ \tau_0 = 4.83\pi \ (\equiv 4.83 \text{ field cycles})$$

$$\omega_1 = 0.793; \ \tau_1 = 1.22\pi$$

$$\omega_2 = 1.207; \ \tau_2 = 0.83\pi$$

The characteristic periods τ_0 and τ_1 are quite evident in the trajectory. Note that although the frequency spectrum is the same for all points on a given iso-β line, the trajectories will differ because of the dependence of the relative weighting factors for the different frequencies on a and q. To take an extreme example, when $q = 0$ only the fundamental frequency of ion motion has a non-zero coefficient and the oscillation is simple harmonic motion.

Now that the basis for understanding ion motion controlled by the Mathieu equation has been presented, we can consider the necessity to establish stable trajectories simultaneously in more than one dimension and we can search for conditions suitable for mass analysis.

B. THE MASS FILTER

(1) *The stability diagram*

For the mass filter, eqns. (2.17) and (2.18) are valid. The condition for simultaneous stability in both x and y directions can therefore be represented as in Fig. 2.9 and in more detail in Fig. 2.10. These composite stability diagrams are arrived at by superimposing two separate stability diagrams (of the type shown in Fig. 2.5) which differ by the factor -1. The y direction is conventionally taken as that for which a is negative. Note that the change of sign of q must be taken into account when considering the "initial phase" in the x and y directions.

There was an early suggestion by Post [1] that the second stable region on the $a = 0$ line near $q = 7.5$ be used for mass analysis. This would seem to offer a resolution of about 100, but there might be complications because of the simultaneous transmission of heavy ions having $q < 0.9$. In any case, calculations indicate large amplitudes of oscillation in the high q regions, so that any such instrument might have a small effective aperture and therefore a very low sensitivity for a given cross-sectional area. However, higher stability areas may have application in special situations and use of the region centred around $q = 3.0$, $a = 2.8$ has been considered [2].

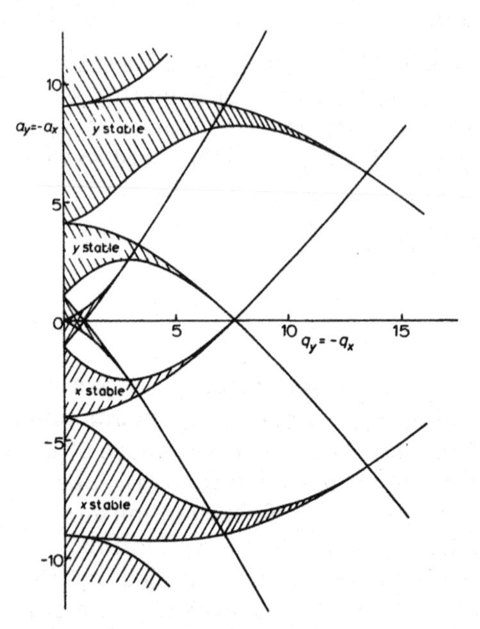

Fig. 2.9. The Mathieu stability diagram for the mass filter showing the regions of simultaneous stability in the x and y directions.

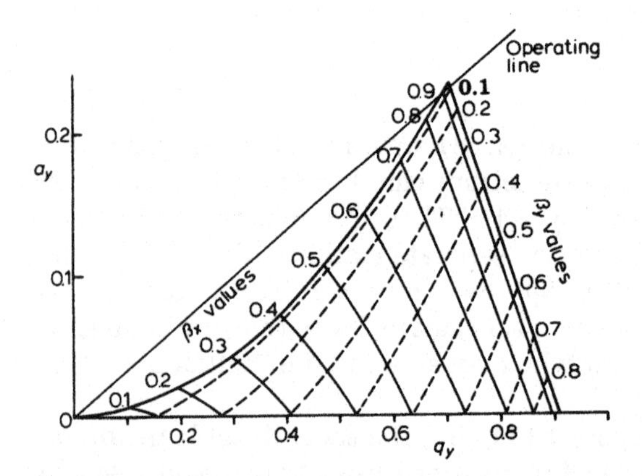

Fig. 2.10. The lower stability region normally used in mass filter operation showing iso-β lines for the x and y directions and a typical operating line.

Paul et al. [3], however, pointed out that it is possible to use the tip of the first stability region, as shown in Fig. 2.10, to obtain mass resolution. For fixed values of r_0, ω, U, and V, all ions of the same m/e have the same operating point (a, q) in the stability diagram. Since a/q is equal to $2U/V$ and does not depend on m/e, the operating points for all ions lie on the same line of constant a/q, passing through the origin of the stability diagram. This is called the mass scan line, mass sampling line, or operating line. Now, when $a \neq 0$, only those ions with operating points lying between the intersections of the mass scan with $\beta_y = 0$ and $\beta_x = 1$ will have stable trajectories in both x and y directions and only those ions will pass through the filter. By increasing the U/V ratio, the mass scan line approaches closer to the tip of the stable region and only a narrow range of m/e values will be associated with stable trajectories. Ions of lower mass will be unstable in the x direction and ions of higher mass unstable in the y direction. The ions with unstable trajectories will strike the hyperbolic electrodes or exit laterally from the field. The mass number corresponding to the stable region can be changed (that is, the mass spectrum can be scanned) by varying the magnitudes of U and V but maintaining their ratio constant in order to maintain a constant mass resolution (defined for this purpose simply as the ratio of the distance of the central point of the stable region from the origin to the width of the stable region measured along the scan line). As is evident from eqns. (2.28) and (2.29), an alternative method of scanning is to maintain U and V constant and to vary the applied frequency ω. This has occasionally been used but is much less common owing to the practical difficulties of frequency sweeping over an extended range.

Two of the attractive features of the quadrupole are evident from the above discussion. First, the resolution can be varied electronically by adjustment of the U/V ratio. Since increasing the resolution generally leads to a decrease in the effective aperture (as will be discussed later), this means that the trade-off between sensitivity and resolution can be made by a simple electronic adjustment exterior to the analyser. This is in contrast with static magnetic deflection instruments where slits on the analyser may have to be changed. The second feature is the simple linear relation between the mass number of the species transmitted and the magnitude of the applied voltages U or V. This enables the ready identification of unknown peaks and facilitates selective multipeak monitoring (see Chapter XIII, p. 310).

At high resolution, the xz trajectories will be similar to that illustrated in Fig. 2.11 (a) and the yz trajectories similar to that in Fig. 2.11 (b). The two are combined to give a composite trajectory in the xy plane in Fig. 2.11 (c).

With the assumption of an analyser of unlimited cross-section and unlimited length, we have the apparent possibility of an infinitely high resolution as the scan line approaches the stability tip at the point $a = 0.23699$, $q = 0.70600$. However, the physical limitations of a real analyser impose limits on both the permitted initial displacement (or instrument aperture) and on the attainable resolution. One has, in fact, a very complex interdependence between the transmissivity and resolution and the initial values of position, velocity, and rf phase.

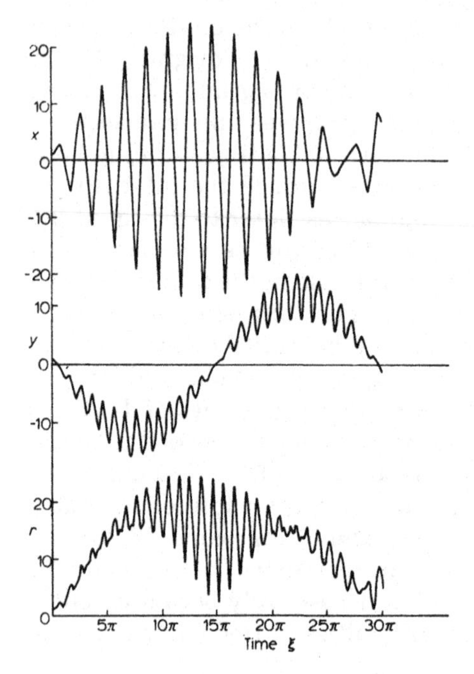

Fig. 2.11. Typical ion trajectories for a point near the tip of the stability diagram on a high-resolution operating line. (a) The trajectory in the xz plane; (b) the trajectory in the yz plane; (c) the radial displacement $r = (x^2 + y^2)^{1/2}$.

(2) *The resolution*

Consider first the limit on resolution imposed by the finite analyser length. One must distinguish short term stability from the "mathematical" stability so far discussed. For example, the analyser must be sufficiently long to enable a distinction to be made between trajectories of the type shown in Fig. 2.6 (a) and those similar to Fig. 2.11 (a); and similarly for trajectories similar to Fig. 2.6 (c) and Fig. 2.11 (b). The time required to make this distinction will obviously depend to some extent on the quarter-wavelength of the ion motion for the stable trajectory; that is $\tau_0^y/4$ or $\tau_1^x/4$. However, it would also depend on the range of initial conditions that are possible. Considering only the first factor, a relationship can be established between the number of rf cycles in a quarter-wavelength (n_0) and the resolution R. By careful calculation of the details of the stability tip, Paul et al. [3] established the following relationships (see Figs. 2.12 and 3.2).

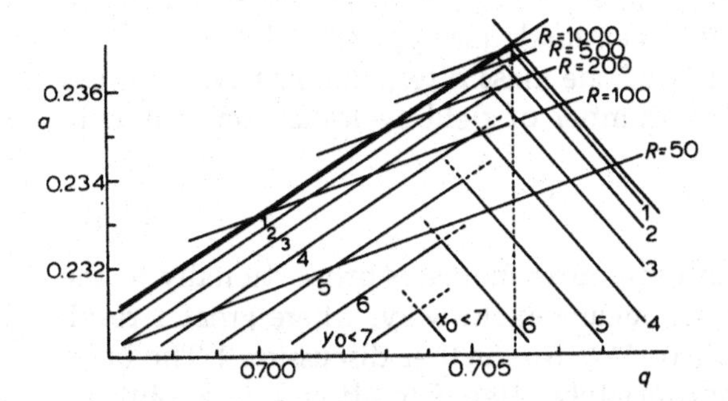

Fig. 2.12. The tip of the stability diagram showing operating lines nominally giving resolutions of 50, 100, 200, 500, and 1000. The lines parallel to the stability boundaries represent the maximum allowable initial x or y displacements, expressed as percentages of r_0, if there is to be 100% ion transmission for all initial rf phases. The calculations [3] were for ion entry parallel to the instrument axis and in the absence of fringing fields.

$$R = \frac{0.178}{(0.23699 - a_{0.706})} \tag{2.35}$$

where R is calculated from the width of the stable region for a particular scan line, and the $a_{0.706}$ is the a value corresponding to $q = 0.706$.

$$\beta_y^2 = \frac{(0.23699 - a_{0.706})}{0.79375} \tag{2.36}$$

$$(1 - \beta_x)^2 = \frac{(0.23699 - a_{0.706})}{1.93750} \tag{2.37}$$

Hence, for the y direction, since $n_0 = 1/2\beta_y$

$$R = \frac{n_0^2}{2.23} \tag{2.38}$$

and for the x direction, since $n_0 = 1/2(1 - \beta_x)$

$$R = \frac{n_0^2}{5.43} \tag{2.39}$$

Therefore one might expect the maximum attainable resolution, R_{max}, to be related to the square of the number of cycles the ions spend in the field n. Experimentally, von Zahn [4] found that $R_{max} = n^2/12.2$ and this type of relationship has been confirmed by other workers (see Chapter VI) although the numerical factor may be even higher. The factor will depend upon the instrument aperture and the field radius, and according to some results of Brubaker [5] seems to depend on whether the correct hyperbolic rods or the conventional round rods are used (see Chapter VI, Fig. 6.5). A definitive understanding of the length limitations of mass filters awaits more detailed computer simulation of their operation, especially in the presence of fringing fields. Some initial calculations have been made (see pp. 33 and 98).

In the normal mode of operation of the mass filter, all ions have a similar energy eE_z in the z direction. The number of field cycles an ion spends in the analyser of length L is

$$n = fL(m/2E_z e)^{1/2} \tag{2.40}$$

The maximum *attainable* resolution therefore increases proportionally to the mass, which is generally, but by no means exclusively, where greater resolution is required. (See Chapter VI p. 123 for further discussion.) The operating line must, of course, be appropriately altered to attain this resolution. Some manufacturers vary the operating line as the spectrum is scanned to take advantage of this feature.

Within the limitations imposed by the degradation of resolution when ions do not remain sufficiently long in the field, an attractive feature of the mass filter is the lack of dependence of the filtering action on the axial ion energy. This is particularly valuable in many special applications, particularly in chemical physics (see Chapter XII).

(3) *The mass filter aperture*

Consider now the limitations imposed by the finite radius of the mass filter. Here the combined effects of ion position and velocity and field phase at ion entry should be considered. Figure 2.13 shows the maximum amplitude/initial displacement ratio for some β_x and β_y values near the tip of the

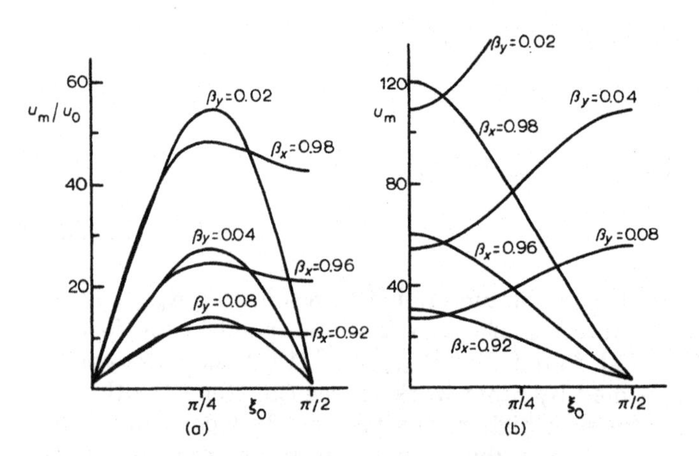

Fig. 2.13. Maximum amplitudes of ion oscillation for various β_x and β_y values near the tip of the stability diagram as a function of the initial phase of the field at ion entry [3]. (a) The initial displacements were unity and the initial velocities zero; (b) the initial displacements were zero and the initial velocities non-zero ($dx/d\xi = dy/d\xi = 2$).

stability diagram as a function of the initial phase [3]. In Fig. 2.13 (a), the initial positions are assumed to be unity and the initial velocities zero, whereas in Fig. 2.13 (b), the initial displacements are zero and the initial velocities non-zero. For most phases, the maximum displacement increases as the stability tip is approached, so that, naturally, increased resolution tends to be associated with decreased useable aperture of the instrument. For zero initial velocity, the zero phase is very favorable for ion transmission, but it is unfavorable if there is an initial velocity in the x direction.

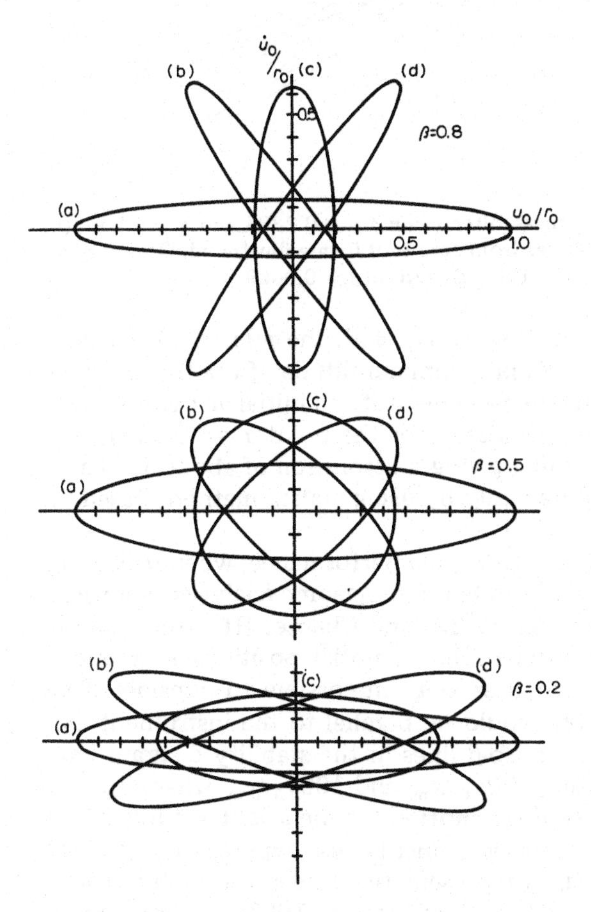

Fig. 2.14. Ellipses connecting equivalent points in $u\dot{u}_0$ phase-space for three different β values and four different initial phases, ξ_0 = (a) 0; (b) $-\pi/4$; (c) $\pi/2$; (d) $\pi/4$. The ellipses connect points which result in the same maximum displacement r_0. In each case $a = 0$.

Combination of position and velocity for particular initial phases can be expressed as families of ellipses connecting those points in the $(u_0, \dot{u}_0)$ plane (or phase-space) which result in the same maximum displacement. Figure 2.14

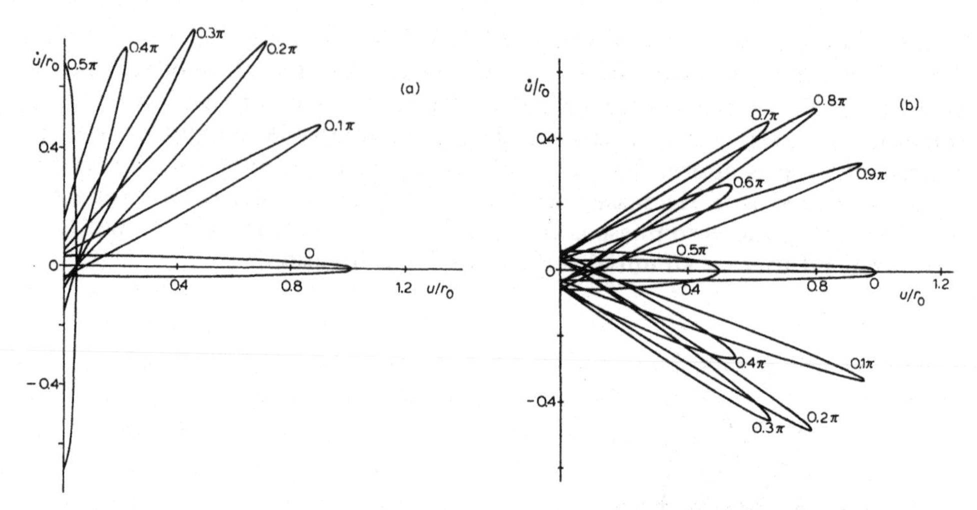

Fig. 2.15. Acceptance ellipses for the mass filter when $a = \pm 0.2334$ and $q = 0.706$. (a) The x direction, $\beta = 0.9569$. Initial phases from zero to 0.5π are illustrated. Other phases are symmetrical about the 0.5π phase. (b) The y direction; $\beta = 0.0669$.

shows some ellipses calculated by Fisher [6] with the maximum displacement taken as one. For $\xi_0 = 0$, the maximum amplitude of the trajectory is, as stated earlier, equal to the initial displacement if the initial velocity is zero. Figure 2.15 shows some ellipses calculated for the point $a = \pm 0.2334$, $q = 0.706$ which lies on an operating line giving a resolution of about 50. These ellipses can now be calculated quite easily by using matrix methods based on phase-space dynamics (see Chapter IV).

In early estimations of probable quadrupole performance, when trajectory calculations were made by using the rather cumbersome analytical solutions to the equations of motion (see eqn. (2.32) and Chapter III), some simplifications were essential, and combinations of initial positions in x and y directions and differing initial velocities were impossible to consider. Paul et al. [3] assumed that ion entry would be parallel to the instrument axis and constructed Fig. 2.12. Here the iso-β lines in the stability diagram have been replaced by lines representing $100x_0/x_m$ and $100y_0/y_m$ where x_m and y_m are the maximum amplitudes of ion motion for the *least* favorable initial rf phase. In other words, since x_m and y_m must be less than r_0 if the ions are to be transmitted, the contour lines represent the percentage of the instrument diameter for which transmission at all phases (100% transmission) is possible. The numbers on each scan line represent the resolution as calculated from the width of the stable region. For a given scan line, if the ion entrance aperture is limited to less than the diameter indicated by the intersection of equal $100x_0/x_m$ and $100y_0/y_m$ contours on that line, then one can expect a flat-topped (trapezoidal) peak. For example, for the $m/\Delta m = 50$ line, the initial x and y displacements must be limited to less than about 5.5% of r_0, if a flat-topped peak is to be obtained. At higher resolutions, such an aperture was expected to give a smaller, perhaps triangular peak, with a decreased percentage transmission. Paul et al., therefore, suggested that there are two operating ranges: (i) region I, low resolution, flat-topped peaks, with 100% ion transmission and intensity independent of resolution and (ii) region II, high resolution, triangular peaks, intensity decreases with resolution. It was expected that the peaks would be asymmetric because of the evident assymetry of Fig. 2.12, but, as we shall see later, detailed calculations do not show this, and assymetric peaks are not usually found in practice.

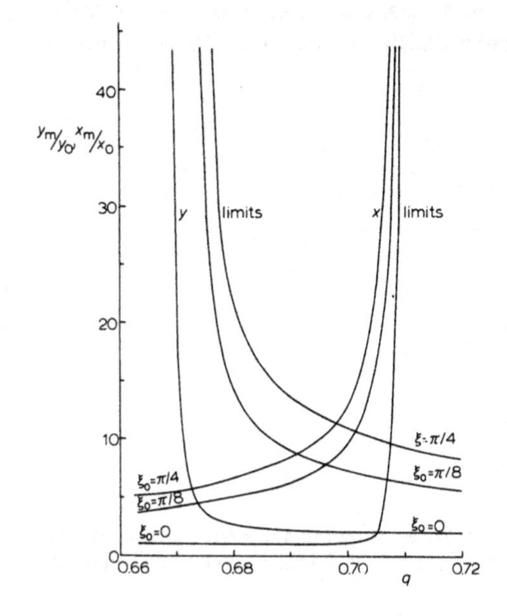

Fig. 2.16. The ratio of the maximum displacement to the initial displacement for various positions along the operating line $q = 3.07a$ when the ion beam enters the mass filter parallel to the z axis. The initial phases illustrated are 0, $\pi/8$, and $\pi/4$ [7].

An alternative representation of y_m/y_0 and x_m/x_0 for parallel ion entry, showing the phase dependence, is that of Fig. 2.16 [7], which is for a scan line $a = q/3.07$. It is evident from this figure that, at high resolution, when the transmission is less than 100% the ion current will be modulated at the frequency of the applied rf. Provided that the z direction energy is well-defined, it has also been suggested that this feature might be used in a "lock in" detection system to improve the signal-to-noise ratio, but the high frequency makes it difficult.

Using the relationships given in the eqns. (2.35)–(2.37) and the approximately linear dependence of x_m and y_m on $1/(1-\beta_x)$ and $1/\beta_y$, respectively (Fig. 2.13), Paul et al. obtained

$$x_m/x_0,\ y_m/y_0 < 1.8(M/\Delta M)^{1/2} \tag{2.41}$$

for 100% transmission at the peak maximum with ions injected parallel to the instrument axis. Or, considering ions entering at the centre of the field with initial x and y velocities, the equivalent condition was

$$(dx/dt)_0,\ (dy/dt)_0 < 0.16\, r_0\, \omega (\Delta M/M)^{1/2} \tag{2.42}$$

Other diagrams similar to Fig. 2.12 have since been constructed for particular combinations of initial displacement and initial velocity [8]. However, use of the phase-space approach gives a more complete and detailed view. This is possible because the ellipses in phase-space have an additional significance besides that of indicating the permissible initial velocity for a given initial displacement.

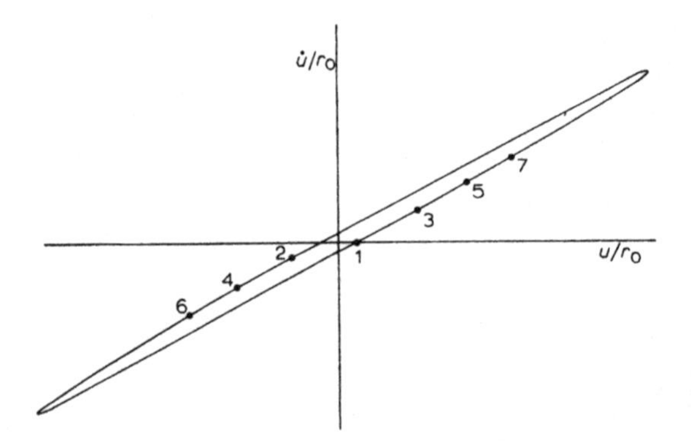

Fig. 2.17. The ellipse for $\xi_0 = 0.9\pi$ from Fig. 2.15(a) showing successive points at this same phase representing ion velocity and position after 1, 2, 3 etc. rf cycles have elapsed. The initial ion velocity was taken as zero.

(4) *The mass filter acceptance*

If any point on an ellipse represents the initial conditions, at that particular rf phase, then the velocity and position N cycles later, where N is an integer, is also given by a point on the ellipse. Figure 2.17 indicates successive points after 1, 2, 3 etc. rf cycles with $\dot{u}_0 = 0$ for one of the ellipses of Fig. 2.15. The ellipses can be represented by the parameters A, B, and Γ, the reduced equation of each ellipse being given by

$$\Gamma u^2 + 2\mathrm{A} u\dot{u} + \mathrm{B}\dot{u}^2 = \epsilon \tag{2.43}$$

where ϵ is called the emittance of the ellipse and is equal to the area of the ellipse divided by π. The variation of Γ, A, and B with the initial phase is illustrated in Fig. 2.18 (a) and (b). This kind of variation is typical, and the parameters are related by

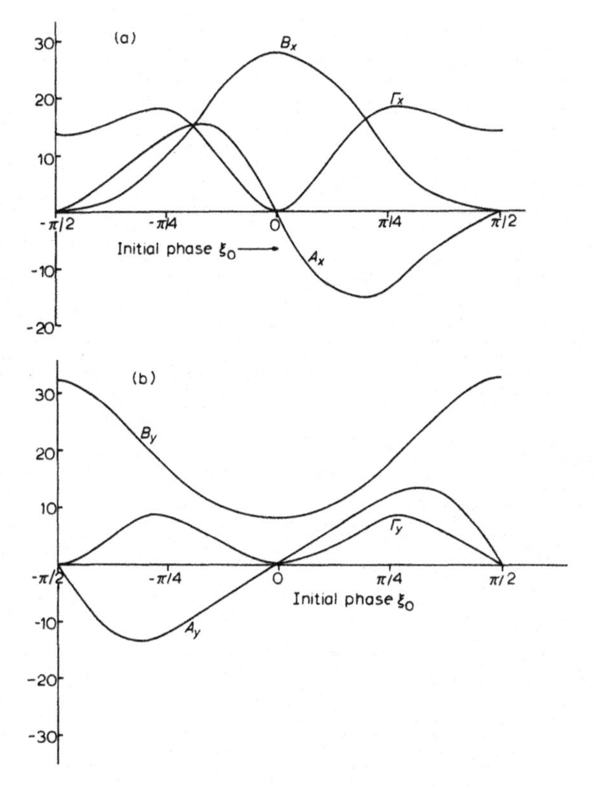

Fig. 2.18. (a) The variation of the ellipse parameters A, B, and Γ with initial phase for the point $a = 0.2334$ $q = 0.706$, i.e. for the x trajectories. (b) The ellipse parameters for the y direction.

$$\mathrm{B}\Gamma - \mathrm{A}^2 = 1 \tag{2.44}$$

One can show (see Chapter IV) that the maximum displacement during a trajectory (with infinite time available) is given by

$$u_{\max} = (\epsilon \mathrm{B}_{\max})^{1/2} \tag{2.45}$$

That is

$$u_{\max} = [\mathrm{B}_{\max}(\Gamma u_0^2 + 2\mathrm{A}u_0\dot{u}_0 + \mathrm{B}\dot{u}_0^2)]^{1/2} \tag{2.46}$$

Thus, the parameter Γ represents the susceptibility of the trajectory to the initial position, B its susceptibility to the initial velocity, and A that to combinations of velocity and position. As one would expect, at certain field phases, a positive initial velocity at a positive initial position may minimize the maximum displacement (when the velocity is "opposed" to the initial field) but at other phases such an initial velocity increases the maximum displacement (change in sign of A). Some examples of A, B, and Γ for (a, q) values important in the mass filter are given in Appendix A.

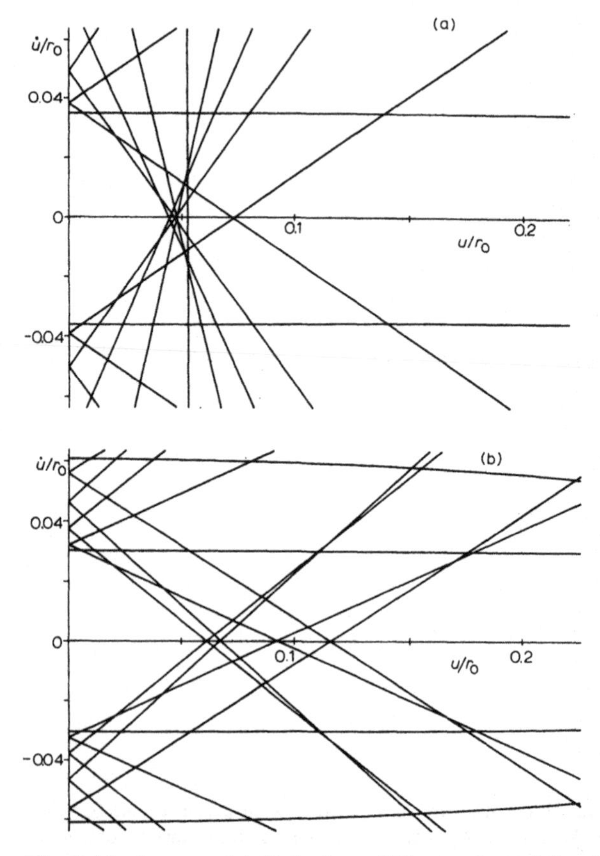

Fig. 2.19. A more detailed view of the overlap of the ellipses of Fig. 2.15. The area at the centre represents those initial conditions appropriate to 100% ion transmission. In the absence of fringing fields, the ion source emittance should be matched to the quadrupole acceptance. (a) x direction; (b) y direction.

Mass filter sensitivity is most properly expressed in terms of the ellipses in phase-space scaled, as in Fig. 2.15, so that the maximum deflection is equal to r_0. For 100% transmission, it is the overlap of the ellipses which is significant. A more detailed illustration of this important region in phase-space is given in Fig. 2.19, again for the point $a = \pm 0.2334$, $q = 0.706$. The area in phase-space is called the *acceptance* of the mass filter. The exact shape of the area is not so significant since an appropriate lens system ahead of the filter can, in principle, be used to transform an area to the appropriate shape. To consider sensitivity only in terms of the intersections of the ellipses with the u axis (the parallel ion entry assumption) can be somewhat misleading. The phase-space *emittance* of the ion source should be transformed by a lens system to match as closely as possible the acceptance of the quadrupole filter [9]. In this process, the area in phase-space is conserved so that the area of ellipse overlap largely determines the overall sensitivity. The acceptance of the filter will generally be modified by the fringing fields (see Chapter V).

(5) *Computed performance*

Using the matrix techniques, the maximum values of the ion displacement have been calculated [10] for several different (a, q) positions along the scan line giving a resolution of 50, for 20 different values of the initial phase. These results were then combined to consider 400 ions entering the mass filter at each phase with a uniform distribution across a square aperture and with no initial x or y velocity. The aperture size could be adjusted so that the ion entrance aperture occupied various fractions of the total square aperture, $4r_0^2$. Any ion exceeding a displacement r_0 in x or y directions was considered as not being transmitted. Note that for a frequency of operation

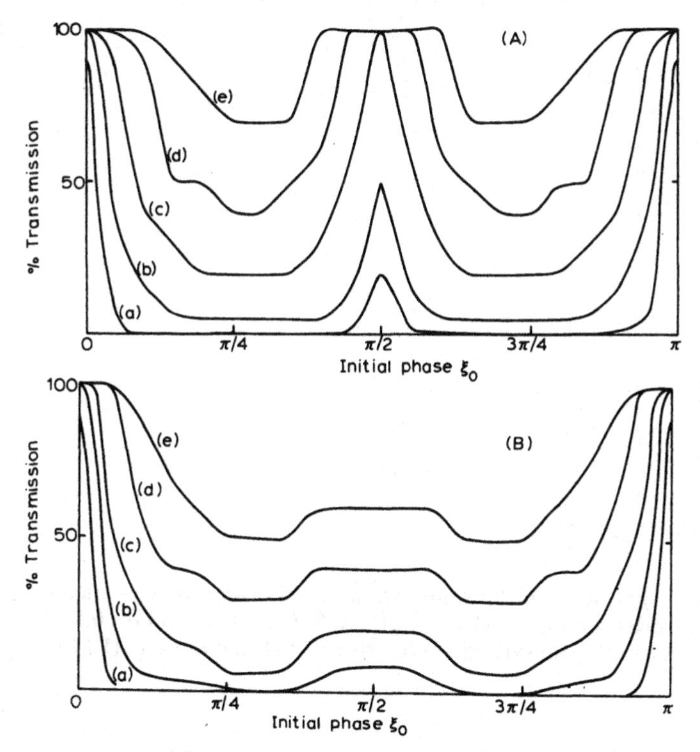

Fig. 2.20. (A) Computed curves of the percentage of ion transmission as a function of initial field phase for the point $a = \pm 0.2301$, $q = 0.696$, which lies on an operating line giving a resolution of 50. The ion entrance apertures given as percentages of the total aperture $4r_0^2$ are (a) 25%, (b) 4%, (c) 1%, (d) 0.25%, and (e) 0.11%. (B) Transmission curves for a point near the x stability boundary, $a = \pm 0.2341$, $q = 0.708$. Apertures as in (A).

of 1 MHz, this computer simulation is equivalent to considering a total ion current of about 10^{-9} A, which is quite realistic. Figure 2.20 (A) shows the calculated ion transmission for several entrance apertures as a function of phase for the point $q = 0.696$, $a = \pm 0.2301$ which is close to the y stability boundary. Figure 2.20 (B) gives similar curves for the point $q = 0.708$, $a = \pm 0.2341$ as one approaches the x stability boundary. Summing the transmission over all phases for various points along a scan line gives the predicted peak shapes as indicated in Fig. 2.21. The aperture areas for 100% transmission agree with the calculations of Paul et al. presented earlier. However, the peaks do not show large asymmetry when the detailed dependence on initial position and initial phase is taken into account as in these simulations. Moreover, when transmission is less than 100%, the peaks do not become triangular but remain somewhat bell-shaped and the half-height resolution is little changed. Paul et al. had suggested a factor of two resolution increase on account of the postulated triangular shape.

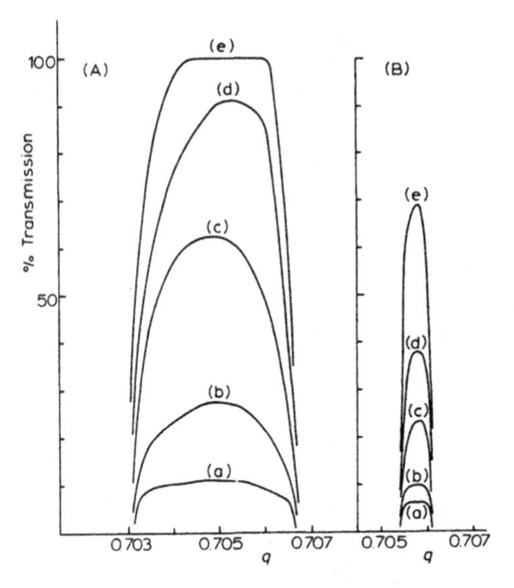

Fig. 2.21. Computed peak shapes obtained by summing ion transmission over all phases. The entrance apertures illustrated are (a) 4%, (b) 1%, (c) 0.25%, (d) 0.11%, and (e) 0.04%. The curves in (A) are for an operating line giving a resolution of 225 and those in (B) have $R = 1100$.

The modulation of the transmitted signal evident in Fig. 2.20 has, in fact, been observed experimentally [11]. For applications where current densities are low, it has been suggested [12] that bunching of ions be used to ensure entry only at the most favorable (zero) phase for transmission. This involves the use of special ion acceleration systems between the ion source and the analyser. One method, unfortunately readily applicable only to one ion of a particular m/e and given axial velocity, is to interpose an equipotential gridded box of thickness, x, whose potential, V, varies with time, t, according to

$$V(t) = mx^2 V_0/[mx^2 + 2eV_0t^2 - 4xt(emV_0/2)^{1/2}]$$

This should give efficient bunching, but would be difficult to apply when mass scanning. A partial experimental verification has been carried out using the approximation of a linear variation in V for a high mass [11].

An alternative, but much less efficient, bunching arrangement [13] is to have an accelerating field between the source and the analyser given by V_{acc} cos ωt. For example, for ions of 63 eV energy emerging continuously from the source, an accelerating region 1 cm long with the potential V_{acc} = 169 V at a frequency of 1 MHz, will bunch about half the ions into less than 1/16th of the rf cycle. These bunching techniques should, of course, be even more valuable at higher resolutions. However, note that once again this discussion assumes that ions enter parallel to the instrument axis. This will not be so, especially because of the presence of fringing fields at the entrance to the quadrupole. It has been shown by calculation that the phase dependence can be completely changed even by very short fringing fields, corresponding, for example, to a time of less than one rf cycle as the ion passes through. The fringes become particularly important as higher performance is demanded, involving the use of low z velocities. This subject is dealt with in detail in Chapter V, as is the "delayed d.c. ramp" technique [14] for modifying ion entrance conditions. Although the theory of fringing fields is poorly developed, there is some experimental justification for the calculations [7, 15] and any theory or calculation which neglects them must be treated with reserve.

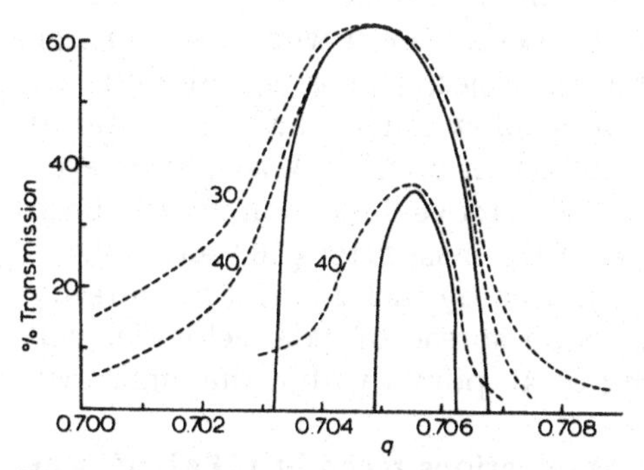

Fig. 2.22. Computed peak shapes for operating lines of nominal resolutions 225 and 600 when the mass filter field is limited in length. The numbers indicate the length in rf cycles. The assumed entrance aperture was 0.25%.

Other departures from ideality considered in Chapter V are (a) errors in the rod geometry, for example, the common substitution of round rods for hyperbolic ones, (b) errors in rod positioning, and (c) errors in the applied waveform. Some experimental observations on these limitations are given in Chapter VI.

The detailed computer simulations using matrix methods have also been used to examine the transmitted ion signal when the field is of limited length. Figure 2.22 shows peak shapes for two different operating lines when the number of rf cycles the ion spends in the field is limited to thirty or forty. The shapes show the kind of asymmetry found by experiment.

(6) *Ion exit*

The ions will exit from a quadrupole with displacements and velocities which vary with the operating point during a scan [6, 16]. Since the principal wavelengths of the ion trajectory are given by $2/\beta_y$ field cycles for the y direction and $2/(1-\beta_x)$ field cycles for the x direction, there will be a return close to the original distributions after these times. Moreover, as discussed later (see exact focussing spectrometers, p. 57), an inverted image would be formed after d cycles whenever $\beta = p/d$ where p and d are integers having no common factors. Considering the y direction, with ions of a fixed axial velocity spending n cycles in the quadrupole field, there will be an imaging effect each time $n/m = d$ where m is an integer. For example, with the scan line giving a resolution of 50 shown in Fig. 2.13 and an ion spending 30 rf cycles in the field, this imaging would take place during a scan of the mass peak when $\beta_y = 1/30$, $1/15$, $1/10$. The x trajectories would be imaged for $\beta_x = 29/30$, $14/15$, and $9/10$. Obviously, for a scan line which coincidentally passes through a point $\beta_y = 1/d$ and $\beta_x = p/d$ when the energy is such that ions spend d cycles in the field, there will be imaging in both directions. This special case might find application for simultaneous mass selection and imaging (for constant energy ions), for a crude form of combined mass and energy selection, or possibly for use of a limiting exit aperture to increase resolution. For general mass selection purposes, it is essential, however, to ensure that all ions are measured by the detector, even when they emerge with large displacements or large lateral velocities. This is usually done by use of a large area collector or electron multiplier or use of a strong field to collect the transmitted ions. If this is not done, poor peak shapes or even mass discrimination effects may be found. The above discussion ignores both the presence of fringing fields at the exit of the mass filter [16] and the distribution of axial velocities likely to be induced by such fields if not already present (see Chapter V). A sufficiently wide spread of axial velocities will give ion exit distributions corresponding to all points within the emittance ellipses such as Fig. 2.15.

There have been some experimental observations recently [16a] of peak shape problems caused by defocussing along certain iso-β lines.

It is usually found in mass filters that there is a background signal which limits the ability to measure trace concentrations. This background originates from excited neutrals which easily pass through the "line-of-sight" analyser. The problem has been solved by off-setting the electron multiplier collector with provision for deflecting the emerging ion beam (see Chapter VI, p. 139). Good design is again important to ensure 100% ion collection under all conditions. Curved quadrupoles [17] or curved sections [18] have also been used to avoid the problem but these forego the mechanical simplicity which is one of the mass filter's attractive features.

When the applied potentials are small (or zero) as at the beginning of a mass scan, ions entering the filter may be transmitted even though their trajectories are mathematically unstable, just because of the weakness of the fields. This gives rise to an output signal at the beginning of a mass scan sometimes called the "zero blast".

Note that an ion which enters the mass filter with no radial velocity at the centre of the field will, in principle, pass through the device undeflected. In practice, since there are always fringing fields, such entry conditions are not very probable. However, to avoid any contribution of such ions to peak "tailing" (or resolution degradation), mechanical cross-wires preventing ion entry along the potential asymptotes ($x = y$) have been used [19].

(7) *The rectangularly driven mass filter*

An interesting variation on the conventional mass filter is a device operated with a square-wave potential instead of a sine wave [20]. As implied in the discussion leading up to eqns. (2.17) and (2.18), any choice of the time-varying term $\Phi_0(t)$ is potentially acceptable provided it is periodic. Some forms may well give rise to anomalous non-linear resonance "lines" of instability within the normally stable region because of resonance interactions between the ion motion and particular sub-frequencies of the applied waveform (see Chapter V). However, Richards et al. [20] have experimentally verified the feasibility of using a rectangular waveform. The equation of motion for the waveform illustrated in Fig. 2.23 (a) is given by

$$d^2u/d\xi^2 + (a + 2q)u = 0; \quad 0 < \xi < \theta \tag{2.48}$$

$$d^2u/d\xi^2 + (a - 2q)u = 0; \quad \theta < \xi < \pi \tag{2.49}$$

with the sign of a and q depending, as usual, on whether the x or y direction is being considered. These equations are forms of the Meissner equation which has been used for many years, e.g. in studying vibrations in locomotives. A departure of the duty cycle, $\delta = \theta/\pi$, from the value of 0.5 is equivalent to applying an additional mean d.c. potential to the device. This is expressed in terms of a stability diagram in Fig. 2.23 (b). With $\delta = 0.39$, one can obtain a high resolution without the application of any conventional d.c. potential. Thus one exchanges the necessity for a very precise control of the

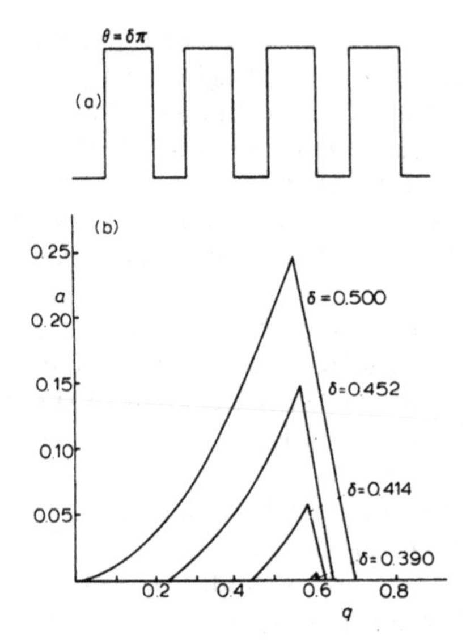

Fig 2.23. (a) The applied potential for a rectangularly driven mass filter. The departure of the duty cycle, δ, from 0.5 is equivalent to the application of a mean d.c. potential and results in greater mass selectivity. (b) The stability diagram for the Meissner equation and its variation as the duty cycle is changed. With $\delta \simeq 0.39$, a high resolution is obtained without applying a conventional d.c. potential.

U/V ratio for high resolution in a conventionally operated quadrupole for a precise and jitter-free control of the duty cycle. The latter appears to offer some technological advantages. There is an additional advantage to the quadrupole designer in that the equations of motion are even more easily solved using matrix methods (see Chapter IV, p. 83) and complete computer simulations can be rapidly carried out for a variety of ion source (or initial ion entrance) conditions. (See Chapter VI, p. 145 for a discussion of the experimental results.)

(8) *A high-resolution device*

Figure 1.1(d) shows another variation on the basic mass filter [21]. The quadrupole field section is operated with $a = 0$, i.e. as a high-pass mass filter. Ions that have q values near the limit of the stability region tend to receive large transverse velocities from the fields at the quadrupole exit. All other ions are rejected by the spherical retarding electrode. At the mass number 500, a resolving power of 1400 was obtained and transmission was about ten times higher than that in the conventional mode of operation. However, the separation from the general background is only efficient at high masses. It may be possible to improve the technique as our understanding of fringing fields improves (see Chapter V).

C. THE MONOPOLE

(1) *Conditions for ion transmission*

The electric field configuration of the monopole (Fig. 2.3) consists of one quadrant of the quadrupole field; it has the negative d.c. potential applied to the rod, i.e. the plane of symmetry of the instrument is the y direction (see p. 11). As shown on p. 14, equations of motion in the x and y directions are again represented by Mathieu equations

$$\frac{d^2u}{d\xi^2} + [a_u - 2q_u \cos 2(\xi - \xi_0)]\, u = 0$$

with $y \geqslant |x|$ and

$$a_u = a_x = -a_y = \frac{8eU}{m\omega^2 r_0^2} \tag{2.50}$$

$$q_u = q_x = -q_y = \frac{4eV}{m\omega^2 r_0^2} \tag{2.51}$$

However, the monopole differs greatly from the mass filter in its mode of operation. Consider, for example, the trajectories for a point near the tip of the stability diagram such as those shown in Fig. 2.11 (a) and (b). The criterion of stability is not sufficient in itself to ensure ion transmission. Obviously, the x deflection must remain small so that $y \geqslant |x|$. Moreover, the ion must spend fewer rf cycles in passing through the instrument than the number of cycles in the characteristic half-beat length in the y direction. For ion entry parallel to the axis, there is then the possibility of ion transmission for the 50% of the initial phases for which y remains positive during the first half-beat length. If the length condition is not fulfilled, the ions will strike the V electrode when y approaches zero. In any case, the ions entering at unfavourable phases are immediately driven towards the V electrode and are lost. The functioning of the monopole thus depends not only on the stability of ion trajectories but also on a partial focussing action in the y direction. A simplified description of how the monopole works is presented at the beginning of Chapter VII.

The necessary (but not sufficient) condition for 50% transmission can be expressed by

$$\frac{1}{\beta_y} > n \tag{2.52}$$

where n is the length of the field in rf cycles. A monopole operated under the conditions giving the trajectories of Fig. 2.11 (a) and (b) would have a very small useful aperture because of limitations on the initial x displacement imposed by the condition $y \geqslant |x|$. Although operation at the tip of the stability diagram has been investigated [22], it is more usual to operate with

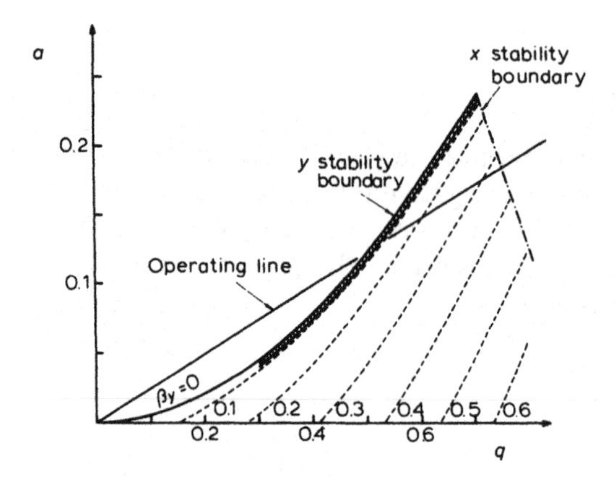

Fig. 2.24. The stability diagram for the monopole spectrometer showing a typical operating line. The shading running alongside the $\beta_y = 0$ boundary represents the possible regions of ion transmission.

a scan line as indicated, for example, in Fig. 2.24 for which $q = 5a$. This not only results in x trajectories of more limited amplitude but also means that the device operates at lower q values and therefore that the voltage requirements are less than in the mass filter [in addition to the factor of 2 expressed in eqns. (2.50) and (2.51), compared with eqns. (2.28) and (2.29)]. In fact, any scan line intersecting the stability triangle can be used, and the conditions for possible ion transmission are roughly represented by the shaded region running alongside the $\beta_y = 0$ boundary. A second advantage over the mass filter is that there is no longer the necessity for extremely accurate control of the U/V ratio at high resolutions, since any scan line is suitable.

Figure 2.25 presents some trajectories in the yz plane for the scan line $q = 5a$ and an ion energy and instrument length such that $n = 12$. Only the 50% of the initial phases for which ions are driven in a positive y direction are shown. Figure 2.25 (a) represents the point of maximum ion transmission, Fig. 2.25 (b) a point during the scan before this maximum is reached, and Fig. 2.25 (c) a point beyond that of maximum transmission. The envelope of the corresponding x trajectories is shown in Fig. 2.26. Contrast the maximum amplitude with that of Fig. 2.11(a).

(2) *Resolution*

Achievement of a good mass resolution will require some limitation of the energy spread of the ion beam entering the analyser. We will reconsider this limitation later and for the moment consider one particular ion energy. As shown in Fig. 2.25, ions of lower masses (higher a, q) than those being transmitted have larger β_y values and will spread out in a spectrum intersecting the V block (at least until the a, q value approaches the x instability

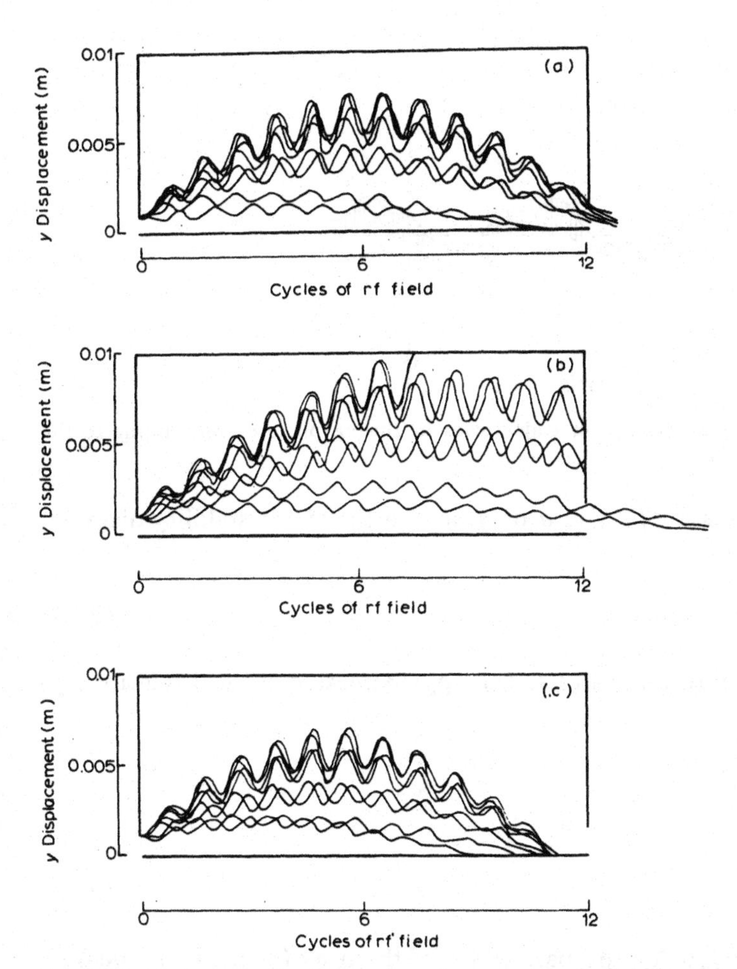

Fig. 2.25. Trajectories in the yz plane for a monopole field in which ions take 12 rf cycles to pass through. The operating line was $q = 5a$. (a) The position of maximum ion transmission $q = 0.425$; (b) for $q = 0.4135$, before the maximum transmission is reached (or for a heavier ion); (c) for $q = 0.4310$, after the point of maximum transmission (or for a lighter ion).

boundary). Ions of higher mass (lower a, q) will have some possibility of being transmitted until the a, q value is near the y stability boundary. To a first approximation, assuming infinite amplitude of oscillation is permissable, one would expect the base width of the peak, and therefore the resolution, to depend on the distance from the y stability boundary to the line $\beta_y = 1/n$. That is

$$R \geqslant \frac{(q_{\beta} + q_{\beta=0})}{2(q_{\beta} - q_{\beta=0})} \tag{2.53}$$

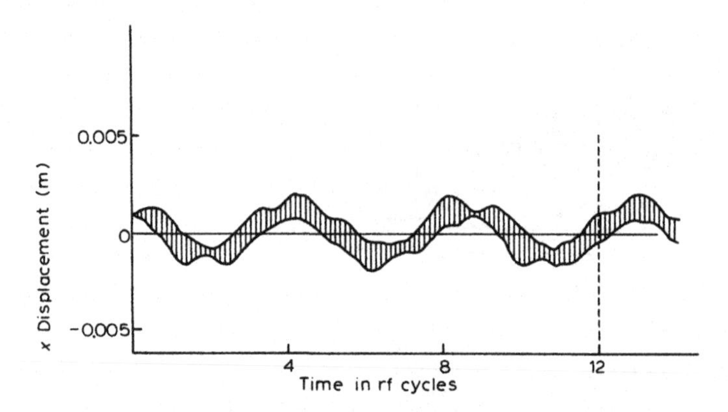

Fig. 2.26. The envelope enclosing ion trajectories in the *xz* plane corresponding to the conditions of Fig. 2.25 (a).

Using eqn. (2.36) relating β_y and a or q, at the tip of the stability diagram, eqn. (2.53) becomes

$$R \geqslant \frac{n^2}{3.3} \tag{2.54}$$

For other scan lines, one can use the approximate empirical relationship [24]

$$q_\beta n^2 = \frac{2}{(q_\beta - q_{\beta=0})} \tag{2.55}$$

to obtain

$$R \geqslant \frac{n^2 q_{\beta=0}^2}{2} \tag{2.56}$$

This gives $R \geqslant n^2/10$ for the base width of the peak for the scan line $q = 5a$. The relationship between R and n has been explored in some detail in computer simulations [23] as we will see shortly (see p. 43). However, some qualitative conclusions can already be drawn. From eqn. (2.56), or from eqn. (2.53) and Fig. 2.24, it is evident that as the slope of the scan line controlled by the U/V ratio is decreased, the resolution for an instrument of a given "electrical length", n, will be expected to deteriorate owing both to the lower q of the boundary and to the greater width between the iso-β lines. A change of scan line can be used to extend the mass range for a power supply of a given maximum output voltage.

(3) *Mass scanning*

As with the mass filter, the monopole can be mass scanned in two ways; amplitude scanning where the frequency is fixed and the magnitudes of U

and V are changed keeping their ratio constant, or frequency scanning using fixed U and V values. The former has been the most commonly used since it is easier to design circuitry for a wide mass range and fast sweep capability. The peaks are approximately equally spaced, with a slight deviation for $M = 1$ and 2 [23]. For a fixed axial energy, the electrical length of the device, n, depends on $M^{1/2}$. Thus, according to eqn. (2.56), $R \propto M$ and the mass spectrum approximates a series of equally spaced peaks of constant half-width. This is useful in applications where simplicity of the output signal is important.

With frequency scanning, the peak positions depend upon 1/(frequency)^2 and, since transmitted ions of all masses remain the same number of rf cycles in the field, the resolution ($M/\Delta M$) is constant, so that the appearance of the spectrum is somewhat analogous to that in a magnetic sector mass spectrometer with magnetic field scanning.

A novel sweep method for extending mass range has been used by Herzog [25]. The first part of the spectrum is scanned in the usual way, varying the magnitude of U and V but maintaining their ratio constant. When the maximum V is reached, it is kept constant while U is decreased, thus changing the mass scan line, and allowing higher masses to be analysed, albeit with a decreased resolution (see Chapter VII).

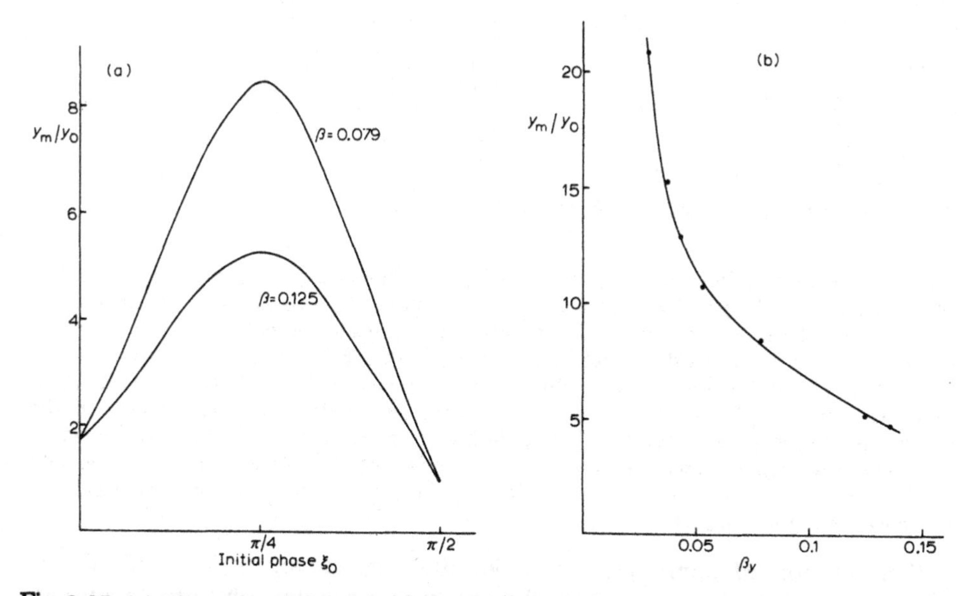

Fig. 2.27. (a) The ratio of the maximum y deflection during one half-beat length to the initial displacement for two points on the $q = 5a$ operating line as a function of the initial phase. (b) The y_m/y_0 ratio at the $\pi/4$ initial phase as a function of β_y for the $q = 5a$ operating line.

(4) *Computed performance*

When one takes into account the finite nature of the allowed y vibrational amplitude, the above discussion has to be modified. Near the stability boundary, where β is small, the maximum y amplitude will tend to be very large for certain initial phases and the usable entrance aperture will be reduced. Ion transmission will be less than 50% even though the ion trajectories are mathematically stable. The important parameter is now y_m/y_0 determined only during the first half-beat length. Figure 2.27 (a) shows two examples of the variation of y_m/y_0 with the initial phase derived from trajectory calculations [23] and Fig. 2.27 (b) shows y_m/y_0 at the $\pi/4$ initial rf phase as a function of β. There is some advantage by comparison with the analogous Fig. 2.13 (a) for the mass filter, partly off-setting the aperture reduction resulting from the use of only one-quarter of the mass filter field.

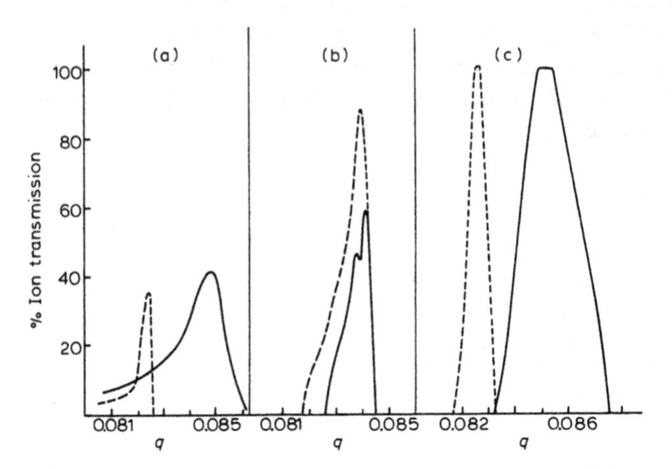

Fig. 2.28. Computed ion transmission curves for the monopole using the $q = 5a$ scan line considering only the yz plane. (a) Ion entry parallel to the axis and exit aperture twice the initial displacement. ———, the ions spend 12 cycles in the field. – – – –, the ions spend 20 cycles in the field. (b) Ion injection at 10^{-3} m but with an initial y velocity of 3×10^{-4} m/rad of the applied field. Ions spend 20 cycles in the field. Assumed exit apertures were ———, 2×10^{-3} m; – – – –, 4×10^{-3} m. (c) Ion injection parallel to the axis but in the presence of a linear fringing field at the analyser entrance. The exit aperture was twice the initial displacement. ———, ions spend 3.5 cycles in the fringing field and 9 cycles in the analyser; – – – –, ions spend 3.5 cycles in the fringing field but 15.75 cycles in the analyser.

Examination of computed ion trajectories, as shown in Fig. 2.25, allows some qualitative conclusions regarding peak shape and resolution. Assuming, as shown in the figure, that ion entry is parallel to the axis and at only one initial y position, the percentage transmission can be determined for a series of points along a scan line for an instrument of a given electrical length, n.

Figure 2.28 shows some examples. The "transmitted" peak shape depends on the limit that is assumed for the maximum possible y displacement or on the aperture that is assumed at the instrument exit. In effect, the finite V block–rod separation serves as a "virtual" exit aperture. As is evident in Fig. 2.25, for ion injection parallel to the instrument axis, Fig. 2.28 (a), one expects a gradual "tailing" of the peak on the low mass side rather than an abrupt cut-off. Furthermore, transmission will be less than 50% and modulated at the frequency of the applied rf. Nevertheless, most monopoles have been operated with apparently parallel ion injection and the tailing has not generally been observed, or has been much less important. Figure 2.28 (b) shows a calculation for injection with a velocity in the y direction of 3×10^{-4} m/radian of the applied field. (That is, 3×10^{-4} m in $1/2\pi f$ sec, where f is the applied frequency in hertz.) This was part of an attempt to solve the dilemma of the computed asymmetric peaks. For a monopole of 0.1 m length, the velocity is equivalent to injection at an angle of 20°. Note that too small an exit aperture may lead to peaks with a central dip.

It is interesting, in the light of Fig 2.28 (a) and (b), that Herzog [26] (see Chapter VII) has found that the sensitivity and background of a monopole can be improved by applying a weak magnetic field near the ion entrance in order to deflect ions away from the central axis.

There still remained to be explained [23] the considerable discrepancy with experimental peak shapes. Measurements of ion transmission as a function of phase [7] have shown a phase-dependent transmission only if the ions spend very little time in passing through the fringing fields. Under the influence of the fringing fields, the phase dependence was no longer observed. This evidence is discussed in Chapter V as is Fig. 2.28 (c) which shows ion transmission calculated taking account of fringing fields, when the anomalous peak "tailing" is no longer predicted.

Although the experimental evidence is as yet fragmentary because of a current lack of research on monopoles, it seems clear that the fringing fields play a vital role in monopole operation. Their presence, while unavoidable to some extent, seems essential to the attainment of high performance.

The influence of geometrical faults in the field is also discussed in Chapter V.

From the calculations of peak shape, one can also examine resolution as a function of n [23]. Only qualitative conclusions are valid because of the crudeness of the model but the relationship $R = n^2/h$ does seem appropriate with h depending on the exit aperture (or rod–V block spacing). The h values are, of course, smaller than the maxima obtained with eqn. (2.56). As expected, for a scan line of lower U/V ratio, h is much larger, corresponding to a degradation of performance. There are different degrees of focussing at the ion exit, depending on β_x and β_y and the instrument length as for the mass filter (see p. 34). The resulting variations in ion transmission efficiencies have been experimentally observed (see Chapter VII).

No detailed computer analysis, taking account of a distribution of initial x and y displacements and simultaneously considering the x and y trajectories, has yet been carried out. Perhaps it would not be useful until a better model of fringing field effects is available (see Chapter V). Furthermore, few detailed experimental studies of monopole performance have been published since von Zahn's original work [22] (see also Chapter VII). The latter did report values of h between 1.5 and 2.25 for scan lines passing near the tip of the stability diagram and an h of about 3.8 for the scan line $q = 4a$. However, other workers [27] have surprisingly reported that a change of the scan line from $q = 5a$ to $q = 12.5a$ produced no loss of resolving power.

Phase-space dynamics techniques have recently been adapted to the calculation of monopole acceptance [27a] and provide a more general view of the instrument's operation.

(5) *Ion energy*

The quadrupole mass filter, as described previously, can accept ions of a wide range of axial energies without degradation of performance, provided that the number of cycles an ion spends in the field, $n, \geqslant h^{1/2}R$. Because of the utilization of the focussing properties as well as the stability properties of the fields in the monopole, it is commonly, and rightly, assumed that the monopole requires a more stringent control of ion energy. However, the requirements are not exacting even for resolution of a few hundred. Again, one has the necessity that $n \geqslant h'^{1/2}R$ where R is the required half-height resolution. In addition, a spread of n values above the minimum will result in some peak broadening. To take a simple example, using Fig. 2.28 (c), the variation in n from 9 to 15.75, an energy variation of a factor of three, would produce a decrease in resolution from about 32 to about 19, assuming all intermediate energies are equally probable. However, a very uneven distribution of energies may produce a peak with "structure" since different energies are transmitted at different points during the scan. Using eqn. (2.55), the above illustration can be expressed more generally. The change in the q value at maximum ion transmission due to a change in ion energy will (near the $\beta = 0$) be approximately given by

$$\Delta q = \frac{2[(1/n_1^2) - (1/n_2^2)]}{q_1} \tag{2.57}$$

where n_1 and n_2 correspond to energies E_1 and E_2. On the other hand, the peak width, corresponding to a single energy E_1 is given by $\Delta q' = hq_1/n_1^2$ from the expression for the resolution. Hence

$$\Delta q/\Delta q' = \frac{2[1 - (n_1^2/n_2^2)]}{hq_1^2} \tag{2.58}$$

For example, for a typical value of h of 4, and the scan line $q = 5a$ for which q_1 is approximately 0.4, this ratio becomes $3[1 - (n_1^2/n_2^2)]$ or $3\Delta E/E$. Thus a 10% spread in energy will produce a broadening of the peaks by about one-third. This is in contrast with a static focussing magnetic sector mass spectrometer where the resolution will be no better than $E/\Delta E$.

(6) *Geometrical variations*

As indicated in Fig. 1.1, there have been several variations proposed on the monopole theme. The most intriguing, involving a better use of the exact focussing capabilities of quadrupole-type fields [24] is described in some detail below. The others are geometric variations rather than different in principle from the simple monopole. There is a "duopole" structure [28] suggested to provide a balanced system of potentials at the output in order to avoid any detector problems due to potentials induced by the oscillating field. There is a proposed four-fold monopole structure, consisting of a central rod and a square-housing, giving four monopoles in parallel [29]. The only practical realization, so far, is the quadruple monopole with a central ground plane cross-electrode and four rods [30], each monopole section being capable of individual tuning for the simultaneous observation different masses (see Fig. 1.1).

(7) *Apparent advantages and disadvantages* (see also Chapter VII, p. 179)

It is perhaps appropriate to finish this section with a summary of the apparent advantages and disadvantages of the monopole in comparison with the quadrupole. "Apparent" because the ultimate achievements of both types of instrument seem to have been so far quite similar. Moreover, the technological effort in monopole design and construction has been quite limited and full measurements of performance in terms of design parameters such as length, power, rod diameter etc. have seldom been made. The monopole is mechanically even simpler than the mass filter, and only a simple single-ended rf power supply is necessary. More important, it is no longer necessary to control the q/a ratio accurately in order to obtain high resolution. A higher q/a ratio may be used, resulting in an extended mass range for a given maximum rf voltage (or, alternatively, giving a reduction in power for a given mass range), in addition to the reduction through the use of a single rod. Furthermore, the monopole may, under well-chosen circumstances, give a higher resolution for a given "electrical" length, n, than the mass filter. The accompanying disadvantages are the potential reduction in aperture by a factor of four (considering a simple geometric factor), probable more stringent restrictions on the initial x conditions to ensure that $|x| \leqslant y$, a possible reduction in maximum ion transmission to 50% of the initial phases (depending on the fringing fields), and a certain dependence of resolution on the initial energy spread. A more serious limitation may be the very strong dependence on fringing fields for the achievement of high performance and the present lack of knowledge of this phenomenon and therefore of its optimum utilization. Since the ion trajectories commence near the V block, one might also suspect a greater sensitivity to electrode contamination, that bane of electrodynamic mass spectrometers. A further limitation, for which evidence is presented in Chapter VII, is the fact that the peak height is not a monotonic function of ion energy and mass peaks may have different ratios if the ion energy is changed. This is presumably due to variations in the quality of focussing at the exit.

D. THE QUADRUPOLE ION TRAP

(1) *Conditions for ion trapping*

The geometry of the quadrupole ion trap, a three-dimensional rotationally symmetric quadrupole field, was shown in Fig. 2.4 and the equipotential lines are illustrated in Fig. 2.29 for the xy and rz planes. The equation for the potential of any point is eqn. (2.12) and the equations of ion motion are eqns. (2.26) and (2.27). One must again consider a combination of ion

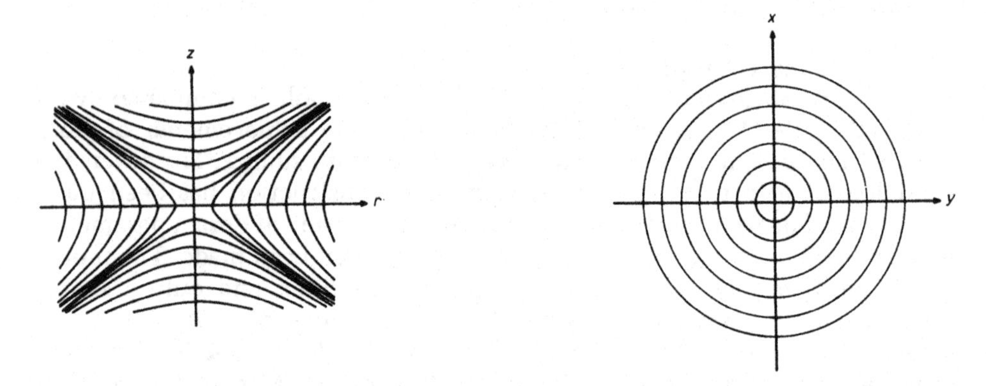

Fig. 2.29. Equipotential contours for the three-dimensional rotationally symmetric quadrupole field. (a) Any rz plane; (b) any xy plane.

motion in two independent directions r and z. The a and q parameters in the Mathieu equation are given by $a_z = -2a_r = -4eU/mz_0^2\omega^2$ [see eqn. (2.28)] and $q_z = -2q_r = -2eV/mz_0^2\omega^2$ [see eqn. (2.29)]. The stability of the trajectory of an ion formed within the ion trap can be readily examined by superimposing two stability diagrams which differ by the factor of -2 and represent the r and z directions. Figure 2.30 (a) shows a general view of the combined stability diagrams and Fig. 2.30 (b) shows in detail the lower region, that usually employed in ion trapping. There is thus a wide range of a, q conditions where ions formed within the quadrupole ion trap might be

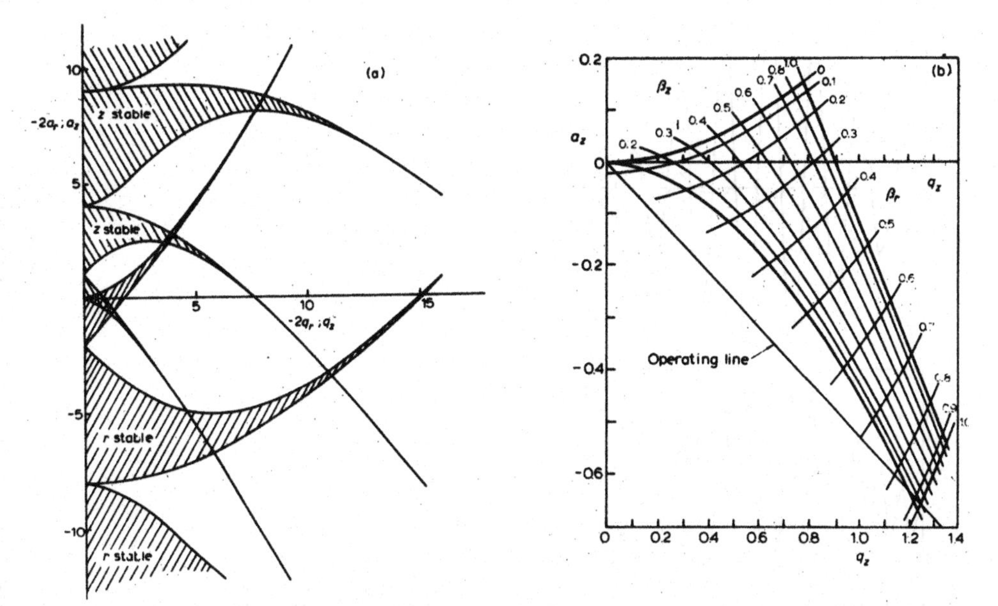

Fig. 2.30. (a) The overall stability diagram for the three-dimensional quadrupole ion trap obtained by superimposing stability diagrams for the r and z directions. (b) The lowest region of stability for the quadrupole ion trap showing some iso-β lines.

expected to have stable trajectories and remain trapped in the device indefinitely unless their trajectories are perturbed by collisions. Of course, as for the mass filter, one must distinguish between the "mathematical" stability as expressed in these figures and the real case where the maximum amplitude of oscillation must remain less than the geometrical boundaries of the device. The significance of stability diagrams and of the β values is discussed earlier in this chapter.

Figure 2.30 (b) is used to analyse the nature of the ion trajectories for any applied voltages U and V, by determining β_r and β_z and considering the motion in the r and z directions independently. However, although this is the commonly presented outline of ion trap operation, the derivation of the equations of motion, eqns. (2.26) and (2.27), involves the tacit assumption [31] that the initial tangential velocity (that is, the initial angular velocity about the axis of symmetry) is zero. To be exact, there are three equations of motion

$$\begin{aligned} \ddot{z} - (2e/mr_0^2)[U - V\cos\omega(t-t_0)]_z &= 0 \\ \ddot{y} + (e/mr_0^2)[U - V\cos\omega(t-t_0)]_y &= 0 \\ \ddot{x} + (e/mr_0^2)[U - V\cos\omega(t-t_0)]_x &= 0 \end{aligned} \tag{2.59}$$

so that one must consider two directions, x and y, which have the same a, q values but different initial conditions. An example is shown in Fig. 2.31 of an ion formed at the position $y = 2$ with the x position of zero and an initial

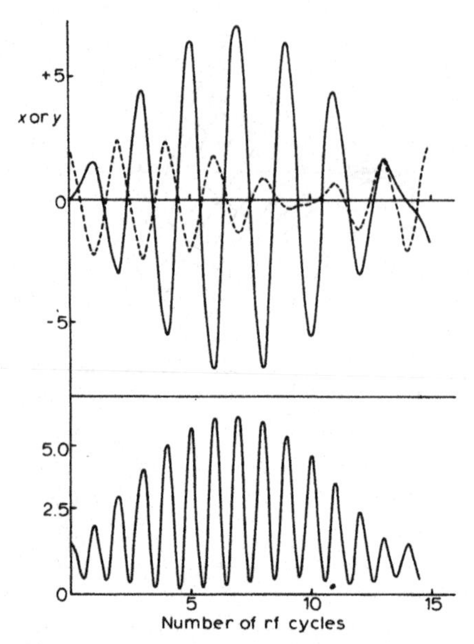

Fig. 2.31. Ion trajectories in the ion trap for $a = -0.65$, $q = 1.25$ to illustrate the case of an initial tangential velocity. The initial y position is 2 units; the initial velocity is 0.5 units/rad. As shown by the calculated radial displacement, r, the ion never passes through the z axis.

x velocity of 0.5 units/radian of the applied field, for the $(a, q)_z$ value of (−0.65, 1.23). The radial displacement $r = (x^2 + y^2)^{1/2}$ is also shown. The physical consequences of a finite tangential velocity become clear on substituting $x = r \sin\theta$ and $y = r\cos\theta$ in eqns. (2.59). After some algebraic manipulations, one obtains

$$\frac{d^2r}{d\xi^2} - r\left(\frac{d\theta}{d\xi}\right)^2 + [a_r - 2q_r \cos 2(\xi - \xi_0)]r = 0 \tag{2.60}$$

and

$$2\left(\frac{dr}{dt}\right)\left(\frac{d\theta}{dt}\right) + r\left(\frac{d^2\theta}{dt^2}\right) = 0 \tag{2.61}$$

Integration of the latter gives $r^2(d\theta/dt) = \text{constant} = \lambda_a$. In other words, this equation represents the conservation of angular momentum λ_a (with respect to the z axis). The extra term in eqn. (2.60) represents the centrifugal force. Equation (2.60) can be re-expressed

$$\frac{d^2r}{d\xi^2} + [a_r - 2q_r \cos 2(\xi - \xi_0)]r = \frac{4\lambda_a}{\omega^2 r^3} \tag{2.62}$$

and it is obvious that if there is an initial velocity in the tangential direction, the ion trajectory *never* passes through the axis of symmetry. However, this digression should not be over-emphasized. For most ionization, the initial velocity may be safely assumed to be close to zero so that we can consider motion in a fixed rz plane. Only for a few fragment ions is the initial velocity appreciable.

The ion trap is an ion source and ion analyser combined. There is no external source. Ions are formed in the trap by an electron (or possibly photon) beam which crosses it. Ion detection has sometimes been external to the device but sometimes intrinsic to it. The different detection methods lead to totally different modes of operation of the ion trap. The external detection is used with mass selective ion storage in the trap. The intrinsic method involves storage of ions of a wide range of masses and a mass-selective detection technique. The two methods are discussed in detail below.

(2) *Mass-selective ion detection*

The original ion detection used by Berkling [32] and Fischer [33] was a resonance absorption technique. Operating conditions were such that several different ionic species were trapped simultaneously but ions of different m/e ratios then have different q values and are oscillating with different characteristic frequencies (see p. 19). The mass-selective detection system uses an auxiliary oscillatory circuit tuned to a chosen fundamental frequency of ion motion. The auxiliary field is usually applied between the two end-caps so that it acts principally in the z direction and is tuned to ω_0^z. This field must be small to prevent ions being too rapidly lost to the walls because of their gain of energy from the field. The power absorption or damping in the auxiliary circuit is a direct measure of the number of ions with the appropriate ω_0^z. The characteristic β_z chosen for ion detection may be in the centre of the stable region where ions are more readily stored.

The equation of ion motion for the z direction in the presence of the auxiliary field is expressed approximately by

$$\frac{d^2 z}{d\xi^2} + [a_z - 2q_z \cos 2(\xi - \xi_0)]z = b \cos(\alpha\xi + \delta) \tag{2.63}$$

where $b = -4V_{RES}/m\omega_0^2 z_0^2$, $\alpha = 2\omega_{RES}/\omega_0$, V_{RES} is the auxiliary voltage of frequency ω_{RES}, and δ is a phase factor. For ion detection, $\alpha = \beta_z$, and the solution of eqn. (2.63) is the unperturbed solution [eqn. (2.32) for example] multiplied by a factor z_{ENV}, which is proportional to the time the resonant field is applied. In fact, the perturbing envelope is given by

$$z_{ENV} = (b/2W)\xi \tag{2.64}$$

where W is the Wronskian determinant for the perturbed Mathieu function. The oscillation therefore has a linearly increasing amplitude owing to the auxiliary field. The resolution capabilities of the detection technique depends upon the behaviour of ions close to, but not at resonance. That is, for $\alpha = \beta$

$\pm(\Delta\beta/2)$ where $\Delta\beta \ll 1$. This results in a beating motion of frequency $\Delta\beta/4\pi$, the trajectory envelope being multiplied by

$$z_{ENV} = (2b/W\Delta\beta)\sin[(\Delta\beta/4)\xi] \tag{2.65}$$

of which eqn. (2.64) is a particular case. The maximum beat amplitude is $2b/W\Delta\beta$. Ions close to resonance may gain sufficient amplitude during the beat to reach the electrodes. Such ions will cause a net absorption of energy from the auxiliary field and will limit the resolving power.

Using the simple assumption that the ions have the maximum amplitudes of their unperturbed trajectories uniformly distributed in the z direction, half the ions will be lost when $z_{ENV} = z_0/2$. The resolution at half-height will correspond to $\Delta\beta_{1/2} = 4b/z_0W$. For ions exactly at resonance and beginning at $z = 0$, one can define a characteristic time for ion loss, τ_1 given by

$$\tau_1 = \frac{4Wz_0}{b\omega} \tag{2.66}$$

Hence

$$\Delta\beta_{1/2} = \frac{16}{\omega\tau_1}$$

and

$$\begin{aligned}\frac{\beta_{RES}}{\Delta\beta_{1/2}} &= \frac{\omega_{RES}\tau_1}{8} \\ &= \frac{\pi}{4}\frac{\tau_1}{\tau_{RES}}\end{aligned} \tag{2.67}$$

where τ_{RES} is the period of the resonating field. The relationship between $\beta/\Delta\beta$ and $M/\Delta M$ depends on the chosen operating point in the stability diagram. For example, Fischer [33] deduced from the stability diagram that for $\beta_z = 0.6$ and $a = 0$, $M/\Delta M = 0.84\ \tau_1/\tau_{RES}$.

The resolving power is thus proportional to the number of resonance periods the ions at resonance can undergo before being lost. If the amplitude of V_{RES} is decreased, the resolving power increases. An upper limit is reached when the loss of ions from the resonating condition is not due to collection at the z electrode but due to collisions. There is no advantage in making τ_1 greater than the mean lifetime of the ions. At higher pressures, the maximum resolution will be limited by ion–neutral collisions, increasing with a decrease in pressure until ion–ion collisions become the limiting factor. (See Chapter VIII for a discussion of ion scattering.)

In Fischer's method of operating the ion trap, the electron beam was in continuous operation so that there was a continuous formation and loss of ions. He chose the detection frequency to give $\beta_{RES} = 0.6$. The mass spectrum was scanned by using a fixed alternating voltage, V, and applying a

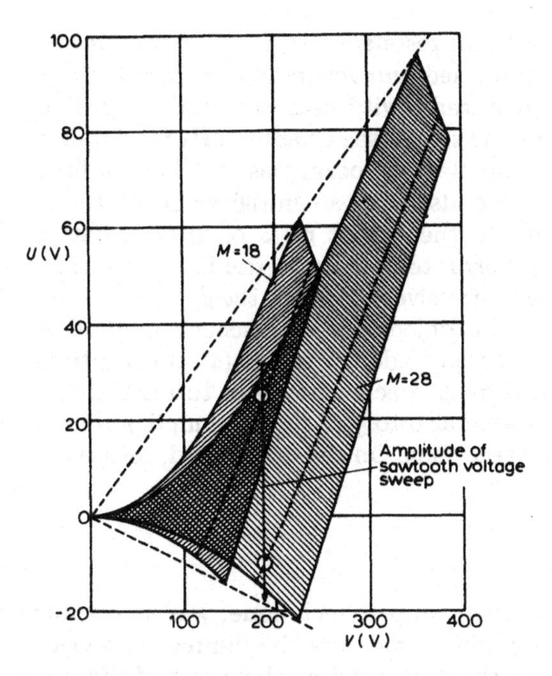

Fig. 2.32. Ion trap stability diagrams for $M = 18$ and $M = 28$ shown in U–V space to illustrate the application of the sawtooth voltage to sweep the different masses through the $\beta_z = 0.6$ line.

slowly varying sawtooth signal as part of the d.c. voltage on the ring electrode. (See Chapter VIII for experimental details.) Figure 2.32 shows two stability diagrams, one for $M = 18$ and one for $M = 28$, given in U–V space, to illustrate the application of the sawtooth voltage to vary U and sweep the different masses through the $\beta_z = 0.6$ line. This scanning technique is of quite limited mass range. A more serious drawback is that conditions at detection are not equivalent for ions of different m/e ratios. The r trajectories will be quite different in the case of $M = 18$ and $M = 28$ as illustrated above. Moreover, the $M = 18$ is not stored during measurements of $M = 28$ but $M = 28$ is stable during the measurement of $M = 18$. Ion–ion interactions (scattering) may influence the observed results or if one species is particularly abundant, the space charge due to its presence may minimize trapping of other ionic species. Despite these limitations, Fischer obtained quite a promising performance from the ion trap (see Chapter VIII). However, this detection method has subsequently been neglected. The resonance technique might well be usefully applied to studying ion–molecule reactions of trapped ions including, for example, the elimination of all but a chosen primary ion.

A somewhat different mass-selective detection technique was later used by Rettinghaus [34]. Ions were detected by the voltage that they induced between the end-cap electrodes at a chosen frequency. These electrodes formed part of a balanced circuit. Ions in resonance at $\beta_z = 0.5$ produced a voltage difference which could be amplified and accurately measured with a phase-sensitive detector. A very great sensitivity was achieved, four ions being claimed as the minimum detectable signal (see Chapter VIII for details). The trap was operated with $a = 0$. The electron beam was switched on for a fixed ionization time, usually five seconds, with an initial value of the rf amplitude, V, chosen to correspond to the lowest mass of interest having $\beta_z = 0.5$. With the electron beam interrupted, the magnitude of V was increased to bring ions of higher mass successively through the $\beta_z = 0.5$ point. This method suffers from many of the objections to Fischer's method as described above. Different ions are stored under different (a, q) conditions and the simultaneous storage of different species may lead to interference effects. The very long storage times tend also to give rise to complications in the spectra because of ion–molecule reactions. On the other hand, sensitivity and resolution were excellent.

(3) *Mass-selective ion storage*

A later development, and yet a much simpler technique, was to operate the ion trap to store mass-selectively and to measure the number of a given species after an appropriate storage period by ejecting them out of the trap through holes in an end-cap electrode into an electron multiplier [35, 36]. The mass-selective storage is entirely analogous to the operation of the mass filter. A scan line passing near the tip of the stability diagram is used (see Fig. 2.30). The more acute tip which corresponds to a positive potential on the ring electrode has generally been used. The spectrum is usually scanned by varying the magnitudes of U and V but keeping their ratio constant. When one ionic species is in the stable region, ions of a lower m/e (higher a, q) have trajectories unstable in the r direction. Meanwhile, ions of higher m/e (lower a, q) have trajectories unstable in the z direction. To avoid complications due to the rejection of these ions from the trap and their measurement in the electron multiplier, the detection circuitry is gated to accept only the signal due to the stored ions pulsed from the trap. Figure 2.33 shows a typical timing diagram. The electron beam, which may be directed radially or axially (the former is preferable at higher pressures), passes through the centre of the trap for an ionization period chosen according to the pressure range of interest. A time of 1 msec would be typical for a pressure in the 10^{-7} torr range. The electron beam is interrupted to allow for completion of the ion sorting process and the consequent diminution of the background signal. After an appropriate delay, usually 25 or 50 μsec, a negative pulse is applied to the end cap nearest the electron multiplier to eject the ions. Slightly later, a pulse in the detection circuitry, opens a "gate" to allow measurement of the magnitude of the amplified ion pulse by a peak-reading circuit (see Chapter VIII).

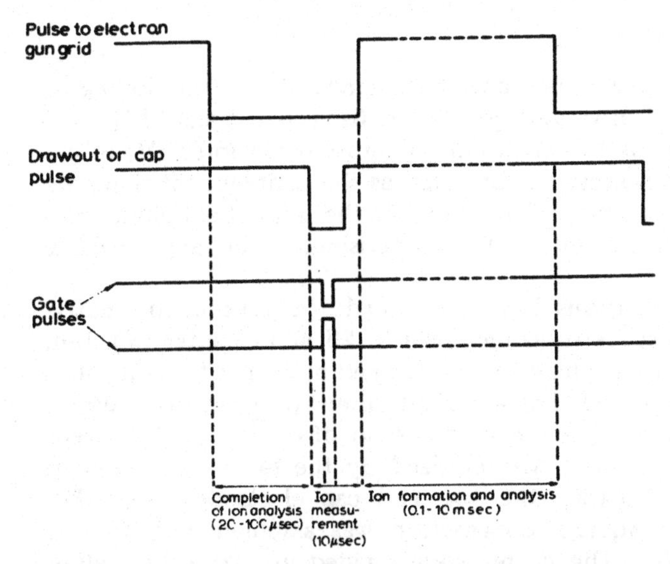

Fig. 2.33. The timing diagram for mass selective operation of the ion trap as a residual gas analyser. The electron beam forms ions and those of a chosen *m/e* are stored. The beam is interrupted to allow ion sorting to be completed. The cap pulse ejects the ions into the detector and the gate pulses open the detection circuitry.

Each mass peak during a scan consists of a train of ion pulses which may be converted to a conventional display by the peak-height reading circuit or by a boxcar detector (see Chapter VIII, p. 189).

The advantages of the techniques are that every ion is treated identically (in terms of the a, q diagram) and interference effects are largely avoided by trapping each ionic species individually. The measurement is facilitated by the integration of the ion formation rate during the storage time and the time of integration can be chosen appropriately to the pressure or the ion formation rate.

The operation of the ion trap has been further refined by synchronizing the ion pulse out [37] and/or the electron beam entrance with the phase of the applied rf field. The latter requires, of course, a very short ionization time [38, 39] and is only feasible when the applied frequency is not too high. However, it is possible to use a lower frequency in the ion trap than in the mass filter because the number of cycles the ions spend in the field is no longer limited by the analyser length. Synchronization of the beginning of the ion pulse-out at the optimum rf phase is essential for reducing noise due to signal variations when this phase is random. When the phase is badly chosen, it has been shown by computer simulation that ions in the trap distant from the multiplier end may leave the trap in the opposite direction to that intended. This odd result has been confirmed by experimental measurements [37].

(4) *Computer simulations*

A number of theoretical analyses have been made of the functioning of the ion trap with mass selective storage. The earlier calculations [40] used the numerical integration of the equations of motion (Chapter IV) to examine individual ion trajectories. Factors such as the containment time, the initial r and z distributions, and the influence of the initial field phase were separately evaluated but could not be considered in any realistic combination.

Recently, detailed calculations have been made [41] using the matrix methods based on phase-space dynamics (see p. 28 and Chapter IV). Ions were considered to be equally likely to be formed at all positions through which the electron beam passed. For a radially directed beam, the assumed beam cross-section was rectangular having a ratio of $\sqrt{2}(= r_0/z_0)$ between the height and the breadth and characterized by the ratio ρ between its height and the distance between the two end cap electrodes ($2z_0$). For axially directed beams,· a square cross-section was assumed with ratio ρ between its width and $2r_0$. The beams were divided into 200 slices along their length and each slice sub-divided into 1600 possible initial positions of ion formation. Initial ion velocity was assumed to be zero. Twenty different initial field phases were examined.

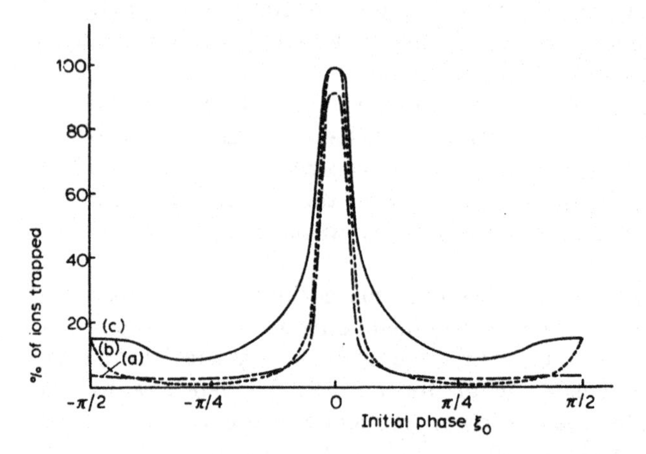

Fig. 2.34. Calculated trapping efficiencies as a function of the initial phase for a radially directed electron beam for three points along the $q = 1.89a$ operating line. (a) $q = 1.247$; (b) $q = 1.204$; (c) $q = 1.215$.

Figure 2.34 shows the trapping efficiencies as a function of the initial phase for three points along the operating line $q = 1.89a$, which gives a mass resolution of 28. The beam with $\rho = 0.05$ was directed radially and because of the wide range of initial r displacements, the trapping efficiency curves

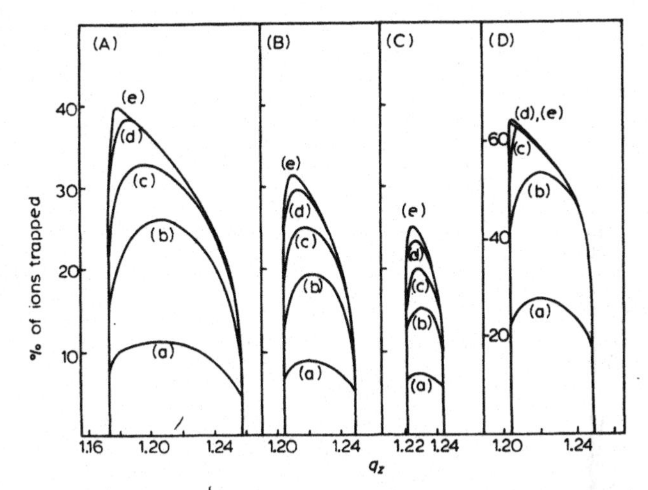

Fig. 2.35. Overall trapping efficiencies for different positions along various operating lines. The assumed radially directed electron beams have widths characterized by ρ equal to (a) 0.5, (b) 0.2, (c) 0.1, (d) 0.05, and (e) 0.033. The electron beams entered only during the positive portion of the rf cycle for (A), (B), and (C) and only during the most favourable one-tenth of the cycle for (D).

largely reflect the characteristic phase dependence of the r trajectories. The importance of the zero phase is more pronounced than in the mass filter (Fig. 2.20). Overall trapping efficiencies can be obtained from the area under the curves. In normal practice, the potentials of electron guns are such that the electron beams enter the trap only during the positive part of the rf cycle. Calculated trapping efficiencies under these conditions are given in Fig. 2.35 (A), (B), and (C) for operating lines that give resolutions of 14, 28, and 54, respectively. The peaks illustrated are for ρ equal to (a) 0.5, (b) 0.2, (c) 0.1, (d) 0.05 and (e) 0.033. At low resolution the peaks exhibit some of the asymmetry suggested by earlier calculations [40]. Evidently, the more the ion formation is confined to the centre of the field, the greater the trapping efficiency. Note, however, that if the electron beam current can be increased as its cross-section is increased, so as to maintain the same beam density, there is no advantage in confining the beam to the centre. In fact, the total number of ions trapped would increase as the beam cross-section was increased (since ρ^2 increases more rapidly than the trapping efficiency decreases). The peaks are also more symmetric with larger diameter beams. Of course, at high pressures, the increased background signal from rejected ions might cause other difficulties. Figure 2.35 (D) shows the trapping efficiency when the electron beam entry is limited to 10% of the rf cycle centred about the favourable zero phase. Only the ions formed at very large radial displacements are then lost from the trap. A comparison for (i) radially, and (ii) axially [41, 42] directed electron beams with ion formation occurring at all initial phases is shown in Fig. 2.36. The peak asymmetries are in opposite

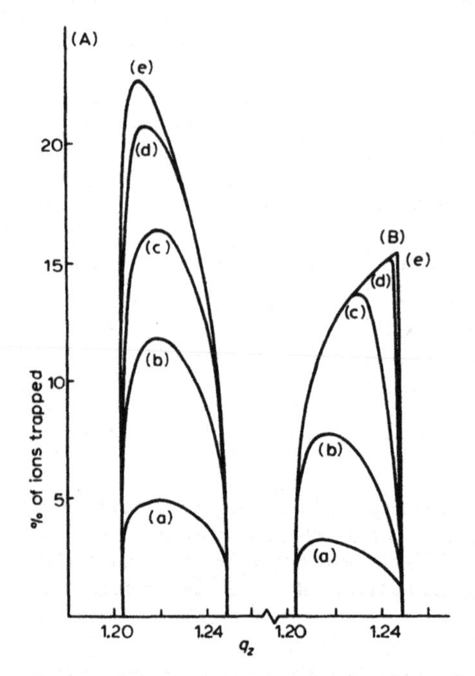

Fig. 2.36. A comparison of (A) radially and (B) axially directed electron beams with ion formation summed over all initial phases. The assumed beam diameters are as given in Fig. 2.35.

directions. The trapping efficiencies are greater for the radially directed beams since $(a, q)_r$ values are smaller and maximum amplitudes of oscillation tend to be smaller in the r direction. A greater fraction of the beam is in a useful region.

Calculations have also been made on the influence of initial ion velocities [41]. These can be the appreciable thermal velocities for light ions such as H_2^+ or He^+ or the initial velocities with which some fragment ions are created. The difficulty of trapping such ions has been invoked to explain some apparent differences in fragmentation patterns [36]. The calculations confirm the likelihood of this explanation in the case of small ion traps used for partial pressure analysis.

Some calculations have also been made for ion storage in cylindrical rather than hyperbolic traps which may have some advantages in certain spectroscopic experiments [43].

(5) *Novel features*

In principle, the advantages that the ion trap offers are (a) the ability to integrate the ion formation rate, (b) the ability to store ions for many rf cycles, thereby avoiding those limitations imposed by length in the quadrupole mass filter. In applications where rf power is important, this feature could be used to reduce the applied rf frequency, thereby greatly reducing rf power, which depends on f^5, and (c) the compactness of the device. These would suggest application to analytical problems requiring high sensitivity and/or high resolution. However, in practice, perhaps because of field imperfections, these hopes have not yet been realized. The success of the ion trap thus far has been in the development of an exceedingly simple device of low performance. In fact, a mesh version is perhaps the simplest imaginable partial pressure analyser [40] as regards device geometry and construction (see Chapter VIII). The ion trap has also found a number of applications in chemical physics (see Chapter XII) but remains relatively unexploited as a ion collection and signal integration device despite ion retention times of several days under appropriate conditions. A further application to mass spectrometry has been as an ion storage source for the mass filter [44] (see Fig. 1.1 and Chapter VIII). The long-term ion storage that is possible with $a = 0$ is used, the ions being subsequently mass-analysed by injection into a mass filter. The ion trap can thus act as a novel chemi-ionization source for the mass filter [45].

E. EXACT FOCUSSING DEVICES

As mentioned earlier (p. 34), quadrupole fields give exact focussing after $2d$ cycles when $\beta = p/d$ where p and d are integers having no common factors. Both the functions in eqn. (2.32) are then periodic. If p is even, the period is πd (in terms of ξ). If p is odd, the functions are inverted after d cycles. For (a, q) values which satisfy the conditions $\beta_{a,q} = p/d$ and $\beta_{-a,-q} = 1/d$, both x and y directions will form exactly focussed inverted images after d rf cycles. Figure 2.37 shows some trajectories calculated by Lever [24] for the point $a = 0.233982$, $q = 0.704396$ which has $\beta_x = 19/20$ and $\beta_y = 1/20$.

The imaging properties of dynamic quadrupole fields have not yet found application in the mass filter geometry (see Sect. G for a static imaging device). The monopole was an attempt to use partial focussing in the y direction and von Zahn [22] mentioned the possibility of using simultaneous focussing in both x and y directions. However, Lever [24] pointed out the difficulties of utilizing this feature in a conventionally operated monopole with ion entry parallel to the z axis since the image is formed with y negative. He suggested using ion entry on the z axis but at an angle to it in both x and y directions so that trajectories would be similar to those in Fig. 2.37. The ions would be injected and would exit through small apertures in the V block [see Fig. 1.1 (f)], which is the ground plane, thereby minimizing fringing field effects.

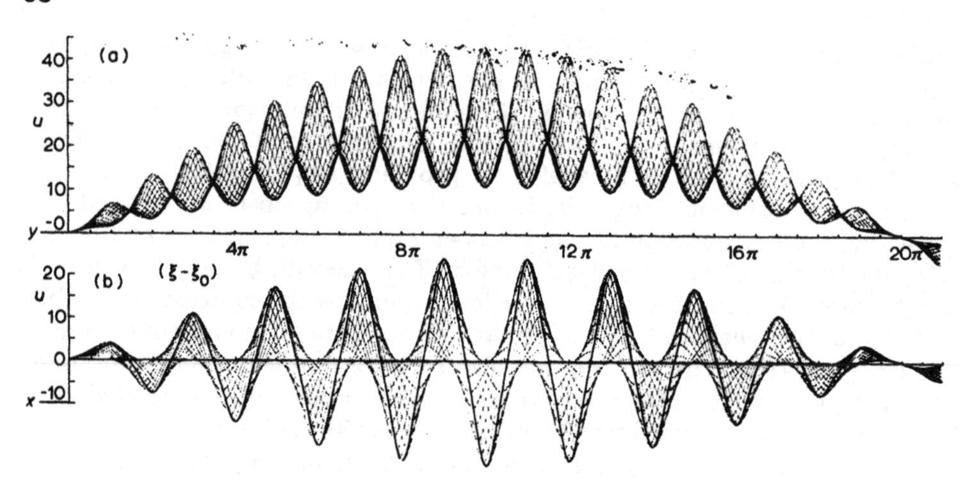

Fig. 2.37. Trajectories for the xz and yz planes in an exact focussing monopole for the point $a = \pm 0.233982$, $q = 0.704396$, where $\beta_x = 19/20$ and $\beta_y = 1/20$. Ion injection is on axis but the initial velocities $dx/d\xi$ are unity.

Lever made a computer study of such a device. The table in Appendix C tabulates his calculated points of double focussing. The angle of injection should be the maximum permitted by the rod—V block spacing. An approximate guide is that the undeflected ion beam would strike the rod about three-quarters of the way down the tube.

Since all ions must spend an exact number of cycles within the field in order to be focussed on the exit, the ions of a particular mass should be monoenergetic and achievement of high resolution might require an energy selector ahead of the spectrometer. To ensure that ions of different masses spend the same number of cycles in the field, frequency scanning would be necessary.

The attainable resolution depends on the variation in the position of focus with a variation in (a, q) along the scan line. The results of some of Lever's calculations are shown in Fig. 2.38 (a) and (b), giving the details of the ion focussing near 10 rf cycles, i.e. for $\beta_y \simeq 1/10$ and $\beta_x \simeq 9/10$. Considering only the y direction, it can be shown that the distance of the focussing point, z_f, for a particular mass and a *fixed* applied frequency, is proportional to the number of cycles n and the square root of the axial energy. However, for a fixed frequency, the energy must be varied with $(m/e)^{1/2}$ (i.e. $q^{1/2}$) in order to keep n constant. Thus $z_f \propto nq^{1/2}$. Lever also established the empirical relationship quoted earlier [eqn. (2.55)], viz.

$$n^2 q = \frac{2}{(q - q_0)}$$

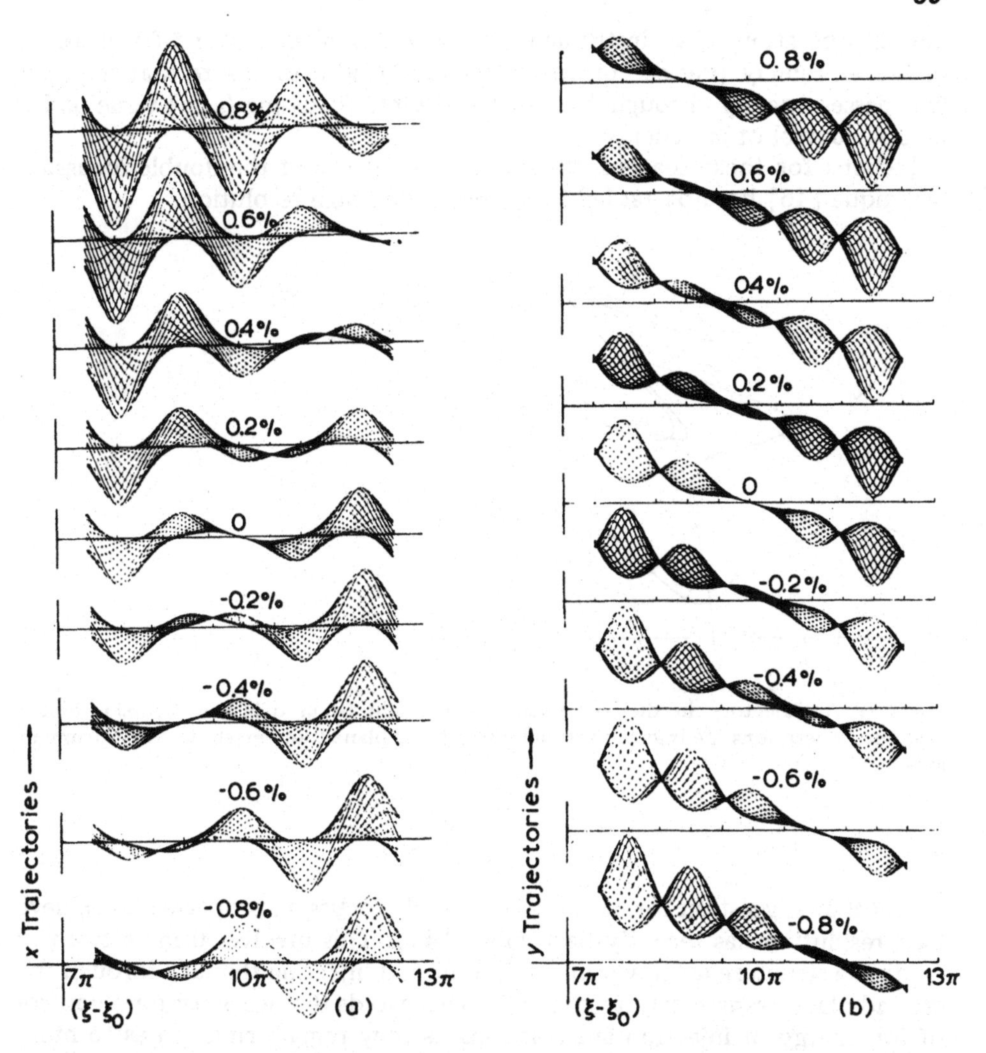

Fig. 2.38. (a) The effect of varying q at constant a/q on the quality and position of focus for the x trajectories. The variations from the double-focussing condition $a = \pm 0.225041$, $q = 0.699745$ are indicated on each curve. (b) Similar results for the y direction.

which, substituting and differentiating, gives

$$\frac{\Delta z_f}{z_f} = \left(\frac{qn}{2}\right)^2 \frac{\Delta M}{M} \tag{2.68}$$

For example, assuming $z_f/\Delta z_f = 100$ is a practical value (giving sufficient instrument aperture for a reasonable sensitivity), and using $p = 19$, $d = 20$ so that $q \simeq 0.7$, then the attainable resolution $(M/\Delta M)$ is 10^4. Thus, the use of the focussing principle seems to offer resolutions so far unattainable in

normal operation of an instrument of conventional size. The difficulties lie in the extremely precise control of the (a, q) ratio to ensure that the scan line passes exactly through the chosen double focussing (a, q) value and a precise control of ion energy.

Perhaps for these reasons, the only reported use of the double-focussing technique [46] has not yet led to the hoped for high resolution.

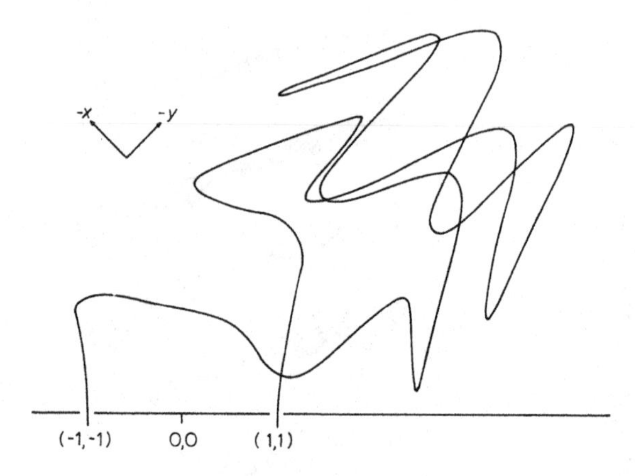

Fig. 2.39. A trajectory in the x–y plane which might be used in the focussing dipole structure where ions are injected through the ground plane transversely to the instrument axis.

Another suggestion for utilizing the double-focussing properties to achieve high resolution has recently been made [47]. This involves using a two-rod or dipole structure as shown in Fig. 1.1(n) and injecting the ions through the ground-plane transversely to the rods. This avoids the necessity for a control of ion energy or injection angle as long as they remain small so as to minimize the maximum ion displacement. Fringing field effects are also largely avoided. Rectangular slits can be used to increase sensitivity. A typical trajectory which might be used is shown in Fig. 2.39, for $\beta_y = 1/10$, $\beta_x = 5/10$. The device has a further potential advantage over conventional devices in that it breaks away from the requirement of very high alternating voltages for high masses. The frequency can be arbitrarily lowered to obtain a convenient voltage for a particular mass. This is impossible in a conventional device since to conserve a given number of cycles within the field, the instrument must be lengthened if the frequency is lowered. Calculations suggest, in principle, the possibility of a spectrometer with a resolving power of 15,000 and mass range up to 5000. The problems are likely to lie in achieving a sufficiently perfect field and a sufficiently exact control of the (a, q) ratio.

F. THE OSCILLATORY FIELD TIME-OF-FLIGHT SPECTROMETER

A time-of-flight mass spectrometer using electrodynamic fields has recently been developed [48, 49] and holds considerable promise for future development. The fields suggested are closely related to those in quadrupole mass spectrometers so that they are appropriately considered in this book. This is done in detail in Chapter IX. Since one is no longer concerned with the stability of ion trajectories but solely with the time-of-flight, the theoretical treatment is essentially independent of that given in the other chapters. Chapter IX is therefore almost complete in itself. Here we will merely outline the general principle.

Figure 9.7 (p. 234) shows a schematic of an apparatus where the principle has been tested and some geometric variations are shown later [46]. The potential distribution in the oscillatory field region approximates the relationship

$$V = -\frac{1}{L^2 - R_1^2/2}(U + V\cos\omega t)\left[z^2 - \frac{(x^2+y^2)}{2}\right] \tag{2.69}$$

Ions are formed by a pulsed electron beam and injected in the z direction at the position of zero field, and then return after a time t_j to be measured by the electron multiplier. The arrival times at $z = 0$ depend on the charge-to-mass ratio.

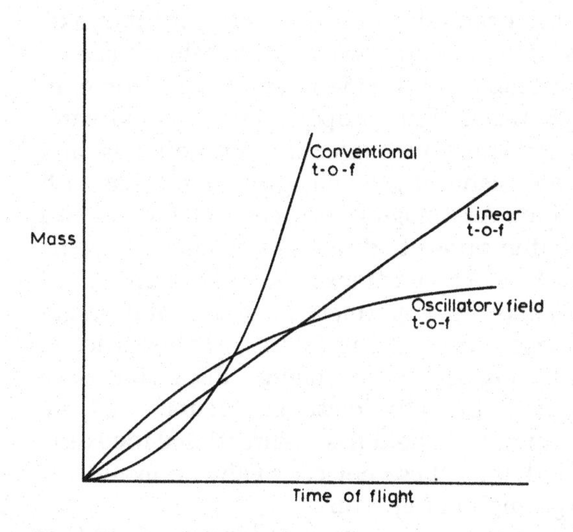

Fig. 2.40. A qualitative comparison of the flight time as a function of m/e for various types of time-of-flight spectrometers.

Considering the z direction motion, a general solution to the equation of motion can be written

$$z = z_0 G_1(a_z, q_z, \xi, \xi_0) + \dot{z}_0 G_2(a_z, q_z, \xi, \xi_0) \tag{2.70}$$

where z_0 and $\dot{z}_0$ are the initial position and velocity. The arrival time for the application we are considering is given by $t_j = 2\xi_j/\omega$ where ξ_j is the value (other than ξ_0) where G_2 first becomes zero. For fixed field parameters and a given phase, t_j depends only on the m/e ratio. It is independent of the initial ion energy, which should be a very valuable property. The general behaviour of the arrival time as a function of m/e is shown in Fig. 2.40 in comparison with other time-of-flight spectrometers [49]. There is potentially a greater mass separation at higher masses than in the conventional instruments.

In choosing the field parameters, a decisive factor is the dependence of G_2 on the initial phase ξ_0. Conditions can be found where ion bunching occurs. That is, the spread in the arrival times for a given ionic species is less than the spread in injection times. Therefore, the peak broadening normally caused by the finite spatial distribution of ions at the time t_0 can be minimized.

Details of the theory of the devices, the search for optimum operating conditions, and the experimental achievements are given in Chapter IX.

G. STATIC QUADRUPOLE DEVICES

Finally, in this section it seems appropriate to outline a few static quadrupole devices. The most intriguing of these is an electrostatic version of the mass filter where the quadrupole has been uniformly twisted about its z axis [50, 51] [see Fig. 1.1(p)]. It has been experimentally verified that it can be used to stably confine charged particles with a wide range of injection conditions and for any charge-to-mass ratio. With properly chosen field parameters, there is an image, reproducing the injection coordinates and momenta. In general, the coordinate plane of the image will be rotated with respect to the xy plane at injection. An imaging lens constructed of two equal sections of quadrupole twisted in opposite directions produces imaging with no rotation. In effect, one has then a beam transportation system.

Static quadrupole fields of the cylindrically symmetric and the rotationally symmetric types have also been suggested as energy filters. An illustration is given in Fig. 1.1 (e). The focussing properties and energy dispersion have been calculated. The cylindrical field has unidirectional focussing of the second order and the rotationally symmetric field has bi-directional focussing of the second order. It is suggested that these devices might be useful for energy selection of beams with large angles of aperture.

Charged particle motion in a static two-dimensional quadrupole field (i.e. cylindrical) with a uniform magnetic field applied in the axial (z) direction has been studied theoretically [53]. No immediate application is evident but the possibility of a mass spectrometer comparable to the trochoidal focussing instrument has been suggested [52]. The static quadrupole trap with a magnetic field applied in the z direction to ensure radial confinement is well-known [54] (see also p. 193). Finally, while considering the use of magnetic fields, ion motion in an oscillating solenoid field [Fig. 1.1(o)] can be represented by the Mathieu equation and could be used in mass analysis [55, 56] but this does not seem advantageous.

REFERENCES

1 R.F. Post, Univ. of Califonria Radiat. Lab. Rep. U.C.R.L. 2209, 1953.
2 P.H. Dawson, J. Vac. Sci. Technol., 11 (1974) 1151.
3 W. Paul, H.P. Reinhard and U. von Zahn, Z. Phys., 152 (1958) 143.
4 U. von Zahn, Diplomarbeit, University of Bonn, 1956, quoted in ref. 3.
5 W.M. Brubaker, 16th Annual Conf. Mass Spectrom., Pittsburgh, 1968 (not generally available).
6 E. Fischer, quoted in ref. 3, p. 148.
7 (a) P.H. Dawson, Int. J. Mass Spectrom. Ion Phys., 6 (1971) 33; (b) P.H. Dawson, J. Vac. Sci. Technol., 9 (1972) 487.
8 R. Nagarajan and P.K. Ghosh, Int. J. Mass Spectrom. Ion Phys., 12 (1973) 79.
9 P.H. Dawson, Int. J. Mass Spectrom. Ion Phys., 17 (1975) 423.
10 P.H. Dawson, Int. J. Mass Spectrom. Ion Phys., 14 (1974) 317.
11 D. Lefaivre, M. Sc. Thesis, Université Laval, 1973; D. Lefaivre and P. Marmet, Rev. Sci. Instrum., 45 (1974) 1134.
12 P. Marmet, J. Vac. Sci. Technol., 8 (1971) 262.
13 P.H. Dawson and N.R. Whetten, Dynamic Mass Spectrometry, Vol. 2, Heyden, London, 1971.
14 W.M. Brubaker, Advan. Mass Spectrom., 4 (1968) 293.
15 A.E. Holme and W.J. Thatcher, Int. J. Mass Spectrom. Ion Phys., 10 (1972/73) 271.
16 M. Mosharrafa and G.M. Wood, 19th Annu. Conf. Mass Spectrom., Atlanta, 1971 (not generally available).
16a P.H. Dawson, to be published.
17 R.W. Fredericks and R.H. Abramson, Final Rep. N66-15618, NASA Contract NASW-1022, 1965.
18 W.M. Brubaker, 19th Annu. Conf. Mass Spectrom., Atlanta, 1971 (not generally available).
19 N. Ierokomos and M.R. Ruecker, NASA Contract. Rep. NAS5-11185, 1971.
20 J.A. Richards, R.M. Huey and J. Hiller, Int. J. Mass Spectrom. Ion Phys., 12 (1973) 317.
21 U. Brinkmann, Int. J. Mass Spectrom. Ion Phys., 9 (1972) 161.
22 U. von Zahn, Rev. Sci. Instrum., 34 (1963) 1.
23 (a) P.H. Dawson and N.R. Whetten, Rev. Sci. Instrum., 39 (1968) 1417; (b) P.H. Dawson and N.R. Whetten, J. Vac. Sci. Technol., 6 (1969) 97.
24 R.F. Lever, IBM J. Res. Develop., 10 (1966) 26.
25 R.F. Herzog, NASA Tech. Brief TSP 70-10057, 1970.
26 R.F. Herzog, Rev. Sci. Instrum., 40 (1969) 1104.
27 R.E. Grande, R.L. Watters and J.B. Hudson, J. Vac. Sci. Technol., 3 (1966) 329.
27a R. Baribeau and P.H. Dawson, to be published.
28 W.M. Brubaker and F.B. Wiens, U.S. Pat. 3, 418, 464 (1968).
29 J.A. Richards, R.M. Huey and J. Hiller, Int. J. Mass Spectrom. Ion Phys., 10 (1973) 486.
30 F. Aulinger and H. Trinks (a) Int. J. Mass Spectrom. Ion Phys., 7 (1971) 92; (b) F. Aulinger and H. Trinks, Int. J. Mass Spectrom. Ion Phys., 10 (1973) 481.
31 M. Baril and A. Septier, Rev. Phys. Appl., 9 (1974) 525.
32 K. Berkling, Diplomarbeit, University of Bonn, 1956.

33 E. Fischer, Z. Phys., 156 (1959) 1.
34 G. Rettinghaus, Z. Angew. Phys., 22 (1967) 321.
35 P.H. Dawson and N.R. Whetten, J. Vac. Sci. Technol., 5 (1968) 11.
36 P.H. Dawson, J. Hedman and N.R. Whetten, Rev. Sci. Instrum., 40 (1969) 1444.
37 P.H. Dawson and C. Lambert, Int. J. Mass Spectrom. Ion Phys., 14 (1974) 339.
38 C.S. Harden and P.E. Wagner, 19th Annu. Conf. Mass Spectrom., Atlanta, 1971 (not generally available).
39 E.P. Sheretov, V.A. Zenkin and V.F. Samodurov, Sov. Phys. Tech. Phys., 18 (1973) 262.
40 P.H. Dawson and N.R. Whetten, J. Vac. Sci. Technol., 5 (1968) 1.
41 P.H. Dawson and C. Lambert, 16 (1975) 269.
42 E.P. Sheretov and V.A. Zenkin, Sov. Phys. Tech. Phys., 17 (1972) 160.
43 M-N. Benilan and C. Audoin, Int. J. Mass Spectrom. Ion Phys., 11 (1973) 421.
44 R.F. Bonner, G. Lawson and J.F.J. Todd, Int. J. Mass Spectrom. Ion Phys., 10 (1972) 000.
45 R.F. Bonner, G. Lawson and J.F.J. Todd, J. Chem. Soc. Commun., (1972) 1179.
46 W. Fock, 1st Australian Conf. Mass Spectrom., 1971 (private communication).
47 P.H. Dawson, Int. J. Mass Spectrom. Ion Phys., 12 (1973) 53.
48 J.P. Carrico, L.D. Ferguson and R.K. Mueller, Appl. Phys. Lett., 17 (1970) 146.
49 J.P. Carrico, Dynamic Mass Spectrometry, Vol. 2, Heyden, London, 1971.
50 E.A. Youssef, F.E. Vermeulen and F.S. Chute, Can. J. Phys., 49 (1971) 2651.
51 F.E. Vermeulen, F.S. Chute and E.A. Youssef, Can. J. Phys., 52 (1974) 379.
52 C. Schmidt, Rev. Sci. Instrum., 41 (1970) 117.
53 H. Portisky and R.P. Jerrard, J. Appl. Phys., 23 (1952) 928.
54 H. Kleinpoppen and J.D. Schumann, Z. Angew. Phys., 22 (1967) 152.
55 P. Kubicek, Czech. J. Phys., B20 (1970) 475.
56 P.H. Dawson, unpublished calculations.

CHAPTER III

ANALYTICAL THEORY

P.H. Dawson

Chapter II was a general introduction to the principles of operation of quadrupole mass spectrometers. It included a brief account of the equations of ion motion and their stability properties. This chapter is intended for those readers who, by necessity or inclination, wish to explore further the mathematical background. There are a number of detailed mathematical texts concerning the Mathieu equation [1–4]. Here the attention is focussed upon the lowest stability region, the only one to have found extensive practical application. Furthermore, after a brief development of the general theory of the Hill equation and the Mathieu equation as a special case, the emphasis will be upon practical aspects such as the computation of stability boundaries, the relationship between a, q, and β, the calculation of maximum amplitudes, and so on.

In the early days of quadrupole development, the computation of such properties by the use of analytical theory was the only approach possible. It is usually quite cumbersome and often involves the summation of a long series of terms. Numerical integration of the equations of ion motion has been a preferred alternative since digital computers became generally available. Lately, in many instances, point-by-point numerical integration has itself been replaced by the use of matrix methods based on phase-space dynamics which enable faster, cheaper computations of greater predictive value. The uses of both numerical integration techniques and phase-space dynamics are discussed in Chapter IV.

Why, then, do we retain an interest in the analytical theory? Undoubtedly it aids our comprehension of the physical processes but, more important, is its usefulness in evaluating new approaches to the utilization of quadrupole fields. An excellent example of this was development of the oscillatory time-of-flight spectrometer described in Chapter IX. The starting point was the analytical theory.

This chapter will first describe the general theory for any periodic waveform applied to a quadrupole structure, but considering only a one-dimensional case since, as described earlier (p. 9), for perfect quadrupole fields the motion in each co-ordinate direction is independent. The derivation of stability criteria is discussed and examples are given of details of the computation of stability diagrams for both sinusoidal and non-sinusoidal waveforms. Harmonics in the waveform are also considered. Various approaches to the calculation of iso-β lines in the a–q diagram are described, including some very simple approximations. Some of the expressions for the calculation of ion motion are reviewed. Finally, special cases are examined, such as the use of an auxiliary field in isotope separation [5] or in oscillation detection [6], the presence of a magnetic field, the presence of drag, and the influence of space charge.

A. THE HILL EQUATION

As explained earlier, quadrupole spectrometers will, at least in principle, operate with any applied waveform provided it is periodic, and the equation of motion is the Hill equation [see eqn. (2.22)]. The general equation might be expressed

$$\frac{d^2u}{d\xi^2} + \left[\theta_0 + \left(\sum_{r=1}^{\infty} 2\theta_r \cos 2r\xi\right)\right] u = 0 \tag{3.1}$$

When $\theta_r = 0$ for $r \geqslant 2$, this is, of course, the Mathieu equation with $\theta_0 = a$ and $\theta_1 = -q$. From Floquet's theorem, one of the solutions of eqn. (3.1) is of the form [4]

$$u = e^{\mu\xi} \sum_{n=-\infty}^{n=\infty} C_{2n}\, e^{2in\xi} \tag{3.2}$$

The equation of motion is invariant under the transformation $\xi \to -\xi$, and the second independent solution is obtained by making this substitution in eqn. (3.2).

Substituting eqn. (3.2) into eqn. (3.1) and removing the common factor $e^{\mu\xi}$ gives an infinite set of simultaneous equations

$$\sum_{n=-\infty}^{\infty} \left\{C_{2n}(\mu + 2in)^2 e^{2in\xi} + \left[\theta_0 + \sum_{r=1}^{\infty} 2\theta_r \cos(2r\xi)\right] C_{2n}\, e^{2in\xi}\right\} = 0 \tag{3.3}$$

Expressing $\cos(2r\xi)$ in terms of $e^{2ir\xi}$ and $e^{-2ir\xi}$ and equating the coefficients of each of the terms $e^{2is\xi}$ to zero (where s is any integer) one obtains a set of recursion relations between the C_{2n} values. That is, if $\theta_{-r} = \theta_r$

$$(\mu + 2in)^2 C_{2n} + \sum_{r=-\infty}^{r=\infty} \theta_r C_{2r+2n} = 0 \tag{3.4}$$

and n takes on all integral values. Thus we see that the values of the constants C_{2n} depend on the values of θ_r [or (a, q) in the case of the pure sinusoidal waveform] and not on the initial conditions (see Chapter II, p. 15).

We will return later to consideration of the determination of the C_{2n} values and describe how the *characteristic exponent*, μ, can be found.

In the region of interest, the complete solution to eqn. (3.1) is made up of the two linearly independent solutions, i.e.

$$u = \alpha' e^{\mu\xi} \sum_{n=-\infty}^{\infty} C_{2n} e^{2in\xi} + \alpha'' e^{-\mu\xi} \sum_{n=-\infty}^{n=\infty} C_{2n} e^{-2in\xi} \tag{3.5}$$

where α' and α'' are integration constants depending on the initial conditions u_0, $\dot{u}_0$, and ξ_0. The significance of the parameter μ has been discussed on p. 17. The stable solutions occur for $\mu = i\beta$ where β is real and non-integral. The boundaries between regions of stability and instability correspond to $\mu = im$ where m is an integer. The solutions are called functions of integral order and the boundaries are called *characteristic curves* or *characteristic values*. For example, in the stability diagram of Fig. 2.5 for a sinusoidally operated device, $0 \leqslant \beta \leqslant 1$ for the first (lower) stable region $1 \leqslant \beta \leqslant 2$ for the second stable region.

If eqns. (3.4) are each divided by the central term $(\mu + 2ni)^2 + \theta_0 = \rho_n$, then the determinant of the coefficients of C_{2n} is convergent and, since $\theta_{-r} = \theta_r$, is given by

$$\Delta(\mu) = \begin{vmatrix} & \vdots & \vdots & \vdots & \vdots & \vdots & \\ \cdots & 1 & \theta_1/\rho_{-2} & \theta_2/\rho_{-2} & \theta_3/\rho_{-2} & \theta_4/\rho_{-2} & \cdots \\ \cdots & \theta_1/\rho_{-1} & 1 & \theta_1/\rho_{-1} & \theta_2/\rho_{-1} & \theta_3/\rho_{-1} & \cdots \\ \cdots & \theta_2/\rho_0 & \theta_1\rho_0 & 1 & \theta_1/\rho_0 & \theta_2/\rho_0 & \cdots \\ \cdots & \theta_3/\rho_1 & \theta_2/\rho_1 & \theta_1/\rho_1 & 1 & \theta_1/\rho_0 & \cdots \\ \cdots & \theta_4/\rho_2 & \theta_3/\rho_2 & \theta_2/\rho_2 & \theta_1/\rho_2 & 1 & \cdots \\ & \vdots & \vdots & \vdots & \vdots & \vdots & \end{vmatrix} \tag{3.6}$$

In order that the set of equations be consistent, this determinant must be zero and it can be shown [4, 7] that this is equivalent to

$$\sin^2[(\pi i\mu)/2] = \Delta(0) \sin^2[(\pi\sqrt{\theta_0})/2] \tag{3.7}$$

That is

$$\cosh(\mu\pi) = 1 - 2\,\Delta(0) \sin^2[(\pi\sqrt{\theta_0})/2] \tag{3.8}$$

This general form for determining the characteristic exponent has been utilized to consider the influence of harmonic composition of the voltage waveform on quadrupole performance [8]. Successive approximations to $\Delta(0)$ can be obtained by considering the determinant to be truncated as the

TABLE 3.1

Values of the characteristic exponent for a Mathieu equation with various percentages of third or fifth harmonic distortion ($\mu = \alpha + i\beta$)

(a)

	Third harmonic		Fifth harmonic	
	α_x	β_x	α_x	β_x
0.1%	0.0	0.94804	0.0	0.94802
1.0%	0.0	0.94823	0.0	0.94802
10.1%	0.0	0.95032	0.0	0.94822

(b)

	Third harmonic		Fifth harmonic	
	α_y	β_y	α_y	β_y
0.1%	0.0	0.03711	0.0	0.03708
1.0%	0.0	0.03741	0.0	0.03708
10.0%	0.0	0.04077	0.0	0.03765

central 3 x 3, 5 x 5, 7 x 7, etc. determinant, as indicated by the broken lines in eqn. (3.6). For points of interest in the sinusoidally operated mass filter, good approximations are obtained even with a 5 x 5 determinant. This implies that higher order harmonics in the waveform which would enter beyond the point of truncation have little influence on the nature of the ion motion. Table 3.1 shows values [8] of the characteristic exponent ($\mu = \alpha + i\beta$) at an operating point for a sinusoidally driven mass filter with in-phase third- or fifth-order harmonics of 0.1%, 1%, and 10%. The operating point lies midway between the stability boundaries on a scan line giving a resolution of about 200. A 15 x 15 determinant was used in the calculation. These results suggested that the important element in stabilizing rf voltage magnitudes for high performance is not to control the peak amplitude of the possibly distorted waveform but to control the fundamental.

Evidently, from eqn. (3.8), the conditions for stability ($0 \leqslant \beta \leqslant 1$) become

$$0 \leqslant \Delta(0) \sin^2 [(\pi\sqrt{\theta_0}/2)] \leqslant 1 \tag{3.9}$$

the equality signs giving the characteristic curves or boundaries in a multi-dimensional θ_r space [7].

The next simplest case, after the Mathieu equation, is $\theta_r = 0$ for $n \geqslant 3$, or

$$\frac{d^2u}{d\xi^2} + (\theta_0 + 2\theta_1 \cos 2\xi + 2\theta_2 \cos 4\xi)u = 0 \tag{3.10}$$

The characteristic boundary surfaces intersect the $\theta_0 - \theta_1$ plane at the usual curves for the Mathieu equation (Chapter II). The intersection with the

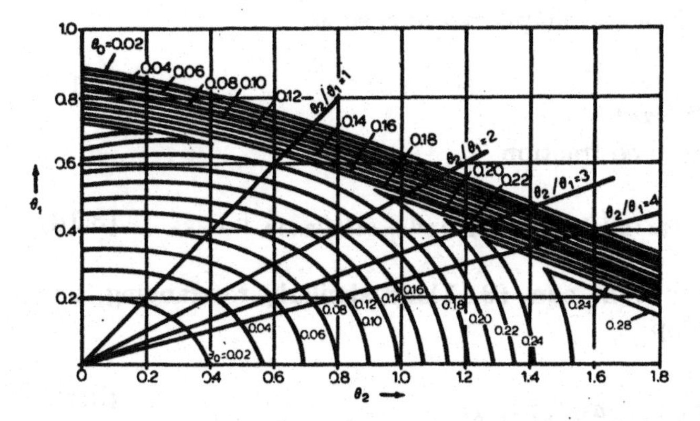

Fig. 3.1. Stability boundaries [9] for the three term Hill equation (3.10): θ_1 versus θ_2 for fixed values of θ_0. The $\theta_2 = 0$ axis represents the Mathieu diagram of Fig. 2.10.

$\theta_0 - \theta_2$ plane is somewhat similar but with $a = \theta_0/4$ and $q = -\theta_2/4$. Figure 3.1 shows calculated boundary surfaces [9] giving the value of θ_0 associated with θ_1 and θ_2 for the mass filter, taking into account the necessity for simultaneous stability in both x and y directions. Thus the $\theta_2 = 0$ axis represents the boundary lines of Fig. 2.10. Again it is interesting to note how little a θ_2/θ_1 ratio of 0.1 changes the stability boundaries. However, such a shift would be evident in high-resolution measurements. The infamous precursor peak sometimes observed in quadrupoles at high resolution (see p. 126) often behaves as if it were controlled by a "shadow" stability diagram slightly displaced from the usual one. This leads to yet another speculation that it could have its origin in random variations in the harmonics of the waveform.

An example of the stability diagram with a rectangularly driven mass filter was given in Fig. 2.23. However, the Meissner equation of motion is much easier to deal with by the matrix techniques described in Chapter IV.

B. THE MATHIEU EQUATION

For the sinusoidally operated instruments, the Mathieu equation applies and the equations are simplified since $\theta_0 = a$, $\theta_1 = -q$, and $\theta_{r \geq 2} = 0$. To calculate the expansion coefficients C_{2n} in the stable region, these θ values are substituted in eqn. (3.4) which, since $\mu = i\beta$, becomes [4, 7]

$$a_{2n} C_{2n-2} + b_{2n} C_{2n} + a_{2n} C_{2n+2} = 0 \tag{3.11}$$

with

$$\begin{aligned} a_{2n} &= \frac{q}{(\beta + 2n)^2} \\ b_{2n} &= 1 - \left[\frac{a}{(\beta + 2n)^2}\right] \end{aligned} \tag{3.12}$$

The set of equations (3.11) can be simply re-arranged as

$$\frac{C_{2n}}{C_{2n-2}} = \frac{-a_{2n}}{(b_{2n} + a_{2n} C_{2n+2}/C_{2n})} \quad (3.13)$$

That is, one obtains a continued fraction

$$\frac{C_{2n}}{C_{2n-2}} = -a_{2n}/(b_{2n} - a_{2n}a_{2n+2}/(b_{2n+2} + a_{2n+2}a_{2n+4}/(b_{2n+4} + \ldots . \quad (3.14)$$

Substituting $(n-1)$ for n in eqn. (3.11)and similarly rearranging, one obtains a second relationship

$$\frac{C_{2n}}{C_{2n-2}} = \frac{b_{2n-2}}{a_{2n-2}} + a_{2n-4}/b_{2n-4} - a_{2n-4}a_{2n-6}/b_{2n-6} - a_{2n-6}a_{2n-8}/\ldots \quad (3.15)$$

Hence, for a given a and q, β can be calculated using eqn. (3.8) with the determinant appropriately simplified in accordance with the Mathieu equation and truncated depending on the accuracy desired. The eqns. (3.14) (for $n \geqslant 1$) and (3.15) (for $n < 0$) are calculated for $|n| = 1, 2, 3$ successively, using sufficient terms that the difference between successive results is less than some specified amount. The nature of the ion motion has then been completely specified from the values of a and q, viz. the fundamental frequency of ion motion, the admixture of higher frequencies, and their relative weighting factors.

There are numerous approximate methods for determining β to be found in the literature. There is little point in repeating them here except for the most commonly used. The simplest [10] is

$$\beta = [a + (q^2/2)]^{1/2} \quad (3.16)$$

(See Chapter VII for an example of its derivation and application.)

A more accurate one [7] is

$$\beta = \left[a - \frac{(a-1)q^2}{2(a-1)^2 - q^2} - \frac{(5a+7)q^4}{32(a-1)^3(a-4)} - \frac{(9a^2+58a+29)q^6}{64(a-1)^5(a-4)(a-9)}\right]^{1/2} \quad (3.17)$$

There are published tables of β values for the lower stability regions [11] and very detailed computations have been made for the area near the tip of the mass filter stability diagram [12] as illustrated in Fig. 3.2(a). The iso-μ lines for (a, q) values near the mass filter operating point but in the unstable region were also computed [13] and an example is shown in Fig. 3.2(b). Such values can be useful when considering [14] the transmission of a "real" mass filter of finite length where some ions with nominally unstable trajectories may be able to pass through the device. However, analytical methods have now generally been superseded by the phase-space dynamics approach described in Chapter IV (see also p. 33).

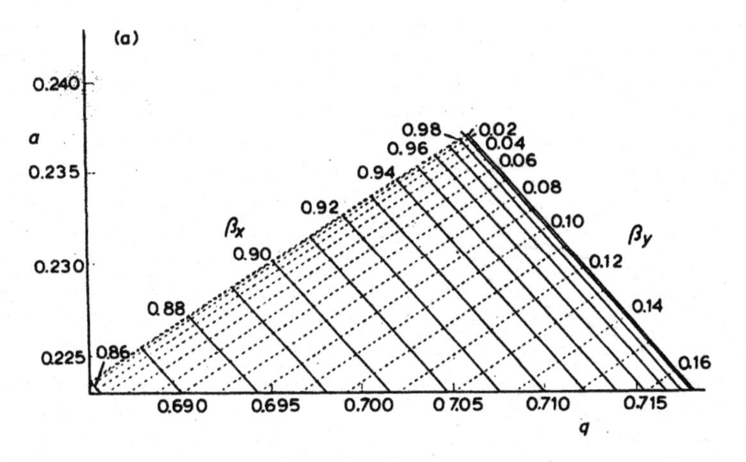

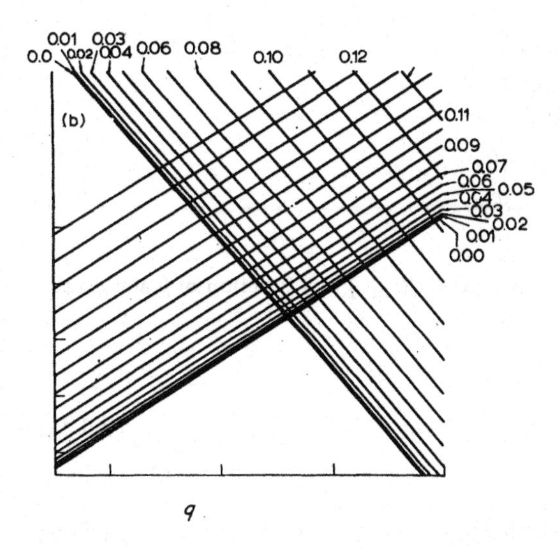

Fig. 3.2. (a) Iso-β lines near the apex of the mass filter stability diagram [12]. (b) Iso-μ lines in the unstable regions near the apex of the mass filter stability diagram [13].

A table in Appendix C gives the (a, q) values for particular double-focussing combinations of β_x and β_y (see Chapter II, p. 57) where $\beta_y = 1/d$ and $\beta_x = p/d$ and p and d are integers having no common factors.

Some values of the coefficients C_{2n} calculated by Berkling [15] using equations such as (3.13) and (3.14) are shown in Fig. 3.3 as a function of β and q for $a = 0$. Values of particular interest in operation of the mass filter [5] are given in Appendix D.

The characteristic values (stability boundaries) for the lowest stable region are given by [3]

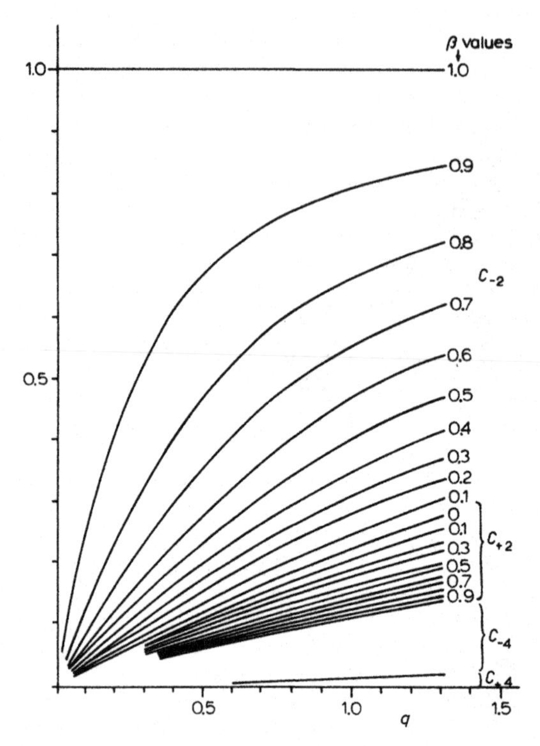

Fig. 3.3. Examples of the ratios of the coefficients C_n/C_{2n-2} as a function of β and q [15] when $a = 0$.

$$a_0 = -\frac{q^2}{2} + \frac{7q^4}{128} - \frac{29q^6}{2304} + \ldots\ldots\ldots \tag{3.18a}$$

$$b_1 = 1 - q - \frac{q^2}{8} + \frac{q^3}{64} - \frac{q^4}{1536} - \frac{11q^5}{36864} + \ldots\ldots \tag{3.18b}$$

C. THE COMPLETE SOLUTION

So far, the nature of the ion motion has been considered, but the integration constants α' and α'' of eqn. (3.5) have been ignored. These depend upon u_0, $\dot{u}_0$, and ξ_0.

If eqn. (3.5) is rewritten as

$$u = \alpha' \mathrm{F}_1(\xi) + \alpha'' \mathrm{F}_2(\xi) \tag{3.19}$$

then

$$\alpha' = \frac{1}{W_0}\left\{u_0 \dot{F}_2(\xi_0) - \dot{u}_0 F_2(\xi_0)\right\}$$

$$\alpha'' = \frac{1}{W_0}\left\{\dot{u} F_1(\xi_0) - u_0 \dot{F}_1(\xi_0)\right\} \quad (3.20)$$

where the Wronksian W_0 is given by

$$W_0 = F_1(\xi_0) \dot{F}_2(\xi_0) - \dot{F}_1(\xi_0) F_2(\xi_0) \quad (3.21)$$

and

$$\dot{F}(\xi_0) = \left[\frac{dF(\xi)}{d\xi}\right]_{\xi=\xi_0}$$

Actually, the Wronksian is independent of ξ_0 [since $(dW_0/d\xi)_{\xi=\xi_0} = 0$] and one obtains

$$W_0 = \sum_{n=-\infty}^{\infty} C_{2n} \sum_{n=-\infty}^{\infty} C_{2n}(\beta + 2n) \quad (3.22)$$

As one would expect, u in eqn. (3.19) scales in accordance with u_0 if $\dot{u}_0$ is zero.

Obviously, the complexity of the complete solution has limited the use of the analytical solutions in particular instrument simulations. However, one important early application [5] was in the estimation of maximum displacement amplitudes. The value of u_{max} obtained by differentiation of eqn. (3.19) is

$$u_{max} = \frac{1}{W_0} \sum_{n=-\infty}^{\infty} C_{2n}\left\{[u_0 \dot{F}_2(\xi_0) - \dot{u}_0 F_2(\xi_0)]^2 + [\dot{u}_0 F_1(\xi_0) - u_0 \dot{F}_1(\xi_0)]^2\right\}^{1/2} \quad (3.23)$$

If u_{max} is then put equal to r_0 (the radius of the quadrupole aperture), then eqn. (3.23) is an ellipse in the $(u_0, \dot{u}_0)$ plane, connecting those points for which the maximum amplitude of oscillation is equal to r_0. There is a different ellipse for each initial phase. Some examples [5] were given in Fig. 2.14. Similar ellipses for points near the tip of the stability diagram (Fig. 2.15) represent the acceptance of the mass filter at each initial phase for an infinitely long device. These ellipses are readily generated [16] by the matrix methods described in Chapter IV or, as illustrated there, can be derived with appropriate scaling from a trajectory plotted in u, $\dot{u}$ space.

There have been some alternative approaches to calculating oscillation amplitudes [17, 18]. One is to consider the ion motion at the apex of the stability diagram as only slightly perturbed from that corresponding to the characteristic curves [17]. The first-order approximation for the x direction is

$$x = (A^2 + B^2)^{1/2}(\cos \xi - f \cos 3\xi) \sin [(1 - \beta_x)\xi + \phi] \quad (3.24)$$

where

$$f = \frac{q}{8[1 - 0.25(1 - \beta_x)^2]}$$

and

$$\tan \phi = \frac{A}{B}$$

For the y direction, the first-order approximation is

$$y = (A^2 + B^2)^{1/2}(1 - 2e \cos 2\xi) \sin (\beta_y \xi + \phi) \quad (3.25)$$

where

$$e = \frac{q}{4(1 - \beta_y^2)}$$

and A and B depend on the initial conditions. Algebraic expressions for the maximum amplitudes can be derived from eqns. (3.24) and (3.25). The quadrupole transmission can then be shown to be approximately proportional to $\beta_y(1 - \beta_x)$.

There have been several attempts [19, 20] to relate the shape of mass peaks analytically to the coupling of ion source and quadrupole field parameters. The complexity of fringing field effects and the wide variation of quadrupole acceptance with the initial phase makes these of doubtful general utility. However, a phase-space dynamics approach to the fringing field problem does now offer the possibility of better ion source design and evaluation (see Chapter V).

Dehmelt [21] has adopted a completely different approach to consider the functioning of the ion trap (see Chapter VIII and p. 77), evaluating ion storage in terms of pseudo-potentials.

D. AUXILIARY FIELDS

Auxiliary fields have been used in the quadrupole mass filter to help achieve high resolution in an isotope separator [5]. In the quadrupole ion trap, ions may be detected by their resonant oscillation [16]. The latter has already been described in Chapter II (see p. 49). The former technique merits further consideration. It was used so that the mass filter could be operated in a high transmission mode with near-zero d.c. applied. Isotope separation was by means of the auxiliary field which increased the amplitude of ion motion for an isotope at resonance producing a dip in ion transmission at the corresponding mass number. This technique has perhaps potential value in increasing resolution in more conventional instruments by selectively diminishing an adjacent peak. However, for two very similar masses a long quadrupole field would be necessary.

In the presence of the auxiliary field, $E' e^{i\omega' t}$, considering it to be applied uniformly in one of the co-ordinate directions, the equation of motion becomes

$$\frac{d^2u}{d\xi^2} + (a - 2q \cos 2\xi)u = \rho e^{2i\omega'\xi/\omega} \tag{3.26}$$

where

$$\rho = \frac{4eE'}{m\omega^2}$$

The general solution of this inhomogeneous Mathieu equation is a linear combination of the solutions of the Mathieu equation discussed earlier with a particular solution of the inhomogeneous equation. A particular solution [5] is

$$u = \frac{\rho}{W_0}\left\{F_2 \sum_{n=-\infty}^{\infty} C_{2n} \frac{\exp[2i(n + \beta/2 + \omega'/\omega)\xi]}{2i(n + \beta/2 + \omega'/\omega)} + F_1 \sum_{n=-\infty}^{\infty} C_{2n} \frac{\exp[-2i(n + \beta/2 - \omega'/\omega)\xi]}{2i(n + \beta/2 - \omega'/\omega)}\right\} \tag{3.27}$$

Now when ω' is chosen to equal one of the components of the frequency spectrum of ion oscillation there is a resonance. At resonance, $\omega' = (s + \beta/2)\omega$ where s is an integer corresponding to the chosen frequency component. Consider a situation close to resonance, defining $\Delta\omega = (s + \beta/2)\omega - \omega'$. Then the dominating term in eqn. (3.27) will be the one in the second summation which has the smallest denominator and corresponds most closely to resonance; so that, after substitution

$$u \simeq \rho C_{2s} F_1 \frac{\exp[-2i(\Delta\omega/\omega)\xi]}{W_0 2i\Delta\omega/\omega} \tag{3.28}$$

At exact resonance, this simplifies to

$$u \simeq \frac{\rho C_{2s} \xi F_1}{W_0} \tag{3.29}$$

This approximation indicates that the ion at resonance has approximately an oscillation with a linearly increasing amplitude. The increase in amplitude will be greatest if the resonance is with the fundamental frequency of ion motion (so that $s = 0$) because C_0 is generally the largest coefficient. Equation (3.28) for the ion at near resonance represents a series of beats of frequency $\Delta\omega$. The amplitude of the beats is proportional to the auxiliary field E' and becomes greater as resonance is approached more closely.

References pp. 77—78

In using an auxiliary field to help separate two adjacent masses, the beat amplitude must be maintained at less than the instrument radius r_0, e.g. by reducing the auxiliary field. The transit time [5] in the device must be about one beat period, $1/\Delta\omega$, in order to completely filter the ions of similar mass.

A more complex suggestion for the use of auxiliary fields [22] would utilize a circuit producing a wide spectrum of frequencies excepting the one corresponding to the transmission of ions of a given m/e.

E. SPECIAL CASES

There are a few special cases which should be mentioned. For example, the presence of a magnetic field, B, in the z direction of the quadrupole produces a coupling of motion in the x and y directions since

$$\frac{d^2x}{d\xi^2} + (a_x - 2q_x \cos 2\xi)x = 2g\frac{dy}{d\xi} \tag{3.30}$$

$$\frac{d^2y}{d\xi^2} + (a_y - 2q_y \cos 2\xi)y = -2g\frac{dx}{d\xi} \tag{3.31}$$

where $g = eB/m\omega$.

The coupling can be removed if one considers a co-ordinate system rotating with frequency g about the z axis [7] but the equivalent a values in the Mathieu equation must be adjusted to $[a + e^2B^2/m^2\omega^2]$. That is, the magnetic field is somewhat equivalent to a change $er_0^2B^2/4m$ in the applied d.c. voltage. Fischer [6] has examined the effect of a magnetic field on the ion trap.

Another modification may be caused by the presence of a viscous "drag" on the particles or ions moving in a quadrupole field. This occurs with macroscopic particles moving through air [23]. The equation of motion is

$$\ddot{u} + 2K\dot{u} + (a - 2q \cos 2\xi)u = 0 \tag{3.32}$$

where $K = 6\pi\eta R/m\omega$, R is the radius of the particle, and η is the viscosity coefficient. By making the substitution $u = u_1 e^{-K\xi}$, the equation becomes

$$\ddot{u}_1 + (\bar{a} - 2q \cos 2\xi)u_1 = 0 \tag{3.33}$$

where $\bar{a} = a - K^2$. Thus one returns to the Mathieu equation, but the solutions involve the term $e^{(\mu - K)}$ and are stable for $K > \mu > 0$, are periodic for $K = \mu > 0$, and unstable for $0 < K < \mu$. The effect of the drag is to enlarge the boundaries of the stable region.

A final modification, which warrants more study, is that due to space charge. Space charge has been considered [5, 6] as producing an outward force on an ion. If this force is proportional to the distance of the ion from the field axis, it is also equivalent to a change in the a_x and a_y values (in opposite directions). In the work on isotope separation, Paul et al. [5] were able to measure the change in the position of resonance for ^{23}Na as the space charge was increased. The change in a is given by

$$\Delta a = \frac{5.7 \times 10^4 I}{f^2 (M V_z)^{1/2}}$$

where I is the ion current in amperes, f is the frequency in MHz, V_z is the ion energy in the z direction, and M is the ion mass in amu. Space charge is of particular importance in the quadrupole trap as the ultimate limitation to ion storage. Fischer [6] estimated the saturated ion density in his trap from the d.c. shift in the stability diagram. Dehmelt [21] has derived an expression for the maximum space charge density in the ion trap assuming that the ions arrange themselves in the bottom of the pseudo-potential well so that they exactly cancel the field. He obtained

$$\rho_{max} = 3eV^2/16\pi m \omega^2 z_0^4$$

(See Chapter VIII, p. 214 for an extended discussion of space charge in the ion trap.)

The influence of collision processes on mass filter operation at high pressures has also been considered theoretically [24].

REFERENCES

1 N.W. McLachlan, Theory and Applications of Mathieu Functions, Oxford University Press, Oxford, 1947.
2 E.L. Ince, Ordinary Differential Equations, Dover Publications, New York, 1956.
3 G. Blanch, in M. Abramowitz and I.A. Stegun (Eds.), Mathieu Functions, Handbook of Mathematical Functions AMS 55, National Bureau of Standards, Washington, 1964.
4 F.M. Arscott, Periodic Differential Equations, Macmillan, New York, 1964.
5 W. Paul, H.P. Reinhard and U. von Zahn, Z. Phys., 152 (1958) 143.
6 E. Fischer, Z. Phys., 156 (1959) 1.
7 J.P. Carrico, Dyn. Mass Spectrom., 3 (1972) 1.
8 J.A. Richards, R.M. Huey and J. Hiller, Int. J. Mass Spectrom. Ion Phys. 13 (1974) 443.
9 L.D. Ferguson, L.H. Krohn, N.O. Tiffany, L.J. Vande Kieft, G.J. O'Halloran, L.M. Sieradski, H.S. Smith and L.W. Walker, Rep. 2462, Bendix Research Laboratories, Southfield, Michigan, 1963.
10 R.F. Wuerker, H. Shelton and R.V. Langmuir, J. Appl. Phys., 30 (1969) 342.
11 T. Tamir, Math. Computations, 16 (1962) 77.
12 K. Maeda, A. Fukada and M. Sakimura, Mass Spectrosc. (Tokyo), 17 (1969) 530.
13 K. Maeda and A. Fukada, Mass Spectrosc. (Tokyo), 18 (1970) 1097.
14 K. Maeda, M. Sakimura and A. Fukada, in K. Ogata and T. Hayakawa (Eds.), Recent Developments in Mass Spectrometry, University Park Press, Baltimore, 1970, p. 39.
15 K. Berkling, Diplomarbeit, Physikalisches Institut der Universität Bonn, 1956.
16 P.H. Dawson, Int. J. Mass Spectrom. Ion Phys., 14 (1974) 317.

17 K. Maeda, M. Sakimura and A. Fukada, in K. Ogata and T. Hayakama (Eds.), Recent Developments in Mass Spectrometry, University Park Press, Baltimore, 1970, p. 273.
18 E.P. Sheretov and B.I. Kolotilin, Sov. Phys. Tech. Phys., 17 (1973) 1547.
19 L.G. Hall and M.R. Ruecker, 17th Annu. Conf. Mass Spectrom. Allied Topics, Montreal, 1964.
20 G.I. Slobdenyuk, Zh. Tekh. Fiz., 37 (1967) 1535.
21 H.G. Dehmelt, Advan. At. Mol. Phys., 3 (1967) 53.
22 R.V. Langmuir, U.S. Pat. 3,334,225 (1967).
23 N.R. Whetten, J. Vac. Sci. Technol., 11 (1974) 515.
24 P. Kubicek and L. Mrazek, Ann. Phys. (Leipzig), 24 (1970) 289.

CHAPTER IV

NUMERICAL CALCULATIONS

P.H. Dawson

While the analytical solutions of the equations of motion can be useful in the effort to develop new general concepts for the application of quadrupole fields, they have been replaced by numerical methods when specific calculations or computer simulations are required. Such methods have involved either numerical integration or the use of matrices.

Numerical calculation of ion trajectories using point-by-point integration of the Mathieu equation has been extensively applied to determining ion trajectories in all types of quadrupole devices and has contributed a great deal to our understanding of their performance. There has been consideration not only of sensitivity and resolution [1–4] but of focusing conditions [5, 6], field errors and distortions [7–9] (see Chap. V), and fringing fields [10–12] at the entrance and exit. Most of these calculations, or computer simulations, have however been of quite an elementary nature. They assume that the electrodynamic fields can be adequately represented in some convenient algebraic form. Recently, there have been some attempts to use more sophisticated techniques based on detailed fields calculated directly from "real" electrode structures. These methods have not so far produced new insights of value but they may be essential for solving the most difficult theoretical problem remaining, that of performing realistic three-dimensional calculations of ion entrance to and exit from quadrupoles (see fringing fields, Chap. V).

One limitation of numerical integration is the necessity to repeat the calculation for each combination of an initial position and an initial velocity. Calculations can be scaled when there is an initial displacement but no initial velocity (or vice versa) as, for example, with an ion beam entering the mass filter parallel to the instrument (z) axis. However, such a beam is unlikely in a real case when fringing fields are present. A second limitation has been the cost of making detailed computer simulations for rf fields consisting of a sufficient number of rf cycles. For these reasons, newly developed techniques using matrix methods with analyses based on phase-space dynamics are now replacing numerical integration.

These new techniques offer the advantage of ready calculation of ion position at any time during a trajectory for any combination of initial velocity and initial displacement. They are an important advance in our ability to make computer simulations of instrument performance under a variety of different conditions. It is also likely that the ideas expressed in phase-space dynamics will open up new concepts in the study and application of quadrupole fields. However, this has not yet occurred and the phase-space dynamics approach is appropriately considered under the heading of numerical calculations. So far, the application of the method has been limited to "perfect" fields in the sense that the ion motion has been assumed independent in each co-ordinate direction (see Chap. II, p. 9). Of course, when x and y motions become interdependent, such as in the presence of field distortion (Chap. V) or in "real" fringing fields, it becomes more difficult to generalize for all input conditions. Step-by-step numerical integration of the equations of motion may still have a role to play in these circumstances.

In this chapter, numerical integration techniques will be discussed first and then the application of phase-space dynamics will be described.

A. NUMERICAL INTEGRATION

The first publication of a detailed study of instrument performance by the digital computation of ion trajectories appeared, appropriately enough, in an IBM journal in 1966 [5]. Lever, while awaiting delivery of one of the recently commercialized monopoles, began a theoretical study of its functioning and finally suggested a new mode of operation utilizing the focusing properties of quadrupole fields (see Chap. II, p. 57). The conventional monopole remained a good subject for trajectory calculations for some years [10, 12–14] because of the mystery of its performance (Chap. II, p. 42). The widespread access to digital computers from the mid-sixties onwards quickly led to other applications of numerical integration to the mass filter [1, 15] and the ion trap [2].

Most of the calculations have been carried out using Cartesian co-ordinates with geometrically perfect fields, motion in each co-ordinate direction being considered independently (Chap. II, p. 9). Occasionally, a system of cylindrical co-ordinates has been adopted [16]. In distorted fields, motion in each co-ordinate direction is no longer independent and simultaneous calculations must be made for all directions [7, 8]. Any of the standard methods of numerical integration may be employed but the most popular has been the fourth-order Runge–Kutta routine. An example of a computer program written in Fortran IVG is given in Appendix E. This calculates the trajectory for two co-ordinate directions simultaneously and prints out the time in rf periods, the two positions and the corresponding velocities. Sometimes, the output is used for automatic trajectory plotting. In this type of program, no effort has been made to increase the efficiency of the calculation. The time step for integration is fixed throughout. The choice of time step depends upon the accuracy required. Generally, a value of $\pi/50$ in ξ (one fiftieth of the rf period) has been found a satisfactory compromise between accuracy and efficiency. A simple improvement would be to have a different time step at different points during the rf cycle. For example, $\delta\xi = A + B \cos^2 2(\xi - \xi_0)$, where A and B are constants, would give [17] a large value of $\delta\xi$ when the rf is at its peak value and a small value of $\delta\xi$ when the rf is rapidly changing. A second weakness of the simple integration program is the lack of any error indication except that obtained by repeating the whole calculation with a smaller time step. These problems have not been serious in the many applications mentioned above, but become important on dealing with "real" electrode structures when the potential can no longer be represented by a convenient geometrical expression.

Trajectory calculation in a device of very imperfect geometry was considered by Chisholm and Stark [17]. They were interested in electron oscillation in a four rod structure where the rod radii were much smaller than the optimum for obtaining quadrupole fields (see p. 117). In such a case, potential distribution must first be computed and then used to determine the charged particle trajectories. Neglecting any space charge, the potential ϕ at any point in the system obeys Laplace's equation

$$\frac{\delta^2\phi}{\delta x^2} + \frac{\delta^2\phi}{\delta y^2} = 0$$

The Laplace equation can be transformed into a finite difference equation which relates the potential at a point P_0 to the potentials at four surrounding points P_{1-4}, equally distributed in space about P_0, viz.

$$\phi_0 = \frac{1}{4}\sum_{i=1}^{4}\phi_i \tag{4.1}$$

The area between the electrodes is divided into a square mesh with several hundred intersecting points. An initial value is guessed for the potential at every point, the boundaries, of course, being known. The relaxation method of solution then recomputes the potential of a point from those of its neighbours as given by eqn. (4.1). This process can be repeated until the difference equation is satisfied to within a specified tolerance. There are over-relaxation methods [18, 19] which improve the speed of convergence.

This calculation of potential is for fixed voltages applied to the electrodes. When an alternating potential is applied to the rods and there is an outer cylindrical container at a fixed potential, the situation is more complex. Let $\phi_B(x, y)$ be the calculated potential with the pairs of rods B at a potential of 1 V and the pair of rods A and the outer cylinder C at zero. Then the distribution with rods A at 1 V and B and C at zero is ϕ_A where

$$\phi_A(x, y) = \phi_B(|y|, |x|) \tag{4.2}$$

With A and B at zero and the cylinder C at 1 V, from the principle of superposition

$$\phi_C(x, y) = 1 - \{\phi_B(x, y) + \phi_A(x, y)\} \tag{4.3}$$

Hence, with potentials $-U - V\cos\omega(t-t_0)$ on B, $U + V\cos\omega(t-t_0)$ on A and V_C on the outer cylinder, the complete distribution is

$$\begin{aligned}\phi(x, y) &= V + V\cos\omega(t-t_0)\{\phi_B(|x|, |y|) - \phi_B(|y|, |x|)\}\\ &\quad + V_C\{1 - \phi_B(|x|, |y|) - \phi_B(|y|, |x|)\}\end{aligned} \tag{4.4}$$

Chisholm and Stark used linear interpolations for positions between mesh points and integrated the equations of motion using a second-order Runge–Kutta method [20]. That is

$$u = u_0 + \delta t \dot{u}_0 + \tfrac{1}{2}(K_0 \delta t) \tag{4.5}$$

$$\dot{u} = \dot{u}_0 + \tfrac{1}{2}(K_0 + K_1) \tag{4.6}$$

where $\ddot{u} = \mathrm{f}(x, y, t) = e/m \operatorname{grad} \phi$, $K_0 = \delta t\ \mathrm{f}(x_0, y_0, t_0)$, and $K_1 = \delta t\ \mathrm{f}(x_0 + \dot{x}_0\delta t, y_0 + \dot{y}_0\delta t, t_0 + \delta t)$.

The upper limit for δt was varied throughout the cycle as suggested above with additional constraints depending on a maximum allowable change in u and $\dot{u}$. The term $\frac{1}{2}(K_0\delta t)$ gives an estimate of the error at each step, and the step was accepted only if this was less than a specified limit. A minimum δt of 10% of the maximum was also prescribed.

This calculation has been described here in some detail, not because of the importance of the results but because of potential future applications of the principles. In fact, Chisholm and Stark found rapid fluctuations in the lifetimes (defined as the time the electrons remained in the system) as potentials were varied, probably related to the resonances that result from imperfect fields (Chap. V).

A more complete and complex analysis of the problem of charged particle trajectory calculations in time-varying Laplacian fields has recently been made [21]. The result is a method said to be an order of magnitude more accurate than that just described and requiring only half the computer time. The potential is again computed for a regular mesh of points. The potential at any other point is then determined with the interpolation formula

$$\phi = \sum_{i=0}^{m} \sum_{j=0}^{m} \alpha_{ij}(x - x_p)^i (y - y_p)^j \tag{4.7}$$

where x_p and y_p are the co-ordinates of the mesh point nearest the particle. A symmetrical group of nine mesh points and a six-coefficient central difference fit was found to give an adequate interpolation. That is, if the mesh point potentials are numbered 0 at the centre and then 1 to 8 in an anticlockwise direction beginning with the point at the right of 0, one has $\alpha_{00} = V_0$, $\alpha_{10} = \frac{1}{2}(V_1 - V_5)$, $\alpha_{01} = \frac{1}{2}(V_3 - V_7)$, $\alpha_{11} = \frac{1}{4}(V_2 + V_6 - V_4 - V_8)$, $\alpha_{20} =$

$\frac{1}{2}(V_1 + V_5 - 2V_0)$, $\alpha_{02} = \frac{1}{2}(V_3 + V_7 - 2V_0)$, and all other values of α are zero. The integration method is based on a Taylor series

$$u = u_0 + \dot{u}_0 \delta t + \ddot{u}_0 \frac{\delta t^2}{2!} + \ldots + \overset{n}{u}_0 \frac{\delta t^n}{n!} \tag{4.8}$$

The accuracy of the second-order interpolation described above is consistent with truncating this series after terms in δt^3. The truncation error at each step in u is $\overset{4}{u}_0(\delta t^4/4!)$ and can be computed before the step itself is executed to select a suitable value of δt. As stated earlier, an interesting but complex application would be the use of analogous techniques in the three-dimensional case of ion entrance and ion exit trajectories.

B. MATRIX METHODS

The original inspiration for quadrupole mass spectrometers came from developments in high-energy accelerators (see Chap. I). In the intervening years, there has been little communication between these diverse fields and only recently has another link been forged with the borrowing of matrix methods based on phase-space dynamics. These concepts have been used for some years in accelerator design and their transfer to quadrupole mass spectrometry came about because of interest in the quadrupole ion trap as a possible source of multiply charged ions for accelerators [22]. The new approach to numerical calculations and computer simulation is being rapidly exploited [23–25]. The representation of ion motion by matrices will first be discussed. This, in itself, can provide an enormous economy in calculation time over the point-by-point integration described earlier. However, the general utility of the matrices is greatly enhanced by the succeeding analysis based on phase-space dynamics.

(1) *Matrix representation*

Matrices were first used to compute ion position and ion velocity (after some given time t) by Richards et al. [26] for a mass filter with a rectangular waveform (see Chap. II, p. 35). The equation of motion is then the Meissner equation [eqns. (2.48) and (2.49)].

During the period 0 to θ, the equation can be represented

$$\frac{d^2u}{d\xi^2} + h_1^2 u = 0 \tag{4.9}$$

The position and velocity ($du/d\xi$) at the end of this period are then given by

$$\begin{bmatrix} u \\ \dot{u} \end{bmatrix}_0 = \begin{bmatrix} \cos h_1\theta & (\sin h_1\theta)/h_1 \\ -h_1 \sin(h_1\theta) & \cos h_1\theta \end{bmatrix} \begin{bmatrix} u_0 \\ \dot{u}_0 \end{bmatrix} = M \begin{bmatrix} u_0 \\ \dot{u}_0 \end{bmatrix} \tag{4.10}$$

For the subsequent interval between θ and π, the equation becomes

$$\frac{d^2u}{d\xi^2} - h_2^2 u = 0 \tag{4.11}$$

and

$$\begin{bmatrix} u \\ \dot{u} \end{bmatrix}_\pi = \begin{bmatrix} \cosh[h_2(\pi-\theta)] & \{\sinh[h_2(\pi-\theta)]\}/h_2 \\ h_2 \sinh[h_2(\pi-\theta)] & \cosh[h_2(\pi-\theta)] \end{bmatrix} \begin{bmatrix} u \\ \dot{u} \end{bmatrix}_\theta = N \begin{bmatrix} u_0 \\ \dot{u}_0 \end{bmatrix} \tag{4.12}$$

Thus, for a complete cycle

$$\begin{bmatrix} u \\ \dot{u} \end{bmatrix}_\pi = NM \begin{bmatrix} u \\ \dot{u} \end{bmatrix}_0 \tag{4.13}$$

Or, to calculate the position and velocity after n complete cycles

$$\begin{bmatrix} u \\ \dot{u} \end{bmatrix}_{n\pi} = (NM)^n \begin{bmatrix} u \\ \dot{u} \end{bmatrix}_0 \tag{4.14}$$

Different initial phases, ξ_0, can be calculated by coupling eqn. (4.14) with an additional matrix covering the period between the $-\xi_0$ and the zero phase. Thus for a rectangular waveform of any duty cycle $\delta(=\theta/\pi)$, it is very simple, given the initial conditions, to calculate the position and velocity at any point in time.

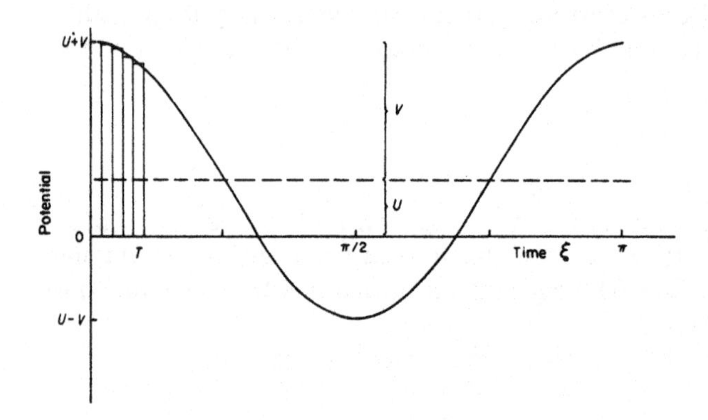

Fig. 4.1. Division of the applied waveform into a large number of narrow rectangular steps of length T in order to simplify the calculation of the matrix representing the action of a complete rf cycle.

In considering other waveforms, the sinusoid being the most important, the simplest approach is that suggested by Pipes [27] and extensively employed by Dawson [23–25]. The waveform is subdivided into a large number of narrow rectangular segments as illustrated in Fig. 4.1. If the interval T is chosen small enough, the representation will become reasonably accurate. Letting M_1 represent the matrix of the form of eqn. (4.10) but for the interval between 0 and T, then

$$\begin{bmatrix} u \\ \dot{u} \end{bmatrix}_T = M_1 \begin{bmatrix} u_0 \\ \dot{u}_0 \end{bmatrix} \tag{4.15}$$

and, after several intervals m of time T [eqn. (4.9) remaining appropriate]

$$\begin{bmatrix} u \\ \dot{u}_0 \end{bmatrix}_{mT} = M_m \cdot M_{m-1} \ldots M_2 \cdot M_1 \begin{bmatrix} u \\ \dot{u}_0 \end{bmatrix} \tag{4.16}$$

When eqn. (4.11) becomes applicable

$$\begin{bmatrix} u \\ \dot{u}_0 \end{bmatrix}_T = N_1 \begin{bmatrix} u \\ \dot{u} \end{bmatrix}_0 \tag{4.17}$$

By the appropriate combination of the required number of matrices M_m and $N_{m'}$, the matrix for a complete cycle, M, is obtained. Partial cycles may be treated the same way. Different initial phases can be treated by combining rectangular segments in the appropriate order, or, as will be seen later, can be deduced indirectly from a calculated trajectory for any given initial phase. Even with a sinusoidal waveform, it becomes a simple matter to calculate the position and velocity at any time without calculating the values at intermediate positions.

In some computer simulations [23, 24] an interval T of $\pi/200$ was used in order to obtain high accuracy. More efficient calculation of the matrices M_m and N_m, allowing larger intervals, is possible using a Runge–Kutta integration or a Taylor series expansion [22]. It has been shown [28] that an interval as large as $\pi/14$ may be adequate for many purposes provided that the height of the steps is chosen so as to minimize higher harmonics in the waveform.

It is evident that the matrix M for the complete cycle, besides being useful for the calculation of position and velocity, must contain information on the nature of ion motion. One factor which immediately emerges from the requirements of stability of the Hill (or general, see Chap. III) equation of motion [26] or from the properties of Twiss matrices [22] is that for stability, the trace of the matrix M must lie between -2 and 2. In fact, if M is represented by $\begin{bmatrix} m_{11} & m_{12} \\ m_{21} & m_{22} \end{bmatrix}$, then

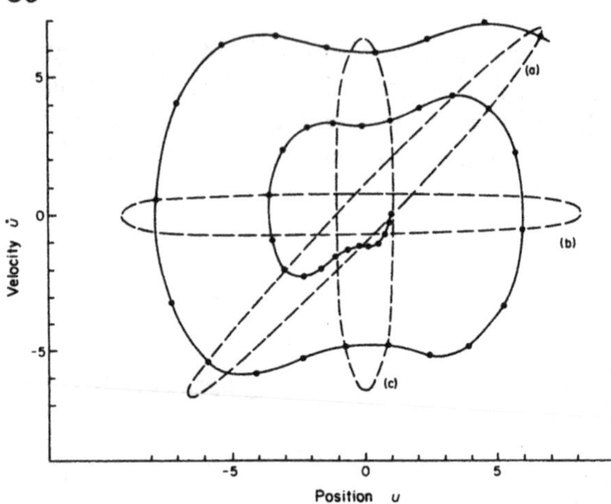

Fig. 4.2. An ion trajectory for the point $a = 0.230114$, $q = 0.696$ plotted in $u\dot{u}$ phase-space. The solid line represents the trajectory during four rf cycles starting at $u_0 = 1$, $\dot{u}_0 = 0$ at the initial phase $\xi_0 = \pi/5$. The points represent the values after each 1/10 of an rf cycle. The ellipses represent the phases (a) $\pi/5$, (b) 0, (c) $\pi/2$.

$$m_{11} + m_{22} = 2 \cos(\pi\beta) \tag{4.18}$$

where β is the well-known parameter (Chap. II, p. 18) characterizing the frequency of ion motion. The calculation of the matrix M gives immediate information on the nature of the ion oscillation and its stability.

(2) *Phase-space dynamics*

The content of the matrix M is even more interesting when an analysis is made using phase-space dynamics [22]. There are numerous texts available on the application of phase-space dynamics to accelerators, lens design, and plasmas [29, 30]. Here the description will be limited to present adaptations to quadrupole mass spectrometry.

Figure 4.2 shows an ion trajectory [23] for the point $a = 0.230114$, $q = 0.696$ plotted in phase-space. The solid line is the trajectory during four rf cycles of a sinusoidally operated mass filter starting at $u_0 = 1$, $\dot{u}_0 = 0$, and the points represent the position and velocity after each one-tenth of an rf cycle. The initial phase ξ_0 was $\pi/5$. The points representing completed cycles of the rf all fall upon the ellipse (a). If β is some rational fraction (see Chap. II, p. 57), the trajectory repeats itself after a certain number of cycles. In that case, only certain points on the ellipse represent positions along the trajectory. On the other hand, when the trajectory is not exactly repetitive, all the points along the ellipse represent points somewhere along the trajectory and all the points are equivalent. Thus the ellipse also represents the combination of different initial conditions which, for the given initial phase, result in equivalent trajectories. If a particular ellipse can be scaled according to the maximum allowable amplitude of oscillation, the ellipse then represents the acceptance of the mass filter for the given initial phase. The ellipses are therefore, with appropriate scaling, equivalent to those of Fischer as illustrated in Fig. 2.14 (see Chap. II, p. 28) and Fig. 2.15.

As can be seen from Fig. 4.2, the ellipses for other rf phases can be obtained from other points along the calculated trajectory. The ellipses have the same area but differ in orientation and eccentricity.

The equation of the ellipses can be written

$$\Gamma u^2 + 2\mathrm{A}u\dot{u} + \mathrm{B}\dot{u}^2 = \epsilon \tag{4.19}$$

where ϵ, the emittance is equal to the area divided by π. In addition

$$\mathrm{B}\Gamma - \mathrm{A}^2 = 1 \tag{4.20}$$

The length of the major axis of the ellipse, L', is given by

$$L' = \sqrt{\epsilon/2}(\sqrt{H+1} + \sqrt{H-1}) \tag{4.21}$$

the minor axis l' by

$$l' = \sqrt{\epsilon/2}(\sqrt{H+1} - \sqrt{H-1}) \tag{4.22}$$

and the angle θ between the major axis and the u axis by

$$\tan 2\theta = -2\mathrm{A}/(\mathrm{B} - \Gamma) \tag{4.23}$$

where

$$H = (\mathrm{B} + \Gamma)/2 \tag{4.24}$$

The parameters of the ellipses depend, of course, on the elements of the matrix M. In fact

$$M = \begin{bmatrix} \cos(\pi\beta) + \mathrm{A}\sin(\pi\beta) & \mathrm{B}\sin(\pi\beta) \\ -\Gamma\sin(\pi\beta) & \cos(\pi\beta) - \mathrm{A}\sin(\pi\beta) \end{bmatrix} \tag{4.25}$$

The calculation of M allows the determination of A, B, and Γ as well as β. Some values of A, B and Γ for a point in (a, q) space typical of mass filter operation [23] were given in Fig. 2.18(a) and (b) and values are tabulated in Appendix A. Some values typical of the second stability region [25], and point $a = 3.15$, $q = 3.24$, are shown in Fig. 4.3. Values typical of ion trap operation with mass analysis are illustrated in Fig. 4.4 [24].

When the (a, q) value corresponds to an unstable trajectory, the representative points in phase-space are situated on a hyperbola. The ellipses can be analysed for their relation to the maximum possible ion displacement. Figure 4.5 shows the critical points on an ellipse expressed in terms of A, B, Γ, and ϵ. The maximum displacement throughout a trajectory u_{max} is given by [22]

$$u_{\mathrm{MAX}} = \sqrt{\epsilon \mathrm{B}_{\mathrm{MAX}}} \tag{4.26}$$

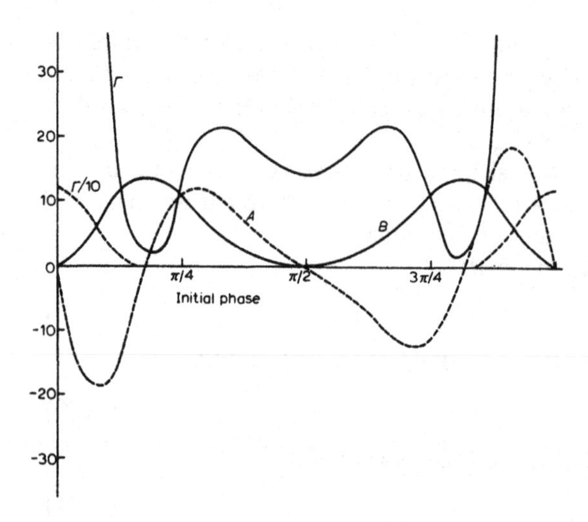

Fig. 4.3. The parameters A, B, and Γ as a function of the initial phase ξ_0 for the point $a = 3.15$, $q = 3.24$ which is in the second stability region.

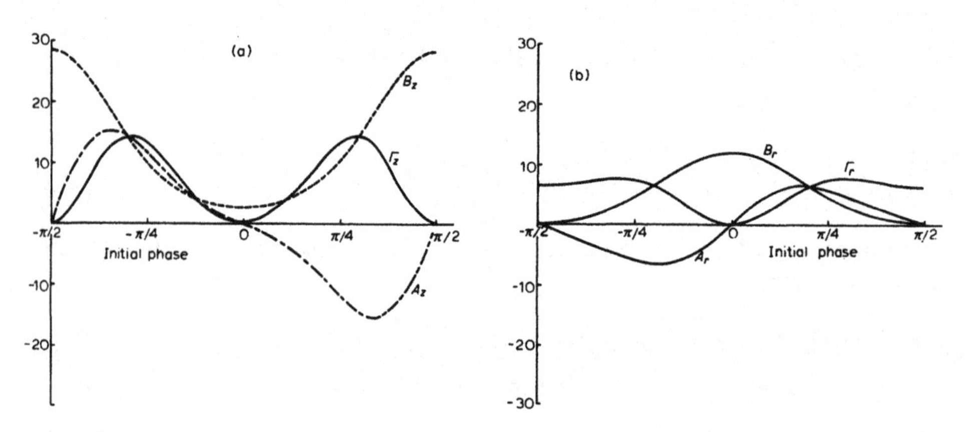

Fig. 4.4. (a) The parameters A, B, and Γ as a function of the initial phase ξ_0 for the point $a = -0.6508$, $q = 1.23$ which could represent the z direction in the quadrupole ion trap. (b) The parameters A, B, and Γ for the corresponding r direction point $a = 0.3254$, $q = 0.615$.

where B_{MAX} is the maximum value of B regardless of the initial phase. The parameter ϵ is determined for a given set of initial conditions using eqn. (4.19). [See eqn. (2.46).] Thus, in a sense, the parameter Γ represents the susceptibility of the trajectory to the initial position, B the susceptibility to the initial velocity, and A is a combination factor. The acceptance ellipses mentioned earlier (see Fig. 2.15) are calculated by introducing a scaling factor obtained by putting $u_{MAX} = r_0$.

Note that an ellipse oriented with θ between 0 and $\pi/2$ represents a

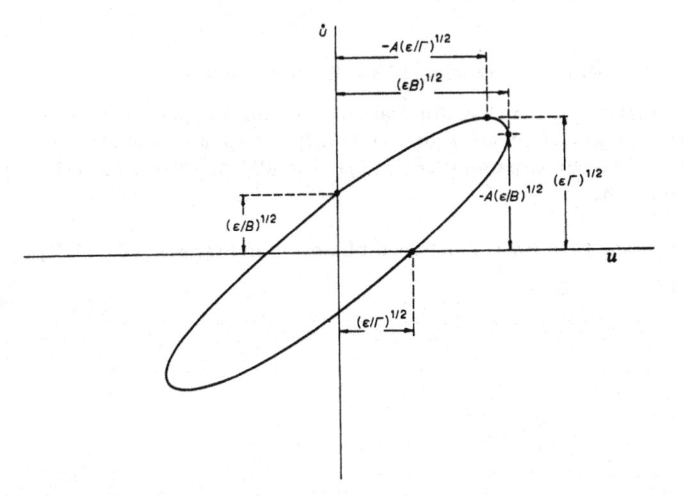

Fig. 4.5. Analysis of a typical emittance ellipse in $u\dot{u}$ phase-space giving four critical points in terms of A, B, Γ, and ε.

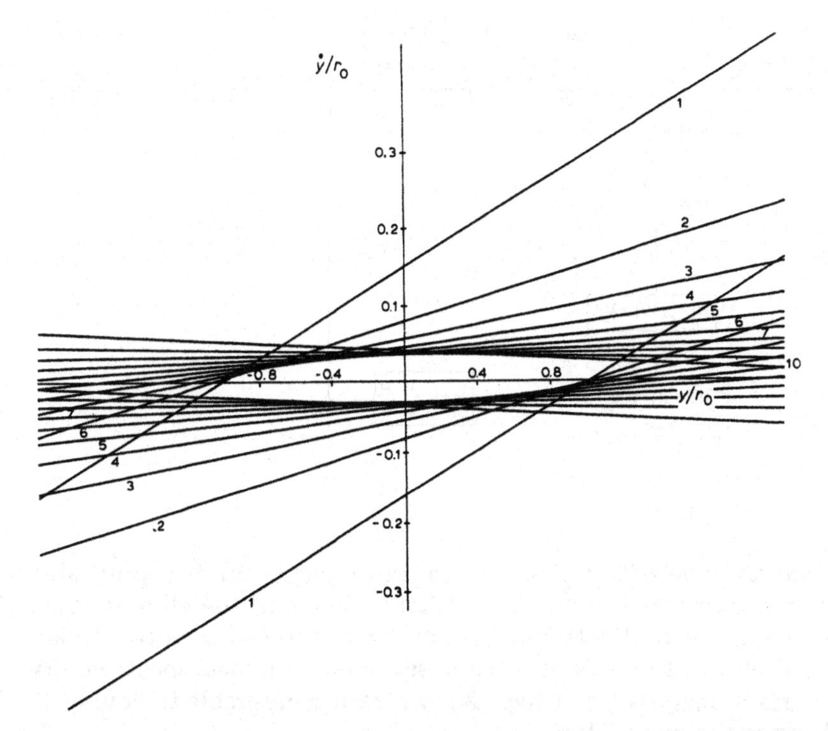

Fig. 4.6. Acceptance of a mass filter of limited length for the y direction with $a = 0.23342$, $q = 0.706$ and the initial phase $\xi_0 = 0.5$. The pairs of parallel lines show how the acceptance is successively restricted if the filter is 1,2,3... etc. cycles long. β_y is approximately 0.067. The ultimate acceptance ellipse is quite well delineated after 7 rf cycles.

References pp. 92—93

TABLE 4.1

The computation of transmission characteristics using phase-space dynamics

There are three steps; (a) calculation of matrices for fractions of the rf cycle, (b) derivation of matrices representing complete rf cycles beginning at different phases and calculation of the corresponding acceptance ellipse parameters, and (c) calculation of transmission from an assumed initial distribution.

(a) $M[1]$ represents the effect of the field on $\begin{bmatrix} u \\ \dot{u} \end{bmatrix}$ between phases ξ_0 and $(\xi_0 + \pi/20)$, $M[2]$ that between $(\xi_0 + \pi/20)$ and $(\xi_0 + \pi/10)$ etc.

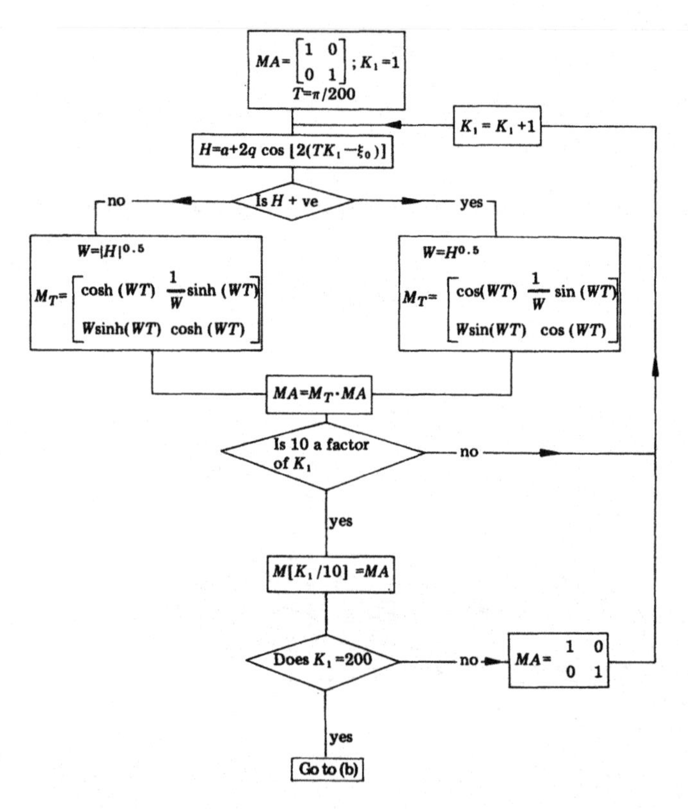

diverging beam and one with $\pi/2 < \theta < \pi$ a converging beam. It is profitable to consider the acceptance of the mass filter in terms of the ellipses when considering fringing field effects and the coupling of ion sources (see Chap. V, p. 104) and also in those cases, such as secondary ion mass spectrometry (SIMS) for surface analysis (see Chap. X), where it is desirable to couple an energy analyser to the mass filter.

Table 4.1 illustrates the use of matrix methods to calculate the percentage of ions transmitted under a given set of initial conditions. The calculation has been simplified to its essential steps and only one coordinate direction is

TABLE 4.1 (*continued*)

(b) M_{0-19} represent the matrices for complete rf cycles beginning at different phases.

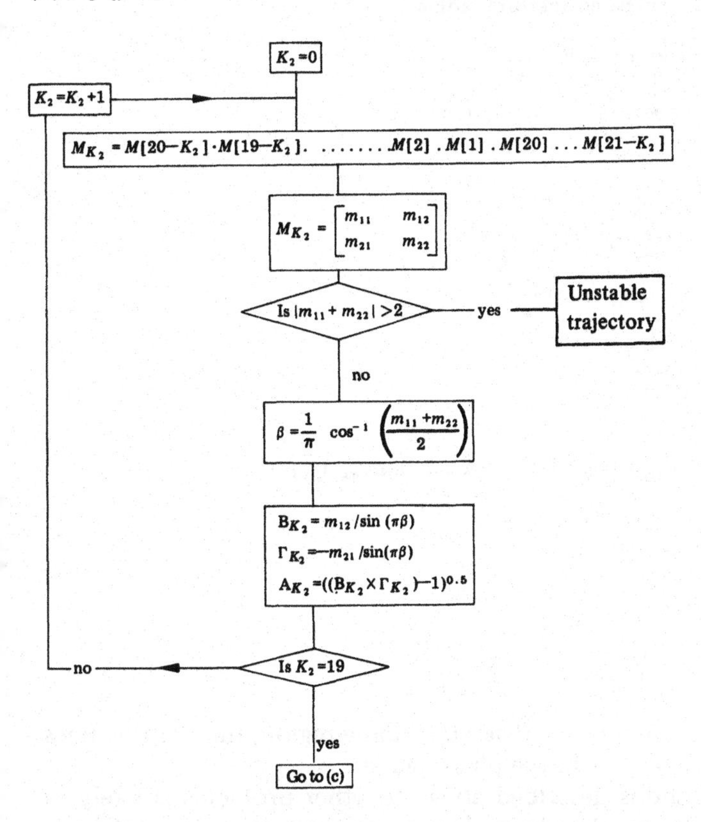

considered. An analogous approach considering two directions simultaneously was used in calculating the results shown in Figs. 2.20–2.22 and Figs. 2.34–2.36.

Thus the matrix approach not only simplifies long-term trajectory calculations but allows rapid computation of device acceptance for any combination of initial position and initial velocity for the "ideal" case of an infinitely long sojourn in the rf field. This has enabled detailed computer simulations of mass filter and ion trap performance [24, 25] which take into account hundreds of different initial positions and velocities.

For a device of limited length, the condition that the point of maximum displacement during each rf cycle be less than r_0 can be represented as a series of pairs of parallel lines in the phase plane. These lines are tangential to the acceptance ellipse as shown in Fig. 4.6 and rotate around the ellipse each period of $\tau_0/2$ [$1/\beta_y$ or $1/(1-\beta_x)$ rf cycles]. The acceptance is that region contained between *all* the lines up to the number n of cycles being considered [31] In the case illustrated, the final acceptance ellipse is quite well delineated

TABLE 4.1 (*continued*)

(c) F represents the distribution of initial ion positions, $\dot{F}$ represents the *corresponding* initial velocities, and r_a is the entrance aperture radius.

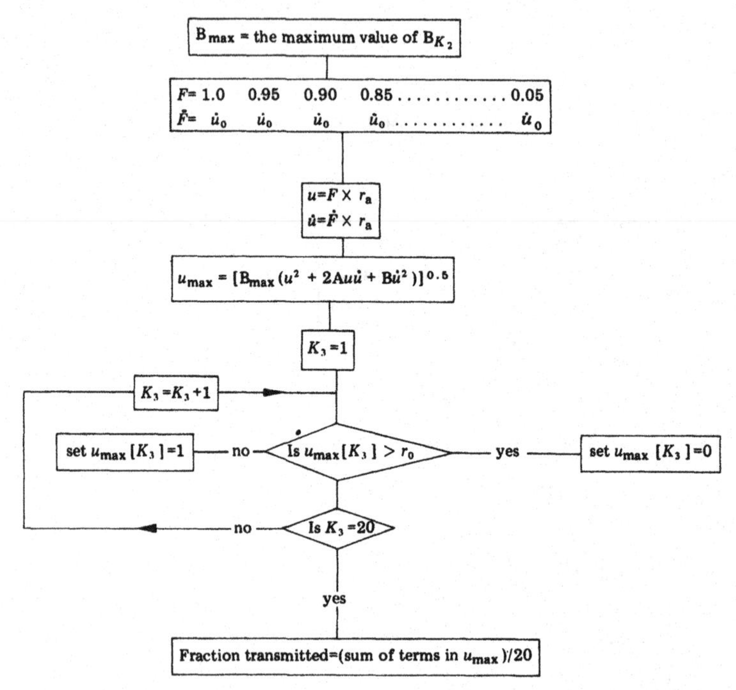

when $n > 1/2\beta_y$. For (a, q) points outside the stable region, the limiting lines are tangential to the hyperbolae in the phase plane.

Extension of the methods described above to other problems is likely in the near future. Studies are already underway on evaluating the optimum storage conditions in the ion trap for the production of highly charged species [22], on calculating ion velocity distributions in the ion trap [32] and on calculating the monopole acceptance [31]. One might also expect the development of matrices larger than the usual $[2 \times 2]$ to represent the fringing field region.

REFERENCES

1 W.M. Brubaker, Advan. Mass Spectrom., 4 (1968) 293.
2 P.H. Dawson and N.R. Whetten, J. Vac. Sci. Technol., 5 (1968) 11.
3 E.P. Sheretov and V.A. Zenkin, Sov. Phys. Tech. Phys., 17 (1972) 160.
4 J.M. Nitschke, Nucl. Instrum. Methods, 78 (1970) 45.
5 R.F. Lever, IBM J. Res. Develop., 10 (1966) 26.
6 P.H. Dawson, Int. J. Mass Spectrom. Ion Phys., 12 (1973) 53.
7 P.H. Dawson and N.R. Whetten, Int. J. Mass Spectrom. Ion Phys., 2 (1969) 45.
8 P.H. Dawson and N.R. Whetten, Int. J. Mass Spectrom. Ion Phys., 3 (1969) 1.

9 P.H. Dawson and N.R. Whetten, J. Vac. Sci. Technol., 7 (1970) 440.
10 P.H. Dawson, J. Vac. Sci. Technol., 8 (1971) 263.
11 P.H. Dawson, Int. J. Mass Spectrom. Ion Phys., 6 (1971) 33.
12 P.H. Dawson, J. Vac. Sci. Technol., 9 (1972) 487.
13 P.H. Dawson and N.R. Whetten, Rev. Sci. Instrum., 39 (1968) 1417.
14 P.H. Dawson and N.R. Whetten, J. Vac. Sci. Technol., 6 (1969) 97.
15 H.M. Powell, Ph.D. Thesis, Vanderbilt University, 1966.
16 M. Mosharrafa and G.M. Wood, 19th Annu. Conf. Mass Spectrom. Allied Topics, Atlanta, Georgia, 1971.
17 T. Chisholm and A.M. Stark, J. Phys. D., 3 (1970) 1717.
18 J.K. Reid, Comput. J., 9 (1966) 200.
19 C. Weber, Philips Res. Rep. Suppl., No. 6 (1967).
20 J.A. Zonneveld, Automatic Numerical Integration, Amsterdam Mathematisch Centrum, 1964.
21 A.B. Birtles, J. Phys. D., 5 (1972) 1396.
22 M. Baril and A. Septier, Rev. Phys. Appl., 9 (1974) 525; C. Schwebel, P.A. Moller and P.T. Manh, Rev. Phys. Appl., 10 (1975) 227.
23 P.H. Dawson, Int. J. Mass Spectrom. Ion Phys., 14 (1974) 317.
24 P.H. Dawson and C. Lambert, Int J. Mass Spectrom. Ion Phys., 16 (1975) 269.
25 P.H. Dawson, J. Vac. Sci. Technol., 11 (1974) 1151.
26 J.A. Richards, R.M. Huey and J. Hiller, Int. J. Mass Spectrom. Ion Phys., 12 (1973) 317.
27 L.A. Pipes, J. Appl. Phys., 24 (1953) 902.
28 J.A. Richards, Int. J. Mass Spectrom. Ion Phys., 17 (1975) 17.
29 A.J. Lichtenberg, Phase-Space Dynamics of Particles, Wiley, New York, 1969.
30 K.G. Wittig, High Energy Beam Optics, Interscience, New York 1965.
31 R. Baribeau and P.H. Dawson, to be published.
32 P.H. Dawson, Int. J. Mass Spectrom. Ion Phys., to be published.

CHAPTER V

FRINGING FIELDS AND OTHER IMPERFECTIONS

P.H. Dawson

In Chapter II, the principles of operation were described for quadrupole mass spectrometers with perfect fields; that is, with perfect geometry, perfectly sinusoidal applied voltages and without fringing fields. Imperfections in the quadrupole fields fall into three categories. The most important are imperfections caused by the presence of fringing fields at the ion entrance and ion exit. For the ion trap, ions are formed directly within the field and it is exempt from this problem. However, in the mass filter and monopole, fringing fields are always present to some extent. Fringing fields at the exit have not been examined in much detail, although special efforts are generally made to ensure collection of all exiting ions even if they are highly defocussed (see p. 34). The exact influence of fringing fields at the ion entrance is the subject of some debate. Some experimental evidence exists but more detailed measurements are needed. The fringing fields certainly become detrimental when the ions take more than three or four cycles to pass through. This is most likely to occur for low-energy and/or high-mass ions when one is attempting to obtain high performance. Various modifications to the ion entrance conditions have been proposed to avoid the problems (see Appendix F), a notable example being the "delayed d.c. ramp" of Brubaker [1] which is described below. Most calculations of fringing fields have been made by the computation of ion trajectories in the xz or yz planes in monotonically increasing fields. Such calculations have proved very useful although they involve some significant approximations. The results suggest that short fringing fields play an important role in the functioning of the monopole and a sometimes beneficial role in normal mass filter operation. The limited experimental evidence supports the general conclusions of the calculations. Recently, fringing fields have been incorporated in a phase-space dynamics approach, which should permit much better ion source evaluation and design.

The second type of imperfection results from systematic field faults which can be represented by modifications to eqn. (2.6) for the potential. These may be geometric faults caused by slight misalignment of the electrodes, by mis-shaping of the electrodes, or by modification of the field due to the electrode housing or to nearby insulators. Systematic faults may also result from harmonics mixed in with the fundamental frequency of the applied rf. Most of the systematic faults become more important as a higher performance is demanded. They take an added significance the longer the time the ions spend within the field and the closer one operates to the limits of the stability diagram. The influence of such field faults has been extensively studied, both by analysis of the equations of motion and by the examination of ion trajectories. The experimental evidence supporting the calculations is quite strong.

The third type of imperfection is a local imperfection of the field, due, for example, to local contamination of an electrode surface or to exposure of some insulator surface. Such surfaces build up charges which perturb the nearby field. Evidently, these localized problems cannot be dealt with in a general way. They should, however, be less important than systematic faults which perturb the ion trajectory throughout its length.

A. FRINGING FIELDS

For the mass filter and the monopole, as for many other types of mass spectrometer, the inevitable presence of fringing fields poses some difficult problems, but also presents opportunities for improving instrument performance. Brubaker was the first to consider fringing fields in detail. Examining the angular acceptance of a mass filter intended for space applications, he found it was particularly poor in the yz plane [2]. While approaching the mass filter entrance, ions are subjected to weaker fields represented, for example, by (a, q) values distributed along the operating line from the origin to the stability tip (see Fig. 2.10). These lower (a, q) values represent regions of incipient instability for trajectories in the yz plane (but not for the xz plane). If the ions spend a large number of rf cycles passing through the fringing fields, the y displacements become very great by the time the ions enter the full field of the mass filter. To avoid the ions passing through the region of instability, Brubaker [1, 3] has developed various modifications of mass filter entrance conditions. They are generally referred to as the "delayed d.c. ramp". The delayed d.c. ramp has been used successfully by Brubaker but has not found general application (see Chapter VI, p. 144).

There are some other more subtle effects of fringing fields. Even very short fringing fields drastically modify the relationship of maximum ion displacement to initial phase (see Chapter II, p. 24), and also change the acceptance ellipses of the mass filter (p. 26). It is this modification that can be beneficial in the case of the mass filter and may be of vital importance in the monopole. The potential advantages of short fringing fields were evident in trajectory calculations by Brukaker [1] and have been further examined by Dawson [4–7]. The phase-space approach [8] is more useful in predicting the mass filter acceptance in the presence of various fringing fields and in evaluating ion source designs for compatibility.

The weaknesses of the calculations lie in the assumptions that in the

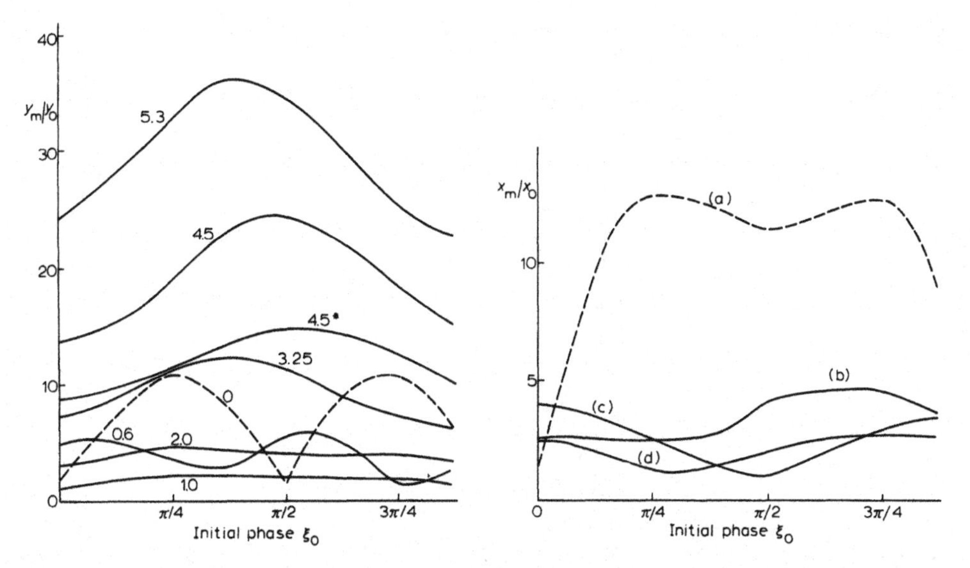

Fig. 5.1. The maximum y displacement in the mass filter as a function of the initial phase for the point $a = 0.228$ $q = 0.7$. The length of the assumed linear fringing field is given on each curve. The curve marked 4.5* was for a quadratically increasing field. The initial ion direction was assumed parallel to the instrument axis.

Fig. 5.2. The maximum x displacement as a function of the initial phase for the point $a = 0.228$ $q = 0.7$ assuming no initial x direction velocity. (a) Zero fringing field; (b) 2 cycle coincident rf and d.c. linear fringing fields; (c) 4.5 cycle coincident rf and d.c. linear fringing fields; (d) 4.5 cycle rf and d.c. fringing fields, the d.c. being delayed by 1.6 cycles.

fringing field region, the motion in the x and y directions remains independent, that the field in the z direction is zero, and that the x and y fields increase in some mathematically convenient way (linearly or quadratically) as one approaches the mass filter entrance. There is, however, a particular form of fringing field which is amenable to the calculation of the coupled trajectories in three dimensions and this has recently been used to establish limits for the validity of the two-dimensional linear approximations [8].

(1) *The mass filter*

The ion trajectories for the yz plane that are illustrated later (Fig. 5.14) for different initial phases of the rf field demonstrate the modification caused by a short fringing field [6] (compare with illustrations in Chapter II). The gradual entry into the field seems to produce a near equivalence for all initial phases. Figure 5.1 shows the effect on the maximum y displacement as a function of initial phase for various linearly increasing entrance "ramps" [6] calculated for an ion beam approaching the mass filter parallel to its axis. The number on each curve indicates the length in rf cycles of the fringing field. Fringing fields as short as 0.6 rf cycles have a profound influence and fields up to about

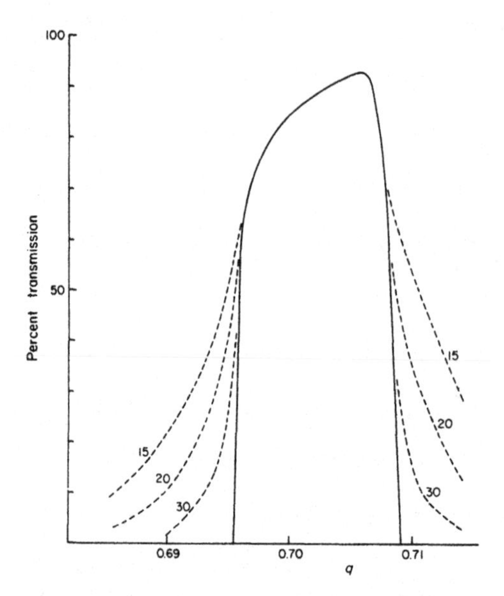

Fig. 5.3. A peak shape by a detailed computer simulation for an ion beam entering parallel to the instrument axis when there is a fringing field through which the ions take two cycles to pass. The broken lines indicate the shapes for fields limited in length to the stated number of rf cycles.

3 cycles long are beneficial. When the fields are longer than 3 cycles, there is a large reduction in instrument aperture (unless a delayed d.c. ramp is used). The curve labelled 4.5* shows a calculation for a ramp which increases quadratically as the ion approaches the mass filter entrance. Similar calculations for the xz plane are illustrated in Fig. 5.2. Averaged over all phases, even long entrance ramps are beneficial in the x direction.

The detailed computer simulations of mass filter performance by matrix methods (see Chapter II, p. 31) have been repeated including the presence of a two-cycle field [7]. Again, a beam entering parallel to the instrument axis was considered. There was an increase in transmission throughout most of the rf cycle and a decrease of the dependence on the zero and $\pi/2$ phases (cf. Fig. 2.20). Figure 5.3 shows some computed peak shapes. A comparison with the peak shapes for "perfect" fields (Fig. 2.21) indicates an increase of usable aperture area by a factor of three or four.

The introduction into the calculations of a realistic length for the mass filter is associated with "tailing" on the side of the peaks, particularly on the low mass side. The decrease of importance of the zero and $\pi/2$ phases in the presence of the fringing fields results in less peak tailing and more symmetrical peaks.

All the illustrations of fringing field effects given above consider the simplified case of ion beams parallel to the instrument axis with no angular divergence. Phase-space dynamics (see Chapter IV) provides a more complete analysis and permits the evaluation of ion source/lens combinations. The acceptance of the mass filter without fringing fields was illustrated in Figs. 2.14 and 2.15. The acceptance ellipses are characterized by the parameters A, B, Γ and ϵ where

$$\Gamma u^2 + 2\mathrm{A}u\dot{u} + \mathrm{B}\dot{u}^2 = \epsilon \tag{5.1}$$

Considering the usual two-dimensional linear approximation with separation of ion motion in the yz and xz planes, the effect of a fringing field can be represented in matrix form by

$$\begin{bmatrix} u \\ \dot{u} \end{bmatrix}_{\mathrm{e}} = \begin{bmatrix} C & S \\ C' & S' \end{bmatrix} \begin{bmatrix} u \\ \dot{u} \end{bmatrix}_{\mathrm{a}} \tag{5.2}$$

where u_e and $\dot{u}_e$ are the position and velocity at the point where the full mass filter field begins and u_a and $\dot{u}_a$ those at the entrance aperture in front of the quadrupole where the fringing field begins. It is then readily shown [9] that the overall acceptance of the mass filter with its fringing field is characterized by $\mathrm{A_a}$, $\mathrm{B_a}$, Γ_a and ϵ where

$$\begin{bmatrix} \mathrm{B_a} \\ \mathrm{A_a} \\ \Gamma_{\mathrm{a}} \end{bmatrix} = \begin{bmatrix} S'^2 & 2SS' & S^2 \\ C'S' & C'S + CS' & CS \\ C'^2 & 2CC' & C^2 \end{bmatrix} \begin{bmatrix} \mathrm{B} \\ \mathrm{A} \\ \Gamma \end{bmatrix} \tag{5.3}$$

The matrix elements C, C', S and S' are determined from two trajectory calculations for each initial phase. Figures 5.4–5.8 show the acceptance ellipses for both x and y directions for a filter with fringing fields of 0.5, 1, 2, 4 and 6 rf cycles for the operating point $a = 0.23342$ and $q = 0.706$. Ten different initial phases are illustrated in each case. The operating point is on a scan line giving a resolution of about 55. For the y direction, the alignment of the ellipses for different phases is remarkable for fringing fields greater than one rf cycle. These calculations are much more complete than the earlier ones, since the allowable transverse velocities are taken into account. Furthermore, for optimum sensitivity, the emittance of an ion source/lens system should be matched to the acceptance of the mass filter. Therefore, for both x and y directions, a converging beam is preferable (positive u associated with a negative $\dot{u}$). Particularly of interest are the areas in phase-space corresponding to 100% or to 50% transmission. These can be approximated by ellipses and the parameters A, B, Γ and ϵ determined (see Fig. 4.5). Appendix B lists some values for 50% transmission. These values can be used in the design of appropriate ion sources or in the evaluation of the mass filter as a detector in ion scattering experiments [10] such as secondary ion mass spectrometry (SIMS) (see Chapter X).

Assuming exact matching of the ion source to the mass filter under all conditions, the optimum relative sensitivities for fringing fields of varying lengths

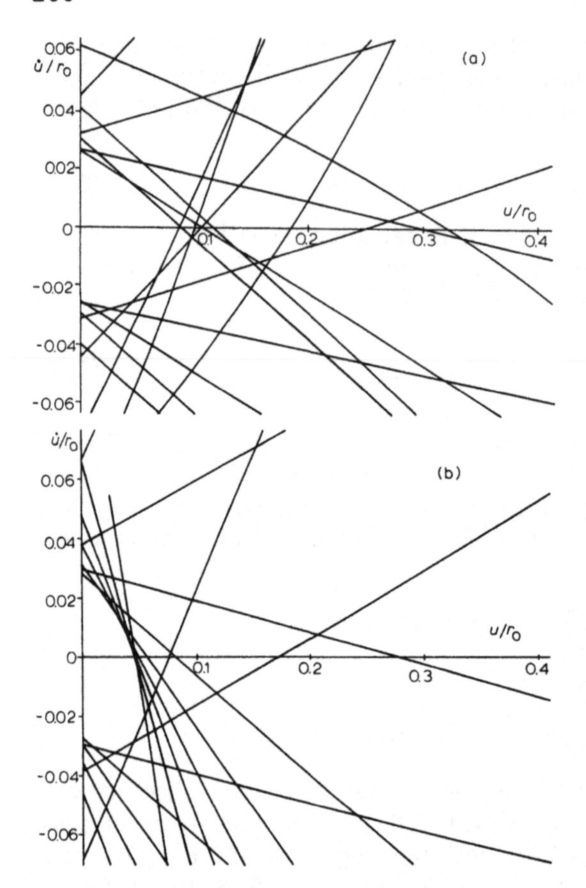

Fig. 5.4. The acceptance ellipses for the mass filter at the operating point $a = 0.2334$ $q = 0.706$ when there is a fringing field of 0.5 rf cycle. Ellipses are shown for 10 different initial phase. (a) The y direction; (b) the x direction.

can be compared by calculating the product of the acceptance areas for 50% or 100% transmission in the x and y planes. There is not a smooth variation of sensitivity with fringing field length but the general variation is shown in Fig. 5.9. A mass scale has been added for the case where all ions have the same axial energy given by $E_z = 15.5r_0^2$ where r_0 is in centimetres and the rf frequency has been assumed to be 2 MHz. The sensitivity variation is very similar to that found experimentally [11]. The use of acceptance area for 100% transmission as a criterion for sensitivity leads to a decrease of sensitivity in proportion to the square of the resolution.

Deceleration of ions in the fringing fields by operation of the mass filter axis off ground is sometimes used to introduce slow ions which have not spent too long in the fringing fields [8].

The validity of the approximations used in obtaining Figs. 5.4–5.9 has been

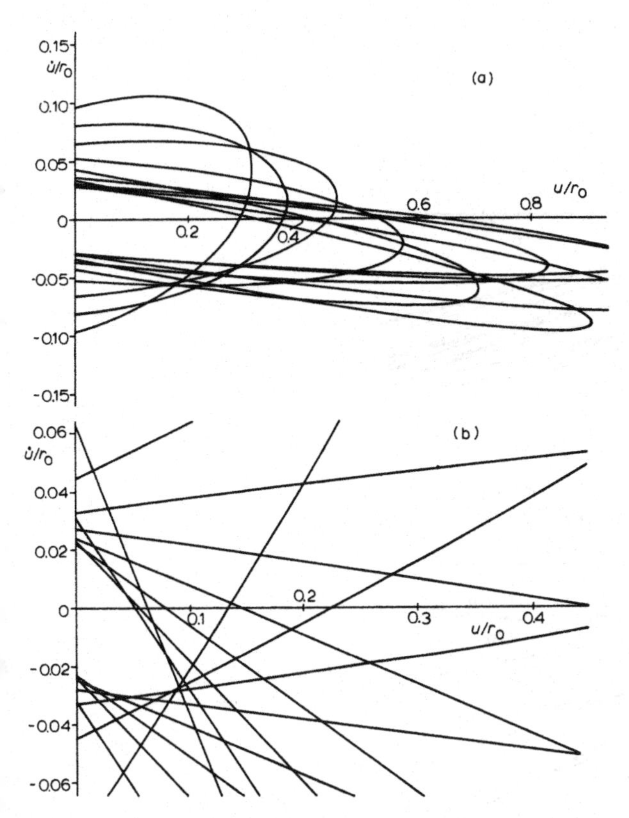

Fig. 5.5. The acceptance ellipses for ten different initial phases at the operating point $a = 0.2334$ $q = 0.706$ when the ions spend one rf cycle in the fringing field. (a) The y direction; (b) the x direction.

partially verified by spot-checks of trajectories calculated using a full three-dimensional field. The assumed potential is of the form

$$\Phi(x, y, z) = \frac{\Phi_0(x^2 - y^2)z}{r_0^3} \tag{5.4}$$

which leads to the equations of ion motion

$$\frac{d^2x}{d\xi^2} + [a - 2q \cos 2(\xi\ \xi_0)]x\frac{z}{r_0} = 0 \tag{5.5}$$

$$\frac{d^2y}{d\xi^2} - [a - 2q \cos 2(\xi - \xi_0)]y\frac{z}{r_0} = 0 \tag{5.6}$$

$$\frac{d^2z}{d\xi^2} + \frac{(x^2 - y^2)[a - 2q \cos 2(\xi - \xi_0)]}{r_0} = 0 \tag{5.7}$$

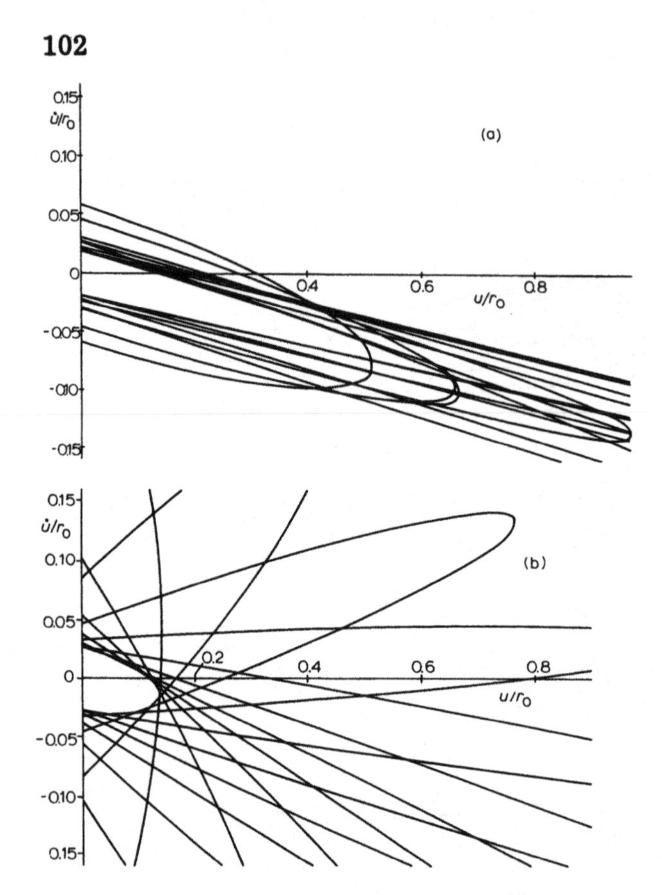

Fig. 5.6. The acceptance ellipses for (a) the y direction and (b) the x direction for the operating point $a = 0.2334$ $q = 0.706$ with a two cycle fringing field.

A similar potential would result from pole pieces rounded at the ends, but it is probably also a reasonable approximation for any case where the fringing field extends over a distance of r_0. When $(x^2 - y^2)$ is small throughout the fringing field trajectory, $dz/d\xi$ is almost constant and eqns. (5.5) and (5.6) become the usual linear approximations. When $(x^2 - y^2)$ becomes appreciable and when the initial value of $dz/d\xi$ is small, the axial velocity can be appreciably modified by the fringing field. Ions tend to be retarded, especially if their trajectory is mainly in the x quadrant, but may be accelerated if it is mainly in the y quadrant. As a result, there will be a distribution of axial velocities in the mass filter which, at high resolutions when slow-moving ions are being used, may limit the attainable resolution or contribute to peak tailing. Under some conditions, ions are reflected at the quadrupole entrance and occasionally may be trapped in the fringing fields for many rf cycles. However, these effects usually only become important for values of $u\dot{u}$ outside the 50% acceptance ellipses. The values given in Appendix B are probably quite good approximations for use in design work.

The potential given in eqn. (5.4) could be used to determine matrices of a

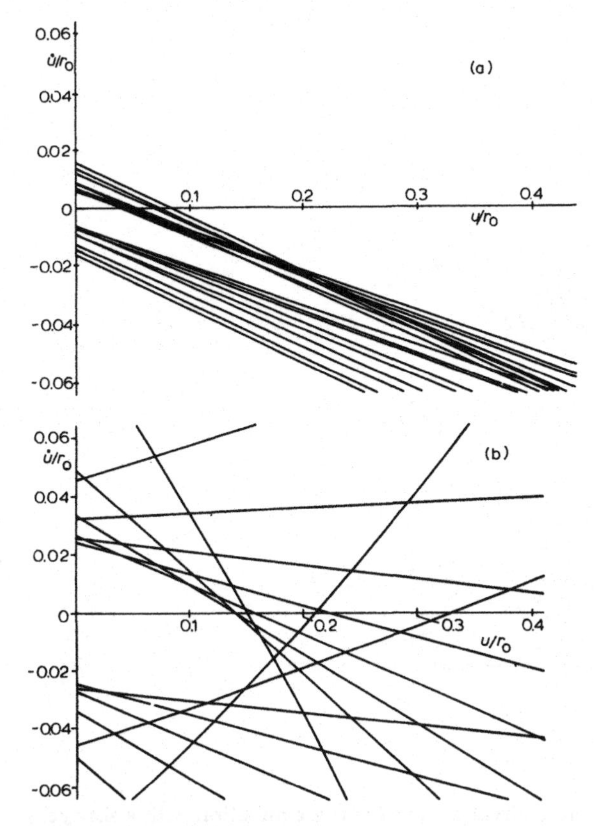

Fig. 5.7. Similar acceptance ellipses for (a) the y direction and (b) the x direction with a four cycle fringing field.

higher order than eqn. (5.2) to represent the fringing field.

An analogous situation exists at the ion exit except that now the emittance ellipses for the mass filter with exit fringing fields correspond to diverging beams (the sign of the parameter A is changed). Ions tend to be accelerated through the fringing exit fields but slow-moving ions emerging with large y values may be trapped within the fringing fields or, at certain phases, even reflected back into the device.

An example of the use of the acceptance parameters of Appendix B is illustrative of their potential value. Since allowable $\dot{u}$ values for a given u are similar even for ions with low axial velocities (Figs. 5.4–5.9), the optimum matching of the ion source requires more convergent beams for lower-velocity ions. This could be obtained with a lens, whose focal length is adjusted as the mass is scanned, between the ion source and the mass filter. A thin lens between the source, s, and the quadrupole aperture, a, gives

$$\begin{bmatrix} u_s \\ \dot{u}_s \end{bmatrix} = \begin{bmatrix} 1 & 0 \\ v_z^2/f_0 & 1 \end{bmatrix} \begin{bmatrix} u_a \\ \dot{u}_a \end{bmatrix} \tag{5.8}$$

References p. 119

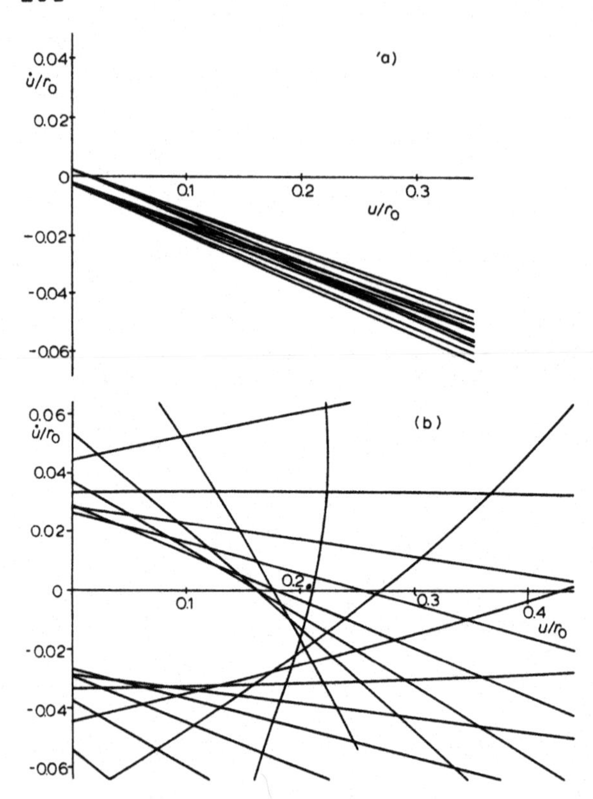

Fig. 5.8. Acceptance ellipses for (a) the y direction and (b) the x direction with a six cycle fringing field.

where f_0 is the focal length of the lens in units of r_0 and v_z is the axial velocity in units of r_0/radian of applied field. Therefore, for the 50% transmission

$$\begin{bmatrix} B_s \\ A_s \\ \Gamma_s \end{bmatrix} = \begin{bmatrix} 1 & 0 & 0 \\ -v_z/f_0 & 1 & 0 \\ v_z^2/f_0^2 & -2v_z/f_0 & 1 \end{bmatrix} \begin{bmatrix} B_a \\ A_a \\ \Gamma_a \end{bmatrix} \tag{5.9}$$

If the source is to be positioned symmetrically about the instrument axis, the axis of the source emittance ellipse should lie along the u direction. That is, $A_s = 0$ (see Chapter IV), and from eqn. (5.9)

$$\frac{1}{f_0} = \frac{A_a}{B_a}\frac{1}{v_z} \tag{5.10}$$

A_a/B_a is roughly constant for fields greater than about 1.5 rf cycles for either x or y directions. Suppose that the lens system is arranged so that v_z/f_0 is constant as was suggested above. A good value would be about 1/6 for the y direction and 1/9 for the x direction. These could be separately adjustable by use

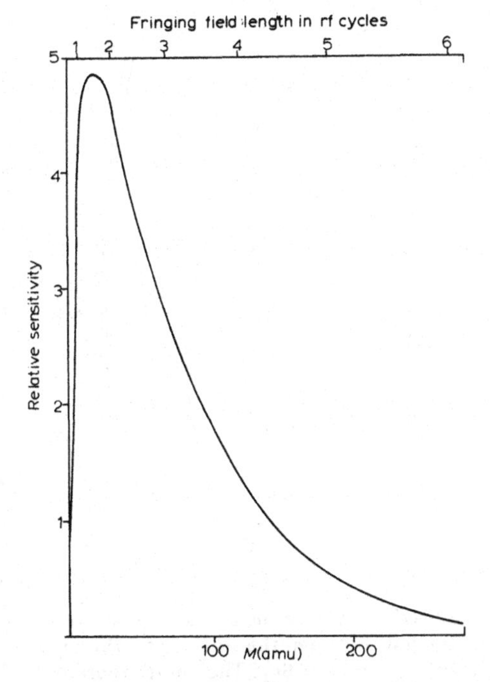

Fig. 5.9. The relative sensitivity of the mass filter as a function of the fringing field length (or of ionic mass if the axial energy is $E_z = 15.5r_0^2$) assuming good matching between the ion source emittance and the mass filter acceptance.

of an appropriate doublet or triplet quadrupole lens. Equation (5.9) can now be used to calculate the overall acceptance of the lens/fringing field/quadrupole system. Some results are shown in Fig. 5.10 which plots acceptable source combinations of u/r_0 and tan θ, the angle of divergence with the axis. This could be used to evaluate ion source designs. This very recent work has yet to be verified experimentally but seems to offer a promising and simple approach to the complex problem of overall quadrupole design.

For very slow ions, such as ions of high mass and low energy, modification of the entrance field may be necessary to obtain good sensitivities. The delayed d.c. ramp pioneered by Brubaker [1] is illustrated in Fig. 1.1(b). The original version consisted of segmented rods with short sections near the mass filter entrance to which only the rf voltages were applied. The delay in the application of the direct voltage means that as ions approach the mass filter, the (a, q) value moves first along the q axis ($a = 0$) and only later does the a value increase until the stability tip is reached when the ions enter the full quadrupole field. In this way, the ion always remains in conditions where the y trajectory is mathematically stable. Another later version, which is simpler to achieve, has four small electrodes ahead of the mass filter with appropriate d.c. voltages applied in order to partially nullify the d.c. fringing fields [3].

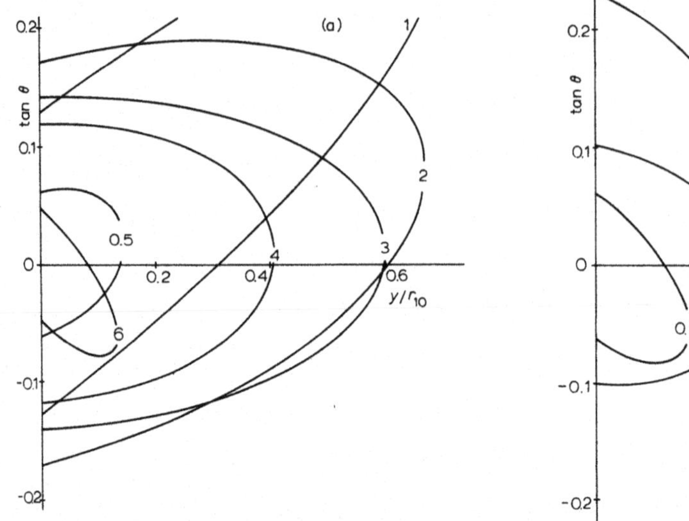

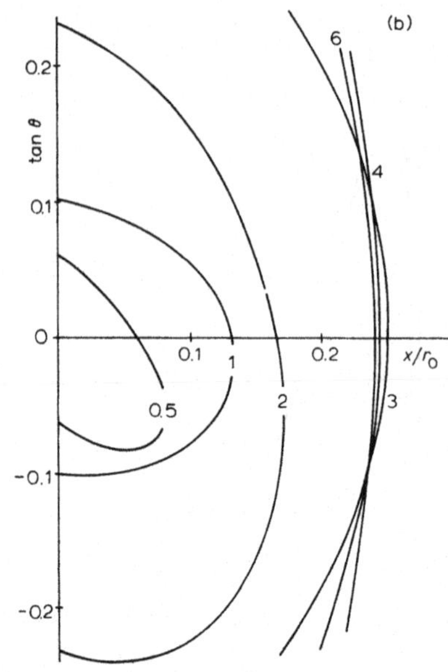

Fig. 5.10. (a) The acceptance of ions from the source given in terms of the angular divergence θ and the initial displacement y/r_0 when the lens between the source and the mass filter is always adjusted so that its focal length is given by $f_0^y = 6v_z$. The curves represent ions of different velocities (or different masses if E_z is fixed, see Fig. 5.9) and each is labelled with the length of the fringing field in rf cycles. (b) Similar acceptance curves for the x direction when the lens focal length is $f_0^x = 9v_z$.

This arrangement has the distinct advantage that the voltages applied can be adjusted empirically to "tailor" the fringing fields for optimum performance.

Figure 5.11 shows calculations of the maximum y displacements for 4.5 cycle entrance ramps with the d.c. being delayed and shortened by the number of cycles shown on each curve. There is a remarkable improvement over the coincident ramps (Fig. 5.1). The improvement in the x direction is also retained as was shown in Fig. 5.2. The ability of the mass filter to accept ions of a large range of energies has been one of its attractive features, but there is always a possibility of discrimination against slow ions if a delayed d.c. ramp is not used. No phase-space analysis of the delayed d.c. ramp mode of operation has so far been made.

The experimental evidence to support the conclusions described above is somewhat fragmentary. The only detailed measurements are those of Holme and Thatcher [12]. By combining time-of-flight, ion counting and signal averaging techniques, they were able to measure the rf modulation of the transmitted ion beam as a function of both the length of the fringing field and the resolution setting of the quadrupole. The fringing field length was varied both by changing the ion energy and by changing the separation of the ion source from the analyser. The experiments were carried out with 15 cm and 5 cm long

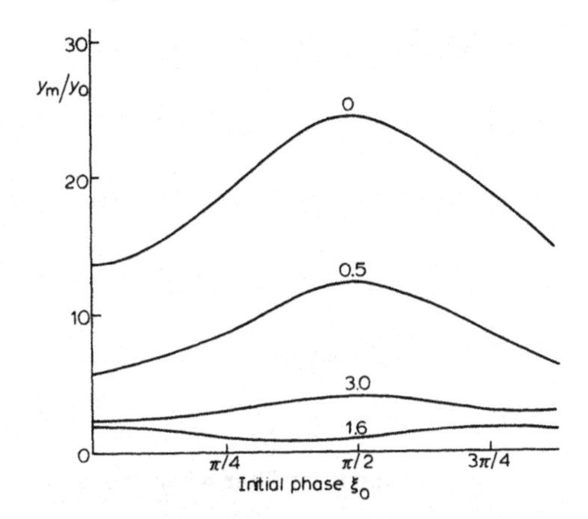

Fig. 5.11. Maximum y displacements as a function of the initial phase for the operating point $a = 0.228$ $q = 0.7$ when a delayed d.c. ramp is used. The fringing fields were 4.5 cycles long but the d.c. component was delayed by the number of cycles indicated on each curve.

analysers with $r_0 = 0.276$ cm. An example of the observed signal modulation for a short fringing field is shown in Fig. 5.12. It was found that the percentage modulation increases sharply as the number of rf cycles spent in the field approached zero. For fringing fields of more than about three rf cycles, the signal modulation was quite small. For a given fringing field length, the modulation increased as the resolution was increased. Holme and Thatcher concluded that, under normal operating conditions of most quadrupole mass filters, the phase dependence is of limited importance except perhaps at high resolution. The importance of the modulation under "normal" conditions has been disputed. Lefaivre [13], who was interested in the advantages of pulsed ion entry at the optimum phase (see Chapter II, p. 33), demonstrated that, for his mass filter, there was a high signal modulation even at low resolutions. Part of the discrepancy is probably accounted for by the difference in quadrupole size. Lefaivre's rods were hyperbolic and 110 cm long with a r_0 of 2 cm.

The other experimental evidence for the mass filter is indirect, i.e. the success of the delayed d.c. ramp. An illustration of its effectiveness, taken from the work of Brubaker [1], is given in Fig. 5.13. The 15 eV ions probably spent less than three cycles in the fringing field, but there is still an improvement at low resolutions. However, the *general* utility of the delayed d.c. ramp is still being debated (see Chapter VI, p. 144).

(2) *The monopole*

Application of the results shown in Figs. 5.1 and 5.2 to the monopole, suggests a dramatic change in instrument performance when there are fringing fields. Normally, only 50% of the ions can be transmitted because, during

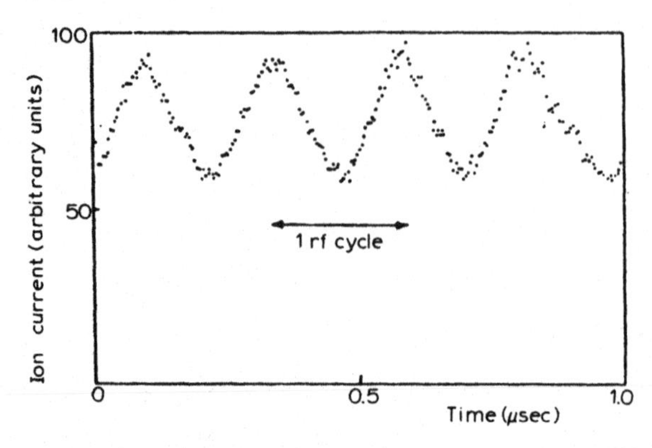

Fig. 5.12. Experimental modulation of the output signal of the mass filter when the fringing field was very short [12].

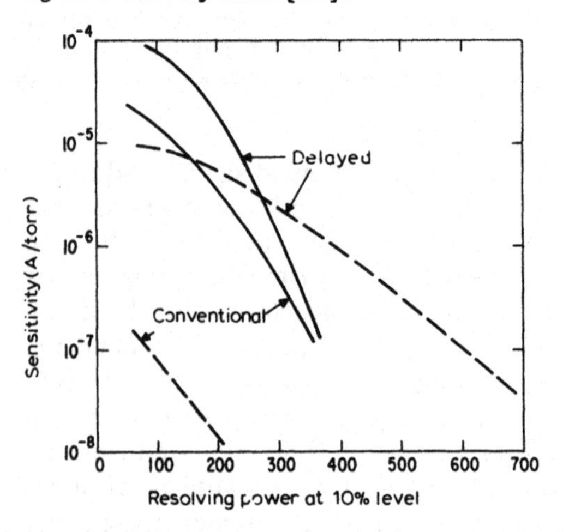

Fig. 5.13. Experimental measurements demonstrating the application of the delayed d.c. ramp technique [1]. The full curves were for 15 eV ions, the broken curves for 4 eV ions. The mass filter radius was 0.75 cm.

half the rf cycle, the ions are pushed towards the V block. Furthermore, the calculated peak shapes are very asymmetric with long "tails" on the low mass side (see Fig. 2.28). The fringing field will, by reducing the x direction ion displacement, facilitate meeting the condition $|x| < y$ and increase instrument aperture. However, more important is the change in the phase dependence of ion transmission. Figure 5.14 illustrates trajectories giving 100% ion transmission [6] for the conditions $a = 0.0825$, $q = 0.4125$ when the ions take 3.5 rf cycles to pass through a linear fringing field and 15.75 cycles to pass through the full analyser field. For clarity, only 4 phases are shown during the first part of the field, but 8 later. Trajectories are quite similar no matter what the initial field phase.

Summing the phases transmitted as one moves along a given q/a scan line gives a peak shape. Of course, this only indicates the influence on peak shape

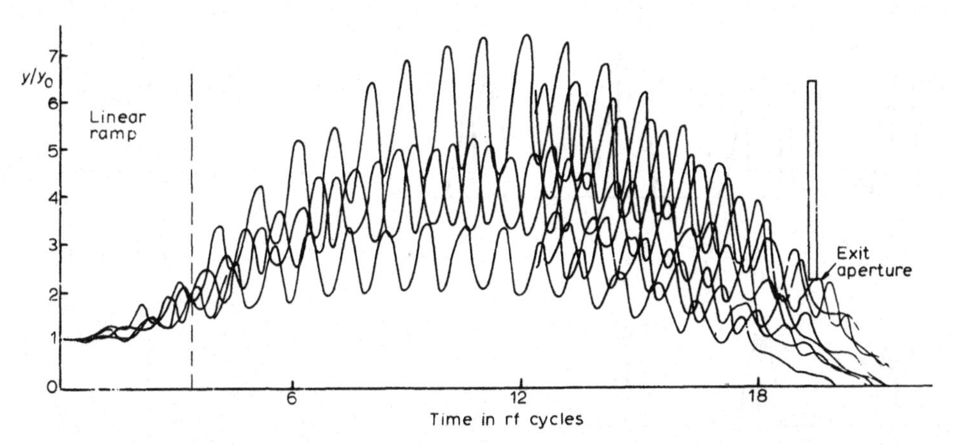

Fig. 5.14. Trajectories in the yz plane in the monopole for the operating point $a = 0.0825$ $q = 0.4125$ when there is a 3.5 cycle fringing field and the full analyzer field is 15.75 cycles long. Ion transmission can be 100%. Several different initial phases spread throughout the rf cycle are shown.

of the initial phase and ignores other factors such as the distribution of initial ion positions. Two results for 3.5 cycle fringes and different field lengths were given in Fig. 2.28(c). It seems likely that these calculations are closer to representing normal monopole performance than those of Fig. 2.28(a) and (b). The calculations show that a fringing field of more than 1.5 rf cycles is necessary for good performance.

A more complete view of monopole operation can be obtained by applying a phase-space analysis [13a].

There is some experimental evidence concerning fringing fields in the monopole, but it is qualitative and incomplete [6]. The output signal of a monopole was observed on an oscilloscope as the axial energy was varied. There was a distinct modulation of the signal at the rf frequency for ion energies corresponding to fringing fields less than about 1.6 rf cycles but no modulation was evident with a 2 cycle fringing field, suggesting that transmission was occurring throughout the rf cycle.

There are many problems in making phase dependence measurements but they could clearly be of considerable diagnostic value in both mass filter and monopole design. In particular, the ease of doing calculations using matrix methods (Chapter IV) provides the possibility of combining detailed performance calculations with quantitative measurements of ion transmission characteristics to arrive at an optimum instrument design.

B. SYSTEMATIC FIELD FAULTS

The question of the influence of field faults on the stability of ion trajectories was examined by von Busch and Paul [14] in the early days of

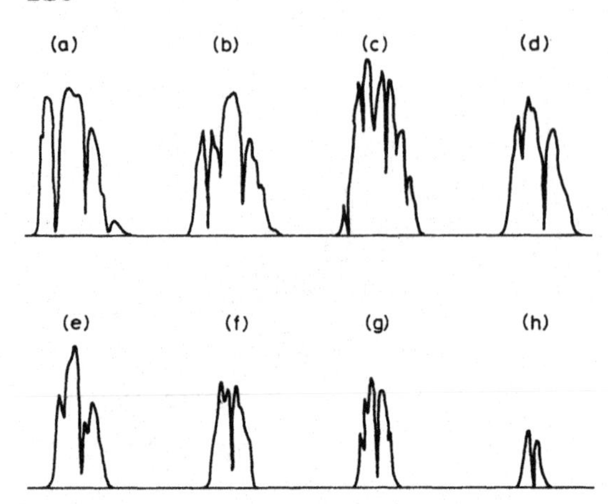

Fig. 5.15. Experimentally observed peaks in a quadrupole ion trap when the ions stayed up to 10^5 cycles in the field. The q/a value of the operating line was decreased from (a) to (h). The half-height resolution for the latter was about 18. The resonance dips can be plotted on the $a-q$ stability diagram.

quadrupole development, but the work was not widely known despite many problems with irregular peak shapes. The field faults can cause a gradual accumulation of "errors" in the ion trajectories for specific (a, q) values, so that a trajectory that would normally be stable becomes unstable or of very large amplitude. The (a, q) values where this occurs will show up in the scan of a peak (in the case of the mass filter or ion trap) as dips or shoulders. The problems will be more severe the greater the number of cycles an ion spends in the rf field and can be very severe in the ion trap operated in the selective storage mode (see Chapter II). Figure 5.15 shows a deliberately exaggerated case [15] for several scan lines of differing resolutions where ions were trapped for about 10^5 field cycles.

Von Busch and Paul showed that the "resonance dips" might be expected to occur along certain "resonance lines" in the a, q stability diagram, the position of the line depending on the type of field fault (see Fig. 5.16). The nature of these resonance lines is discussed next, using the mass filter as an example and later considering the ion trap. The extensive computer simulations are described and the experimental evidence is examined. The monopole is considered later as a special case.

(1) *The analytical approach*

In the most general case, where all types of field faults are present, the potential in the mass filter is given not by eqn. (2.9), but by a more general expression

$$\Phi = \sum_N A_N (r/r_0)^N \cos N(\phi - \phi_N) \left[U - \sum_n b_n V \cos (n\omega/2)(t - t_n) \right] \quad (5.11)$$

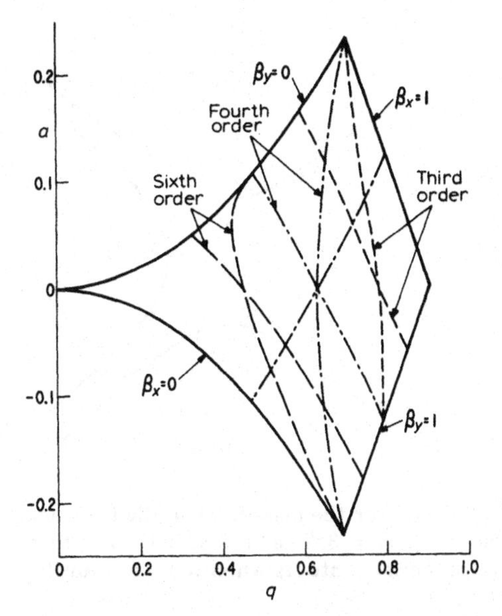

Fig. 5.16. Stability diagram for the mass filter showing the resonance lines caused by third, fourth- and sixth-order field distortions. Only one of the two possible third-order distortions is shown.

where r and ϕ are polar co-ordinates, A_N and b_n are weighting factors and the t_n are phase factors. For the ideal field, the only terms are those with $N = n = 2$ and eqn. (5.1) transforms into eqn. (2.9). Terms that have both $N \neq 2$ and $n \neq 2$ can be neglected. Consider first the case $n = 2$, that is, the time-varying term is purely sinusoidal and the field faults are geometric. In addition to the usual second-order term with $N = 2$, for which we take $A_2 = b_2 = 1$, there are higher-order terms. An N of three gives a term representing a hexapole field. Converting to Cartesian co-ordinates

$$\Phi_3 = \frac{A_3}{r_0^3}(3yx^2 - y^3)[U - V\cos\omega(t - t_2)] \tag{5.12}$$

Note that Laplace's equation is still obeyed but the restoring force on an ion is no longer linearly dependent on its displacement from the centre. There is also a coupling of motion in the x and y directions. The third order term is asymmetric with respect to the x and y directions. There are two possible third-order distortions. Figure 5.17(a) shows how the equipotential lines would be modified by the presence of a large third-order distortion with $A_3 = 3r_0$. A contribution of the third-order type would evidently be produced by an asymmetric positioning of one of the electrodes.

A fourth-order, or octopole, term with $N = 4$, is given by

$$\Phi_4 = \frac{A_4}{r_0^4}(x^4 - 6x^2y^2 + y^4)[U - V\cos\omega(t - t_2)] \tag{5.13}$$

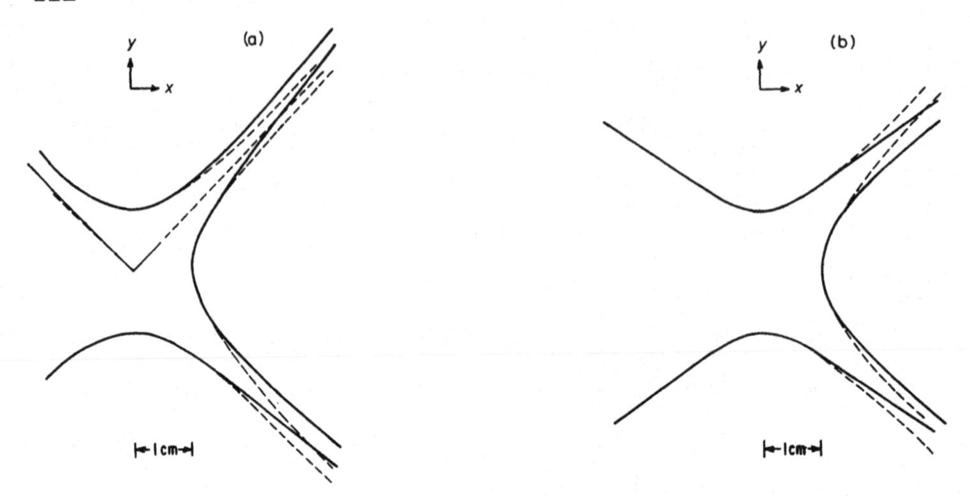

Fig. 5.17. (a) The solid lines are equipotential contours for the mass filter in the presence of a third-order distortion with a weighting factor $A_3/r_0 = 3$. The broken lines are equivalent contours with no field distortion. (b) Equipotential contours when there is a fourth-order distortion with a weighting factor $A_4/r_0^2 = 100$.

and Fig. 5.17(b) illustrates the kind of geometric error that is involved.

Higher-order terms are generally less significant but one of potential importance is the sixth-order distortion. This is the principal fault caused by the common use of round instead of hyperbolic rods. The sixth-order term is

$$\Phi_6 = \frac{A_6}{r_0^6}(x^6 - 15x^4y^2 + 15x^2y^4 - y^6)[U - V\cos\omega(t - t_2)] \tag{5.14}$$

Figure 5.18 shows how the round rod structure can be well represented by the addition of an appropriate sixth-order term.

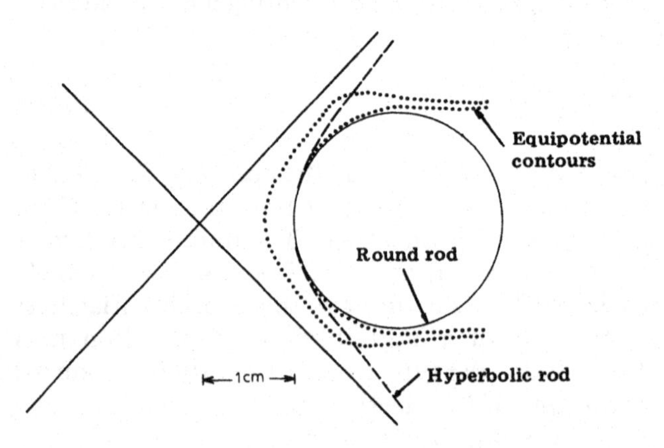

Fig. 5.18. Equipotential contours showing how the use of a round rod instead of a hyperbolic rod can be well represented by including a sixth-order term with a weighting factor $A_6/r_0^4 = 10^6$.

In the ion trap, the third- and fourth-order terms are similarly given [15] by

$$\Phi_3 = \frac{A_3}{4z_0^3}(3r^2z - 2z^3)[U - V\cos\omega(t-t_2)] \tag{5.15}$$

$$\Phi_4 = \frac{A_4}{4z_0^4}(r^4 + \tfrac{8}{3}z^4 - 8r^2z^2)[U - V\cos\omega(t-t_2)] \tag{5.16}$$

From the general theory of harmonic oscillators [14], one can deduce that the higher-order geometric terms produce sum resonances at those conditions where (for the mass filter)

$$\frac{\beta_x}{2}K + (N-K)\frac{\beta_y}{2} = 1 \tag{5.17}$$

where K can have the values N, N—2, N—4, . . . Thus the third-order resonance lines occur at

$$\beta_x = \frac{2}{3} \quad \text{and} \quad \frac{\beta_x}{2} + \beta_y = 1 \tag{5.18}$$

and evidently since the distortion is asymmetric, there is the other possibility

$$\beta_y = \frac{2}{3}, \quad \frac{\beta_y}{2} + \beta_x = 1 \tag{5.19}$$

The fourth-order resonances are at

$$\beta_x = \tfrac{1}{2}, \quad \beta_x + \beta_y = 1 \quad \text{and} \quad \beta_y = \tfrac{1}{2} \tag{5.20}$$

and the sixth-order distortion produces resonances at

$$\beta_x = \tfrac{1}{3}, \quad \beta_y = \tfrac{1}{3}, \quad 2\beta_x + \beta_y = 1, \quad \beta_x + 2\beta_y = 1 \tag{5.21}$$

Some of the resonance lines were shown in Fig. 5.16 for the mass filter and some corresponding lines for the ion trap are given in Fig. 5.19. The lines that pass through the tip of the stability diagram are particularly important.

The analytic theory gives the location of the resonance lines but does not indicate how severe the effects might be for a given amount of distortion. This can, however, be estimated by computer simulation.

The terms in eqn. (5.11) with $N = 2$, $n > 2$ representing harmonics in the rf supply are generally not very important. If the additional terms are sufficiently large, there will be a modification in the position of the zone of stability (see Chapter III). However, odd harmonics of the sub-frequency $\omega/2$, which may be present in circuits which use frequency doubling [14], might cause resonance lines at $\beta = 1/2$.

(2) *Computer simulation*

The technique of computer simulation of ion trajectories by numerical integration of the equations of motion (Chapter IV) has been applied to dis-

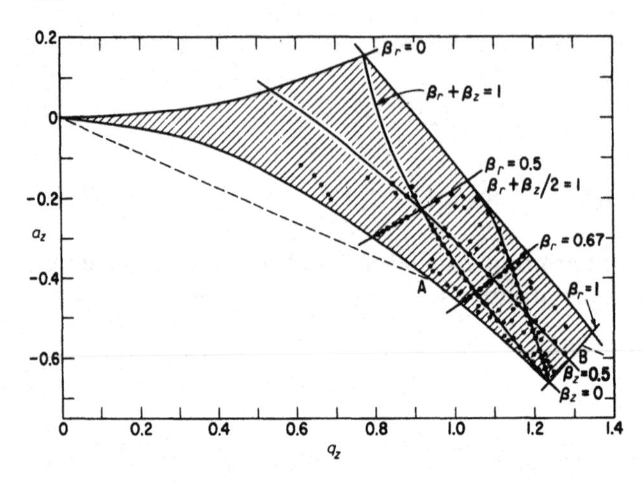

Fig. 5.19. Stability diagram for the ion trap showing the resonance lines caused by third- and fourth-order distortion. The points represent experimental measurements. The principal resonance dips fall on the theoretically predicted curves.

torted fields. The calculations are more complex since x and y motions are no longer independent. They have been carried out in detail only for one particular initial phase of the rf field. Figure 5.20 shows the peak shapes that are theoretically obtained with various scan lines for the ion trap operated in the selective trapping mode (see Chapter II), assuming that a maximum displacement of ten times the initial value is permissible before an ion is lost from the trap. Ions were considered to be accumulated during $200/\pi$ rf cycles and there was a third-order field distortion with a weighting factor A_3 equal to $10z_0/3$. The peak splitting becomes more severe as one approaches the tip of the stability diagram. If the positions of the principal resonance dips are plotted on an $a-q$ diagram, one can see a general agreement with the analytical theory. The field distortions also appear to produce a "blunting" of the tip of the stability diagram which would limit the attainable resolution. Such a limit is experimentally observed with ion traps of approximate geometry. Similar calculations have been made for both third- and fourth-order distortions in the mass filter [16].

The degree of field imperfection that can be tolerated in obtaining a given performance can be estimated from these calculations recalling, however, the limitation that only one initial phase has so far been considered. The results are presented in Fig. 5.21(a)–(c), as the number of rf cycles before an ion at the (a, q) position of the main resonance dip will exceed ten times its initial displacement versus the appropriate distortion weighting factor. The requirements are more severe the higher the resolution. Figure 5.21(a) is for a third-order distortion in the ion trap. At low pressures, the accumulation time in the ion trap is often very long (see Chapter VIII) even at low resolution and geometric distortions are very important. Figure 5.21(b) and (c) are for dis-

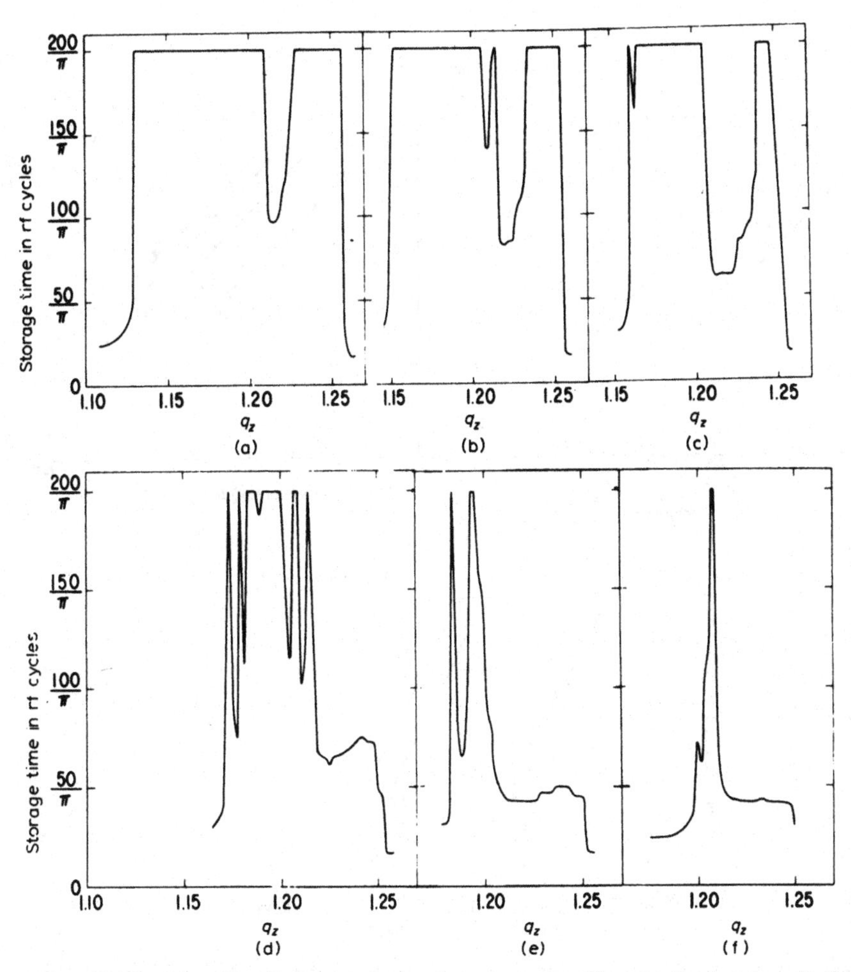

Fig. 5.20. Computed peak shapes for various operating lines in the ion trap in the presence of a third-order distortion with a weighting factor $A_3 = 10z_0/3$. The operating lines had the following q/a ratios: (a) 2.025; (b) 2.0; (c) 1.9875; (d) 1.975; (e) 1.9625; and (f) 1.95.

tortions in the mass filter. The broken lines represent von Zahn and co-workers' [17] criterion for an optimum quadrupole performance; that is, the resolution R is limited to $n^2/12.25$ where n is the number of cycles within the field. Using this criterion, which is more favourable than that often found (see Chapter VI, p. 123), the intersection of the lines indicates the degree of distortion that can be tolerated. The results are summarized in Fig. 5.22. For low performance, gross distortions are permissible, but conditions become stringent at high resolution [18] due to the combination of closer approach to the limit of the stability diagram and the necessity for more rf cycles in the field. Experimental evidence is presented in Chapter VI, Fig. 6.3, which is in accord with the theory.

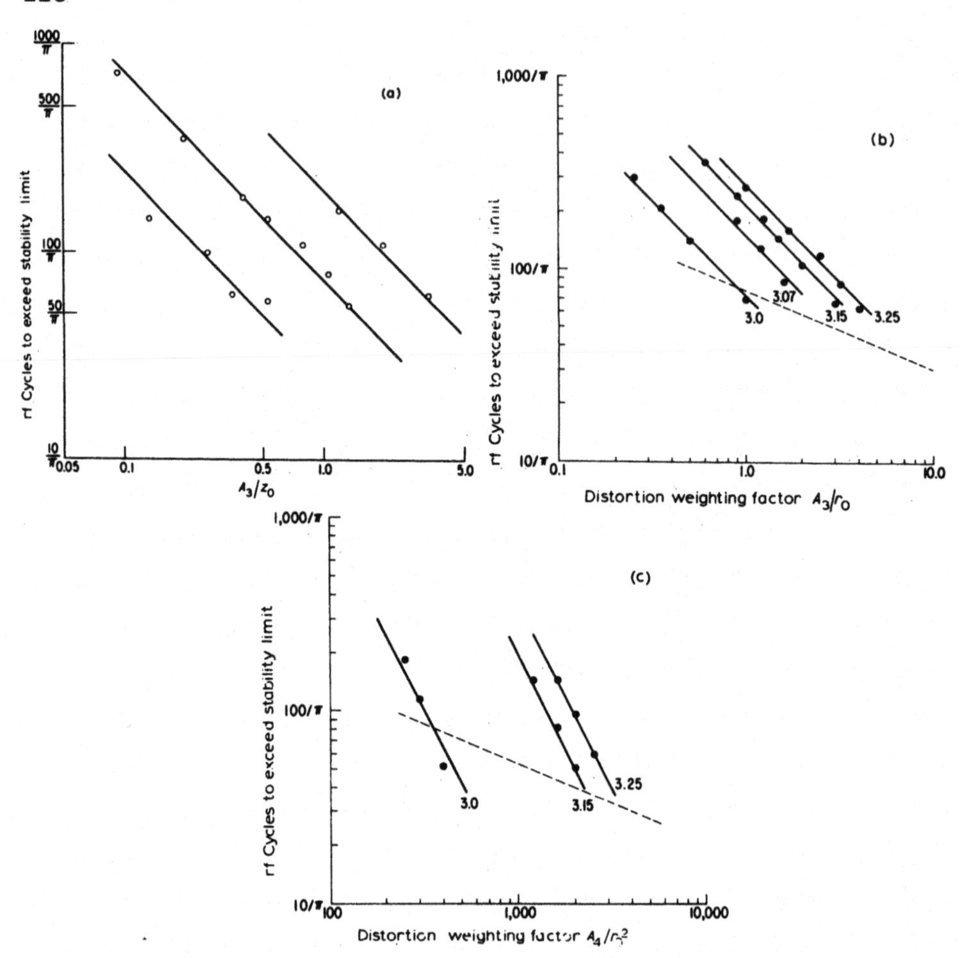

Fig. 5.21. (a) A log–log plot of the number of rf cycles needed before the stability limit is exceeded in the ion trap at the a, q position of the third-order resonance dip versus the distortion weighting factor. The curves are for operating lines having nominal resolutions of 10, 26 and 52. (b) Similar curves for a mass filter with third-order distortions. The q/a ratios are given on each curve and correspond to nominal resolutions of 54, 20, 12 and 8. The broken line gives von Zahn's criterion for maximum quadrupole performance, i.e. $R = n^2/12.25$. (c) A similar plot for a fourth-order distortion in the mass filter.

(3) *Experimental evidence*

Some good "bad" examples of split peaks for the ion trap have already been shown in Fig. 5.15. If the positions of dips are plotted on a stability diagram for many different scan lines, Fig. 5.19 is obtained and the most pronounced resonance dips fall on the expected resonance lines [19]. The origin of the resonance lines as third- and fourth-order distortions has been confirmed by changing the spacing of the end-cap electrodes in an asymmetric or

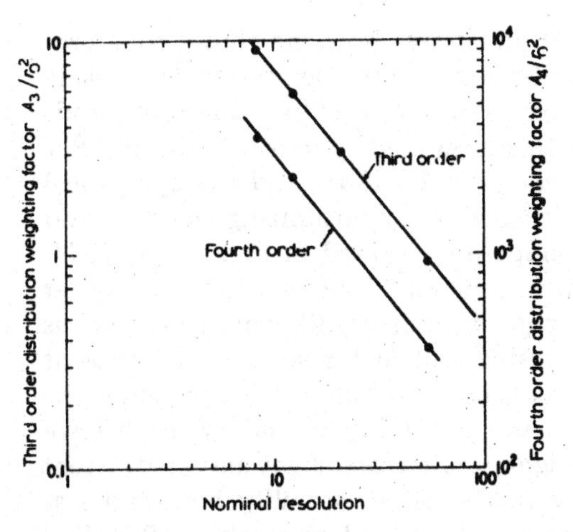

Fig. 5.22. Using von Zahn's criterion for optimum quadrupole performance, the maximum tolerable field distortions are plotted as a function of the resolution. Note the different ordinates for third- and fourth-order distortions.

symmetric way and observing the changes in the importance of the resonance dips.

Plotting the experimentally observed resonance dips on the stability diagram can be used as a diagnostic technique for determining which field faults are present.

For the ion trap, it was observed empirically [20] that a small bias applied between the end-caps (usually a few per cent of the d.c. voltage applied to the ring electrode) produced an improved peak shape. This was explained as being due to a slight shift in the stability diagram to higher q values, with little change in the positions of the resonance lines. The hypothesis has been confirmed [16] both by experimental observation and by computer simulation. Less bias is required at high resolutions since a very small shift removes the tip of the stability diagram from the position of the resonance lines. The computer simulations suggest some loss of sensitivity accompanying the use of a bias voltage. In experiments with various mesh ion traps [21] of very approximate geometry, it is found that good performance can sometimes be obtained without application of a bias between the end caps and this seems to be also associated with resonance lines that no longer pass exactly through the tip of the stability diagram. The geometric origins of such a shift are not clear.

(4) *Round rods*

In most quadrupole mass filters, round rods have been used instead of the correct hyperbolic rods. The optimum size for the rods is the radius $r = 1.148r_0$ where r_0 is the field radius [22, 23]. A value of $r = 1.16r_0$ has often been misquoted and calculations show that the larger rods may reduce the useful

References p. 119

aperture of the instrument. Figure 5.18 illustrates how the round rods produce a sixth-order (and smaller tenth-order) distortion. Denison [23] has suggested that the sixth-order distortion can be minimized by using round rods of appropriate radius in conjunction with a grounded cylindrical housing. For example, the distortion is near zero when $r = 1.1468r_0$ and the cylindrical housing radius $R = 3.54r_0$. Similar calculations for optimizing the r/r_0 ratio have been made for the four-fold monopole illustrated in Fig. 1.1(h) [24].

Dawson and Whetten [25] estimated from Fig. 5.22 that the sixth-order resonance lines caused by round rods might become significant at resolutions greater than a few hundred. They suggested that such resonance might lie at the origin of the pre-cursor peaks often observed at high resolution, since the resonance line $\beta_x + 2\beta_y = 1$ is close to the y stability boundary. High resolution without pre-cursor problems might result from slight and inadvertent asymmetries between opposite rods which shift the stability boundaries as described above for the ion trap. However, Leck and co-workers [26] find that the occurrence of the precursor peak is not systematic enough to be explained in this way.

The only reported experiments comparing directly the performance achieved using round and hyperbolic rods in the same mass filters are by Brubaker [27, 28]. These are difficult experiments because of the variations that sometimes occur even when a particular mass filter is dismantled and reassembled. Brubaker's results are illustrated in Fig. 6.4. The delayed d.c. ramp (see p. 105) was used in both cases. For a given sensitivity, the resolution can be augmented by a factor of two by the use of hyperbolic rods. Undoubtedly, several manufacturers have experimented with hyperbolic rods, but it seems that in instruments of moderate performance, any gains do not offset the costs of increased complexity in manufacture (see Chapter VI, p. 129 for further discussion).

(5) *Field distortions and the monopole*

Since the scan line for the monopole does not have to pass through the tip of the stability diagram (Fig. 2.24), resonance lines are not important. The q/a ratio for the scan line can be chosen to avoid all resonance lines in the neighbourhood of the y stability boundary. However, the field distortions will affect the focussing qualities of the monopole [3] and of the instruments which utilize the focussing properties of quadrupole fields (see Chapter II). In the absence of fringing fields, the monopole, as described in Chapter II, has rather poor focussing in the yz plane. At the position of optimum focussing (Fig. 2.25), the trajectories for those initial rf phases which give the maximum displacement are also farthest from the axis at the exit aperture. A slight third-order distortion ($A_3/r_0 \leqslant +0.5$) caused by a closer positioning of the rod to the V block will give a stronger than usual field for the larger ion displacements and actually improve the focussing. However, monopole operation is so little documented in detail, and apparently so dependent on fringing fields, that these theoretical subtleties are best left in abeyance.

REFERENCES

1 W.M. Brubaker, Advan. Mass Spectrom., 4 (1968) 293.
2 W.M. Brubaker, Proc. 5th Int. Instrum. Conf., Stockholm, 1960.
3 W.M. Brubaker, J.Vac. Sci. Technol. 10 (1973) 291. (Abstract)
4 P.H. Dawson, J. Vac. Sci. Technol., 8 (1971) 263.
5 P.H. Dawson, Int. J. Mass Spectrom. Ion Phys., 6 (1971) 34.
6 P.H. Dawson, J. Vac. Sci. Technol., 9 (1972) 487.
7 P.H. Dawson, Int. J. Mass Spectrom. Ion Phys., 14 (1974) 317.
8 P.H. Dawson, Int. J. Mass Spectrom. Ion Phys., 17 (1975) 423.
9 K.G. Steffen, High Energy Beam Optics, Interscience, New York, 1965.
10 P.H. Dawson, Int. J. Mass Spectrom. Ion Phys., 17 (1975) 447.
11 T.C. Ehlert, J. Phys. E, 3 (1971) 237.
12 A.E. Holme and W.J. Thatcher, Int. J. Mass Spectrom. Ion Phys., 10 (1972/3) 271.
13 D. Lafaivre, M.Sc. Thesis, Laval University, 1973.
13a R. Baribeau and P.H. Dawson, to be published.
14 F. von Busch and W. Paul, Z. Phys., 164 (1961) 581.
15 P.H. Dawson and N.R. Whetten, Int. J. Mass Spectrom. Ion Phys., 2 (1969) 45.
16 P.H. Dawson and N.R. Whetten, Int. J. Mass Spectrom. Ion Phys., 3 (1969) 1.
17 W. Paul, H.P. Reinhard and U. von Zahn, Z. Phys., 152 (1958) 143.
18 M.S. Story, J. Vac. Sci. Technol., 4 (1967) 326. (Abstract)
19 N.R. Whetten and P.H. Dawson, J. Vac. Sci. Technol., 6 (1969) 100.
20 P.H. Dawson and N.R. Whetten, J. Vac. Sci. Technol., 5 (1968) 11.
21 P.H. Dawson and C. Lambert, unpublished results.
22 I.E. Dayton, F.C. Shoemaker and R.F. Mozley, Rev. Sci. Instrum., 25 (1954) 485.
23 D.R. Denison, J. Vac. Sci. Technol., 8 (1971) 266.
24 A.B. Birtles and D.J. Mellor. J. Phys. E., 5 (1972) 1203.
25 P.H. Dawson and N.R. Whetten, J. Vac. Sci. Technol., 7 (1970) 440.
26 A.E. Holme, W.J. Thatcher and J.H. Leck, J. Phys. E., 5 (1972) 429.
27 W.M. Brubaker, 16th Annu. Conf. Mass Spectrom. Allied Topics, Pittsburg, 1968.
28 W.M. Brubaker, J. Vac. Sci. Technol., 4 (1967) 326. (Abstract)

CHAPTER VI

THE MASS FILTER: DESIGN AND PERFORMANCE

W.E. Austin, A.E. Holme and J.H. Leck

A. BASIC CONSIDERATIONS OF MASS RANGE AND RESOLUTION

The two important operating characteristics of a quadrupole, the mass range and the maximum resolution are dependent upon five basic parameters. These are the length and diameter of the rods, the maximum supply voltage to the rods, the rf supply frequency and the ion injection energy. The interdependence of mass range and maximum resolution is shown in Fig. 6.1, which emphasizes the point that these two basic operating characteristics cannot easily be varied independently in a given instrument. The mass range, or rather the maximum mass for which an instrument can be tuned, is calculated from the fundamental theory developed in Chapter II. This is the characteristic normally determined first when designing or setting up a particular instrument. The maximum mass is given by the simple relation

$$M_m = \frac{7 \times 10^6 V_m}{f^2 r_0^2} \tag{6.1}$$

where $V_m \cos 2\pi ft$ is the rf voltage applied between adjacent rods, r_0 (meters) is the inscribed radius of the rods and M_m, the maximum mass, is measured in amu.

For the majority of instruments designed and operated to the present date, the available choice of V_m and r_0 has been relatively limited. V_m is of the order of 3000 V and r_0 has acceptable values in the range 3–10 mm. V_m is

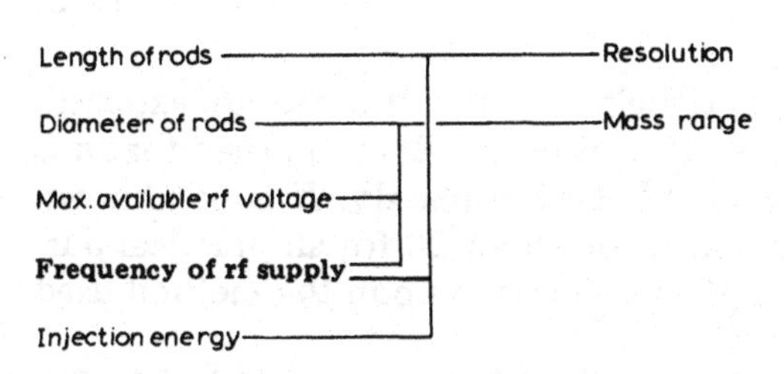

Fig. 6.1. Relationship of resolution and mass range to the fundamental instrument parameters. [Reference eqns. (6.1) and (6.3).]

limited by the difficulties inherent in the design of very stable high-voltage a.c. generators and r_0 by the problem of maintaining a high mechanical accuracy in manufacture of the rod assembly at a reasonable cost with both very large and very small instruments. There is no practical lower limit to the frequency, f, so that by adjustment of this parameter, the mass range can be extended indefinitely. In fact, f is normally set in the bracket 500 kHz to 6 MHz, which is adequate to cover all masses extending up to the order of 1000 amu. The values of mass range that can be obtained in practice for a quadrupole with relatively small cross-section are illustrated in Table 6.1. For a larger instrument with r_0 increased to 10 mm, all the quoted values are reduced by a factor of about twelve.

TABLE 6.1

The instrument operating range expressed in amu shown as a function of the a.c. frequency (MHz) and the maximum a.c. voltage (V_m) applied between adjacent rods. For this instrument, rod diameters of 0.25 in ($r_0 = 2.76$ mm) have been chosen.

	f			
V_m	1	2	4	6
1000	900	225	55	25
2000	1800	450	110	50
3000	2700	675	165	75
4000	3600	900	220	100

The selection of the mass range sets a limit to the resolution that can be obtained throughout the operating range. As has been pointed out in Chapter II, the finite length of the quadrupole electrodes limits the time spent by the ions in the focussing field and hence limits the resolution that can be obtained. It is well established that the resolution limit is governed by the number of cycles of rf field to which the ions are exposed. Although a precise relationship cannot be quoted, there is strong evidence from the various experimental studies that a good representation is given by

$$\frac{M}{\Delta M} = \frac{1}{K} N^n \tag{6.2}$$

where N is the number of cycles of the rf field to which the ions are exposed, ΔM is the width of the "peak" at mass M. It is now well established that n is either exactly or closely equal to 2 [1, 2]. Unfortunately, K is not known with such certainty but can be assumed to be about 20 for all practical purposes (obviously the precise value of K must depend upon the method used to define ΔM).

The experimental evidence supporting eqn. (6.2) is presented in Fig. 6.2 [3, 4]. This shows, in the context of the design of practical instruments, a

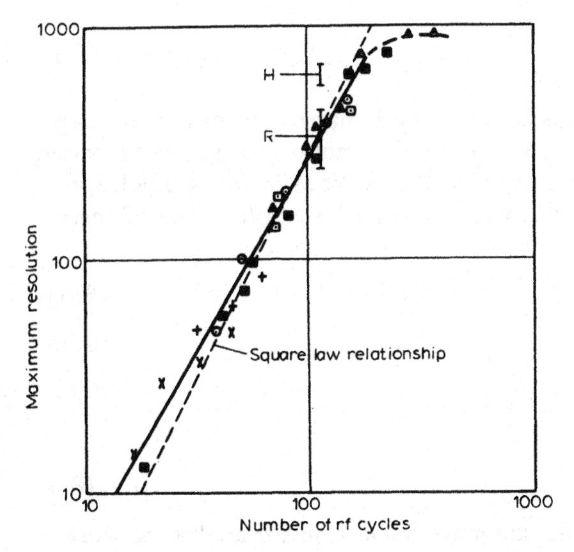

Fig. 6.2. The dependence of maximum resolution (ΔM measured at 10% peak height) on the number of rf oscillations. Data obtained by Brubaker lie within the bars for both round (R) and hyperbolic (H) rods. △, Xe^+; ■, Ar^+; ⊙, Ar^{2+}; ⊡, Kr^+; +, He^+; ×, H_2^+.

consistent performance. It is noticeable that agreement becomes worse with increasing resolution. This is inevitable because eqn. (6.2) does not take account of the dispersions caused by imperfections in the electric field. These, always present to some degree, become of increasing importance at high resolution. In fact, the point of "break" from the straight line is a measure of the accuracy of construction of a particular instrument.

In a given instrument, the number of cycles, N, and hence the limiting resolution, can easily be determined in terms of the following basic operating parameters: (1) the length of the quadrupole rods, L; (2) the rf frequency, f; and (3) the ion injection energy, V_z. Equation (6.2) can be re-written

$$\frac{m}{\Delta m} = 0.05\left\{fL\sqrt{\frac{m}{2eV_z}}\right\}^2$$

where e is the electron charge in coulomb, m the atomic mass and Δm are measured in kg, $K = 20$ and $n = 2$.

$$\frac{m}{\Delta m} = \frac{0.05 f^2 L^2 m}{2eV_z} \tag{6.3}$$

or

$$\Delta m = \frac{40eV_z}{f^2 L^2} \tag{6.4}$$

Simplifying and expressing ΔM in amu gives

$$\Delta M = \frac{4 \times 10^9 V_z}{f^2 L^2} \tag{6.5}$$

The range of values of ion injection energy is narrow in practical instruments. The lower limit, of the order of 2 eV, cannot easily be extended because of the difficulties of efficient injection of ions into the quadrupole.

The following values can be taken as typical of a whole range of instruments.

$$L = 0.2\,\text{m} \tag{6.6a}$$

$$f = 2.0\,\text{MHz} \tag{6.6b}$$

$$V_z = 5\,\text{eV} \tag{6.6c}$$

Equation (6.5) now simplifies to

$$\Delta M = 0.125\,\text{amu} \tag{6.7}$$

Equation (6.7) indicates that an adequate resolution for analytical work is possible, certainly up to a mass of the order of several hundred amu. It also emphasizes that, neglecting second-order effects such as field imperfections, the minimum attainable peak width, ΔM, is independent of ion mass in a given operation (see also Chapter II, p. 24).

The interrelation between "mass range" and "resolution" can be appreciated by rearranging the terms of eqns. (6.1) and (6.5). Eliminating f (the independent variable normally chosen in a given instrument to select mass range) gives the following relationship between ΔM and M_m.

$$\Delta M = \frac{570 r_0^2 V_z M_m}{L^2 V_m} \tag{6.8}$$

Clearly, the peak width ΔM is directly proportional to M_m, which means that in a given instrument the theoretical maximum resolution, given by

$$\frac{M_m}{\Delta M} = \frac{L^2 V_m}{570\, V_z r_0^2} \tag{6.9}$$

is independent of the operating frequency.

For the particular instrument described by eqn. (6.7) with V_m set at 3000 V and $r_0 = 3$ mm, the maximum resolution is 4700.

It is important to recognize that the above results represent the maximum possible performance and have been obtained making no allowance for instrument imperfections. In practice, the optimum resolution cannot be obtained. Factors such as misalignment of the rods must detract from the resolution. This extremely important aspect of the quadrupole design and operation is discussed in detail in the following sections. Selective multipeak monitoring is discussed in Chapter XIII (see p. 310).

B. FIELD IMPERFECTIONS AND THEIR EFFECT ON PERFORMANCE

It is easy to identify the three most important factors which detract from the ideal performance of a given instrument due to the degradation of the hyperbolic fields. These are (1) mechanical misalignments of the rods, (2) contamination of the rods, and (3) use of circular rods to approximate a hyperbolic field.

Unfortunately, there is a lack of data from which to make a critical quantitative analysis and to draw firm conclusions regarding the relative importance of these three factors. However, general trends can be discerned with sufficient certainty to make worthwhile conclusions possible. The individual factors do not, in general, interact and are therefore considered separately in the following sections.

(1) *Mechanical misalignments of the rods*

The requirement for the analyser field is an assembly of four circular rods held straight and parallel to each other in a square array. The accuracy must not deteriorate even after the instrument has been baked for long periods at 300–350°C. To obtain a measure of the accuracy of construction necessary to achieve a given performance, the relevant data from a number of reports [3, 5–8] have been collected together and presented in graphical form in Fig.

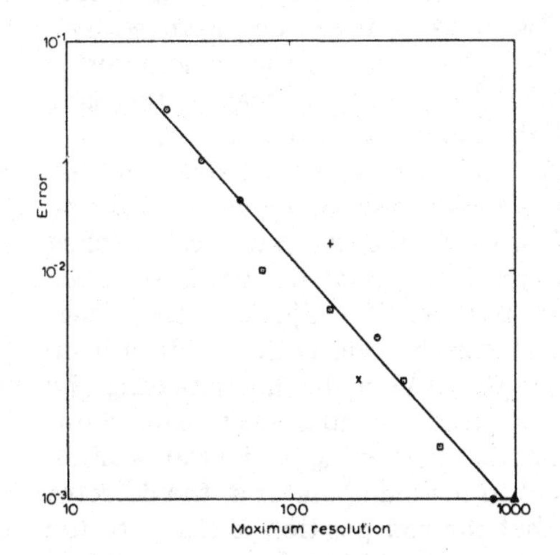

Fig. 6.3. The relationship between maximum resolution and the mechanical error in construction for a practical quadrupole. □, Arnold; △, Story; ×, Munro; +, Fairburn; ○, Liverpool University. In this graph, the limits are the imperfections in the mechanical construction and not the number of cycles of rf seen by the ions.

6.3. The parameter considered is the maximum resolution obtainable with a given instrument. In every case it is recognized that the limitations are the imperfections rather than the fundamental factors expressed in eqn. (6.9). (Although sensitivity will be adversely affected by inaccurate construction, this depends upon so many other design aspects that it is impracticable to compare data obtained from different instruments.) The constructional error has been arbitrarily defined as $(2\epsilon + \epsilon')/D$ where ϵ is the general manufacturing tolerance specified, ϵ' is any additional error deliberately introduced and D is the diameter of the circular electrodes.

It can be seen from Fig. 6.3 that there is a relatively large scatter in the individual points. This is not surprising because, in addition to the difficulty in estimating the errors of construction, the errors appear in many different forms. Furthermore, in a given instrument they may either accumulate or cancel. The data are, however, sufficiently consistent to indicate a relationship of the form

$$\text{maximum resolution} \propto (\text{error})^{-n} \tag{6.10}$$

where n is approximately 1.3. In Chapter V, it was shown theoretically that asymptotic imperfections are the most significant and would be expected to cause an error of the form shown in eqn. (6.10) with $n = 1.23$. Thus practical experience, although limited, is in good agreement with theory.

It is interesting to consider the practical importance of eqn. (6.10). Consider, for example, an instrument having rods 0.25 in. in diameter ($r_0 = 2.76$ mm) each aligned to an accuracy of ± 0.0005 in. Rod imperfections would limit the resolution to approximately 750. This limit can be improved by a factor of two by either (a) keeping r_0 fixed and reducing the manufacturing and assembly tolerances by a factor of $2^{1.3} = 2.5$ or (b) keeping the same tolerance and increasing the diameter of the rods by the same factor.

Experimental evidence showing directly the effect of rod misalignment has been obtained by Story [9], who made a quadrupole in which the position of one rod could be altered by means of two micrometer screws, one attached to each end of the rod. These screws could be operated from outside the vacuum so that each end of the movable rod could be adjusted during operation. The resolution was to about 1 μm. Figure 6.4 shows the result of scanning m/e 502 of perflurotributylamine (PFTBA) while manipulating the position of the rod. Because of linkages involved, the numbers in parenthesis are an arbitrary distance scale, the first number pertaining to the rod position at the ion source rod end and the second to the manipulator at the detector rod end. Increasing numbers indicate that the rod position is closer to the center. As can be seen, a "precursor" peak can be produced by such manipulations. Note that sensitivity is also affected by these field perturbations. (Time-of-flight data from a pulsed ion source on this type of precursor ion indicate that it is faster than the ions that make up the mass peak.)

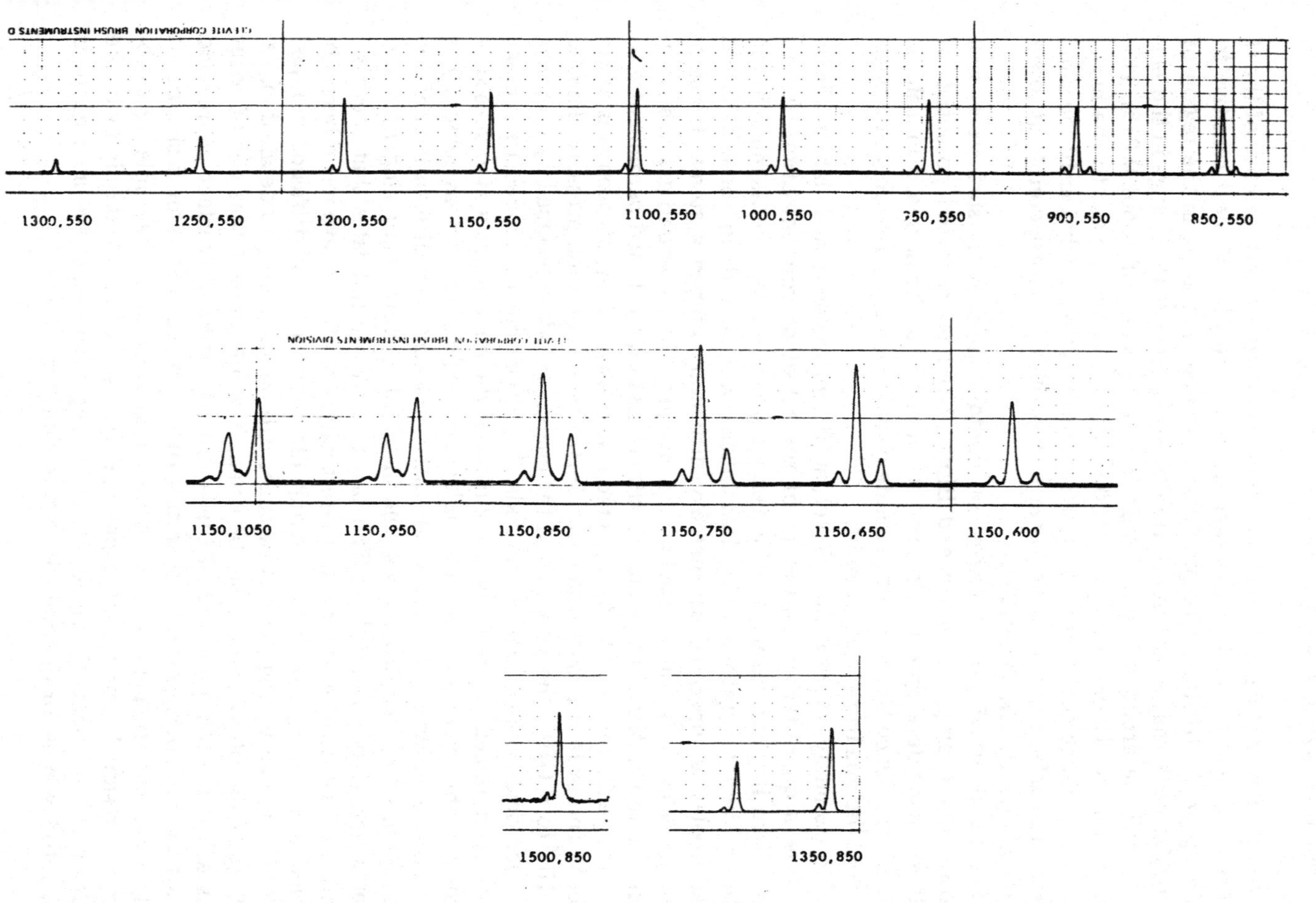

Fig. 6.4. The change of peak shape resulting from moving one rod towards the centre of the structure.

(2) *Contamination of the rods*

Owing to its critical dependence upon the accuracy and stability of the rf and d.c. voltages applied to the analyser rods, the filter is most sensitive to build up of electrostatic charge. Typically, when operating in the region of triangular peak shapes at a resolution of about 100, a 1% change in the sensitivity will be produced by a change in the a/q value of about 0.01%. This corresponds to 10 mV at $V_{d.c.} = 100$ V, the latter being near to the maximum d.c. voltage supplied to the rods in most systems. Obviously the same change in surface potential at lower values of $V_{d.c.}$ (for example at lower mass) will be even more serious.

The build up of surface charge on contaminated metal surfaces is well known. For example, Scruton and Blott [10] have reported potentials of 400 mV being developed on fingerprint impurities under normal atmospheric conditions. Hence the need for absolute cleanliness in the assembly and operation of the quadrupole is essential.

Unfortunately, there is not, as yet, sufficient experimental data available to allow a quantitative assessment of the effect of contamination. In this connection it is interesting to summarize a series of experiments made in the authors' own laboratory [11]. Instabilities were noted on three different quadrupoles all operated on mercury diffusion pumped systems. In every case, the sensitivity for a particular gas changed when a second gas was introduced to the system. For example, with an instrument tuned permanently to the Ar^+ peak, the sensitivity for argon decreased when nitrogen was introduced. The order of magnitude was 5–10% when the total pressure changed by 10^{-5} torr. Often the direction of the change could be reversed by reversing the electrical connections to the rods. The phenomenon tended to become more pronounced if either the resolution of the instrument was increased or measurements were made on low molecular weight gases such as helium. The instruments became perfectly stable after they had been dismantled and the rods cleaned by electropolishing, the instabilities slowly returning, however, over a period of several days. This last observation indicates that the basic problem is centred on the rods. Interestingly, the effects have never been observed when these and similar instruments operated on oil diffusion pumped systems over very long periods of time. Although, as yet, no detailed explanation of the phenomenon can be given, it would appear that out of focus ions affect the d.c. potential of the rods by introducing surface charge and hence the sensitivity/resolution setting of the instrument. This is significant only when the stainless steel has been contaminated, for example by exposure to a residual mercury atmosphere. Potential changes need only be of the order of 10 mV and, of course, are more important when the instrument is tuned to low mass or operated at high resolution.

(3) *The use of circular rods to approximate hyperbolic fields*

Because of the difficulty of machining hyperbolic surfaces, circular rods are used to form the electrode surfaces in the vast majority of analysers. By positioning the circular electrodes (each of radius r), in a square array such that $r = 1.148\, r_0$, the best approximation to the hyperbolic field is achieved. There are, however, imperfections which in the limit impair the performance of the instrument. Dawson and Whetten [12], for example, have calculated that when operating with resolutions in excess of 200, small "precursors" are to be expected adjacent to the main peaks. Although these have not been positively identified, there must, undoubtedly, be some broadening of the main peaks.

The only detailed experimental comparison between round and hyperbolic rods has been made by Brubaker [4]. In an extensive series of experiments, he investigated this problem with both types of electrode mounted in the same instrument. Figure 6.5, reproduced from his publication, is a typical result. Brubaker found, quite consistently through his work, that there was an improvement of a factor of two in resolution (measured at the same sensitivity) when hyperbolic replaced circular surfaces. To allow a general comparison, some of the data reported by Brubaker for both hyperbolic and circular surfaces is included in Fig. 6.2. It is obvious that his hyperbolic surfaces gave an improvement in resolution of about 50% over the best reported by other workers. Thus, bearing in mind the large error of measurement for these parameters, the claim of a factor of two improvement by Brubaker appears reasonable.

In spite of the improved resolution, the value in practice of the hyperbolic surfaces is questionable because the greater difficulties in machining and mounting these electrodes are likely to give both large and asymmetrical misalignment errors. This is likely to offset the advantage of the theoretically more perfect field boundary conditions especially in high-resolution instruments.

A more likely application for hyperbolic surfaces may come in special applications where economy in the provision of power supplies is important. For, as pointed out by Brubaker [4], the radius r_0 must be increased by a factor of about two when circular rather than hyperbolic surfaces are used in order to keep the same performance. This arises because now the field is a satisfactory approximation to the ideal only over the central region. Since, for a given field, the rf power varies as the fourth power of r_0, the difference is significant and may be important in, for example, space applications.

It is interesting to note that satisfactory approximations to the hyperbolic field can be achieved by various unusual electrode configurations. Hayashi and Sakudo [13] showed that the circular concave structure illustrated in Fig. 6.6 forms the basis of a viable filter. The electrodes were formed by plating copper onto an accurately machined aluminium cylinder. Four

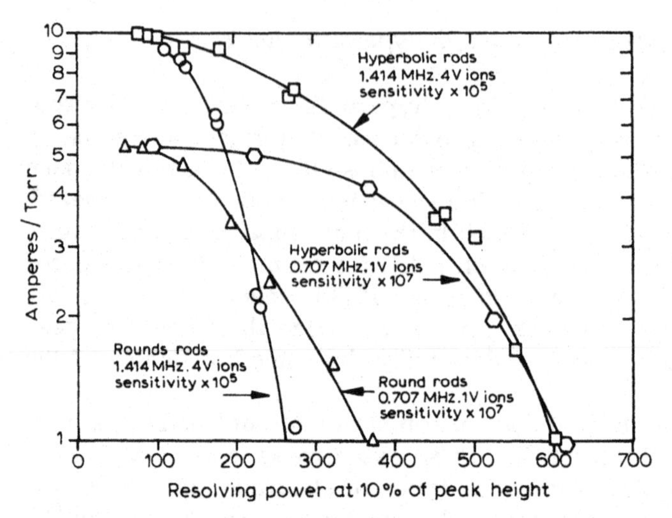

Fig. 6.5. Sensitivities of round and hyperbolic quadrupoles as functions of resolution, as measured by Brubaker. (The entrance aperture was 0.05 in.)

horizontal slots were made in the plated cylinder and after fixing glass supporting rings to the outside of the cylinder, the aluminium core was dissolved in hydrochloric acid. The inside diameter of the electrodes was 10 mm and their thickness 0.8 mm. For these dimensions, the optimum angle between the electrodes, 2γ, was calculated, and confirmed experimentally, to be 40°. The advantage claimed for this novel geometry was the ease of construction, which minimizes the all important asymmetrical errors. In a very brief report, Hayashi and Sakudo indicated that the initial experiments with the new structure gave a performance not greatly inferior to that of the conventional analyser.

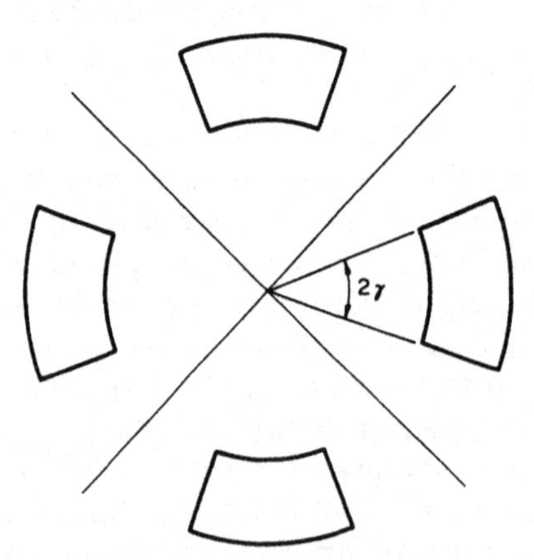

Fig. 6.6. Electrode structure proposed by Hayashi and Sakudo to approximate a hyperbolic field.

C. THE ION SOURCE

Because the quadrupole mass filter depends upon the "path stability" criterion rather than on "exact focussing" properties, the performance is not critically dependent on the energy of the ion beam. For this reason it has been used to analyse ions originating from a wide variety of sources. In addition to its basic role as a gas analyser, the filter has been successful in sampling ions, from gas discharges [14], from the upper atmosphere [15], from field desorption sources [16], and from surfaces following the impacts of ions (SIMS) [17] or electrons (ESD) [18] (see Chapter X).

(1) *Design and construction*

With the exception of these and similar specialized uses, ions are normally formed directly from the gas under analysis by electron impact as in the conventional magnetic deflection mass spectrometer and thermionic cathode ionization gauge. Such a source, typical of many, and illustrating all the fundamental requirements, has been described in detail by Brubaker [4]. It is illustrated in the schematic diagram of Fig. 6.7 which is self-explanatory. A proportion of the ions formed inside the cylindrical grid cage are extracted and focussed into an approximately parallel beam by a lens system. Many varied designs of lens are used, some complex, some extremely simple. As it is desirable to direct the maximum ion flux into the quadrupole field, the lens aperture is usually made as large as possible, consistent with the geometry of the rod assembly. Typically, the final lens aperture is designed to have a radius approximately equal to $r_0/2$. Although there is some evidence [19] to suggest that for the highest resolutions the apertures should be very small, this is by no means conclusive. Undoubtedly, the full importance of the ion

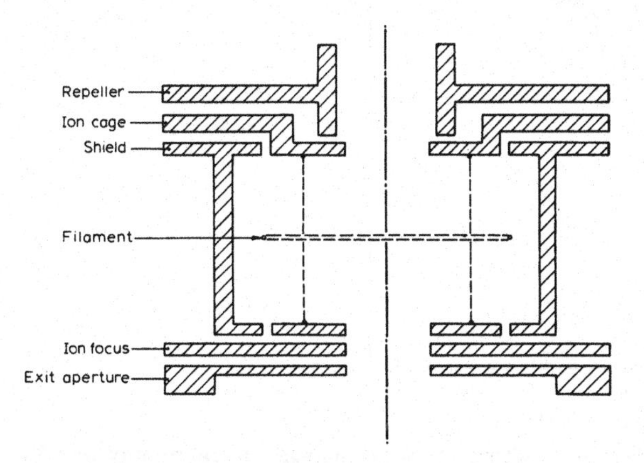

Fig. 6.7. Ion source with circular geometry.

injection characteristics is not yet fully understood. The behaviour of a given source obviously depends not only upon the size of the exit but also upon the divergence of the ion beam as it enters the fringing fields of the quadrupole. (See also Chapter V.)

Story [20], in a time-of-flight experiment, measured the increase in energy of the ions as they entered the quadrupole analyser from the fringing fields. Figure 6.8 shows that ions of all masses experience a gain in average energy for initial energies from 2 to 12 eV. For initial energies up to 9.5 eV, the lighter ions gain more energy but above this value the average emerging energy was less for higher mass ions. This may be due to the higher amplitude of oscillation of the larger ions. Story [20] points out that this increase in ion energy is due to fringing fields at the entrance. He also showed, in another series of experiments, the effect of injecting ions at different positions with respect to the central axis. Figure 6.9 shows peak shapes resulting from traversing a small ion beam (0.025 cm) across the *y* axis of the entrance of a misaligned quadrupole (0.63 cm rod diameter) while scanning $m/e = 219$ of PFTBA. Note that under some injection conditions the anomalies in peak shape do not exhibit themselves and that if this structure were used with a larger aperture, the resulting peak shape would be a weighted average of these shapes depending on the number of ions entering at any particular position.

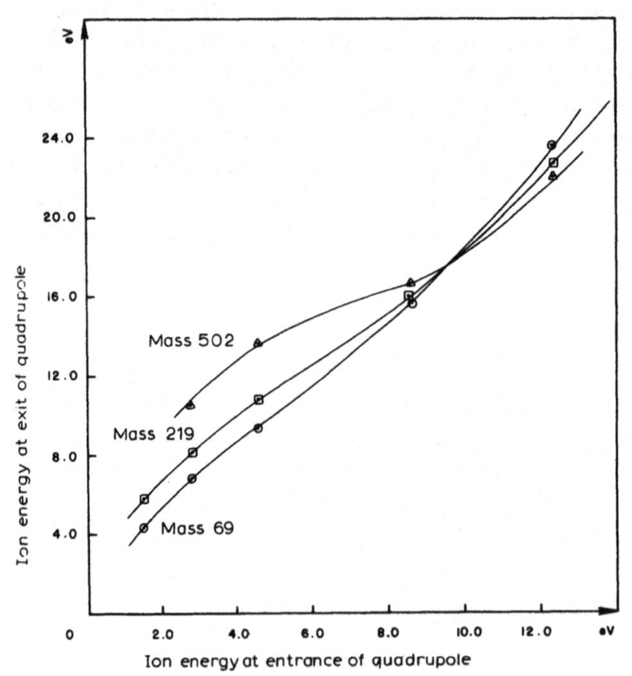

Fig. 6.8. Experimental results obtained by Story showing increase in ion energy due to transmission through the quadrupole filter.

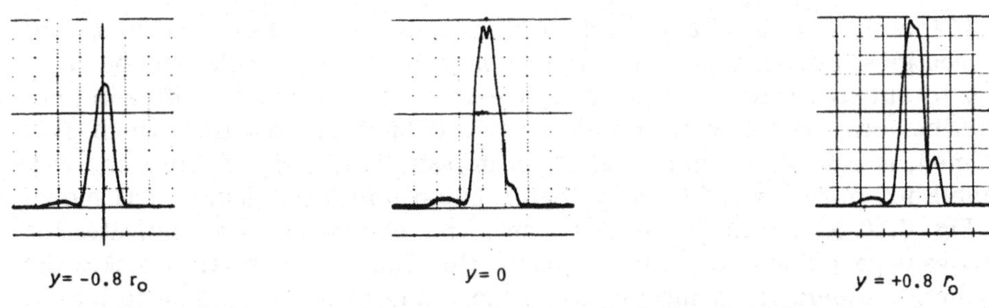

Fig. 6.9. Effect of ion entry position on peak shape of misaligned quadrupole as observed by Story.

In a very commonly used modification to the design shown in Fig. 6.7, the filament is mounted parallel to the principal axis of the quadrupole outside, but close to, the grid cage. This design has considerable merit because it allows two separate filaments to be mounted in the same assembly although it loses the symmetry of the circular filament. It is common practice to design the source so that it can easily be dismantled and reassembled for cleaning. Advantage is usually taken of this to make replacement of the filament assembly a quick and simple operation.

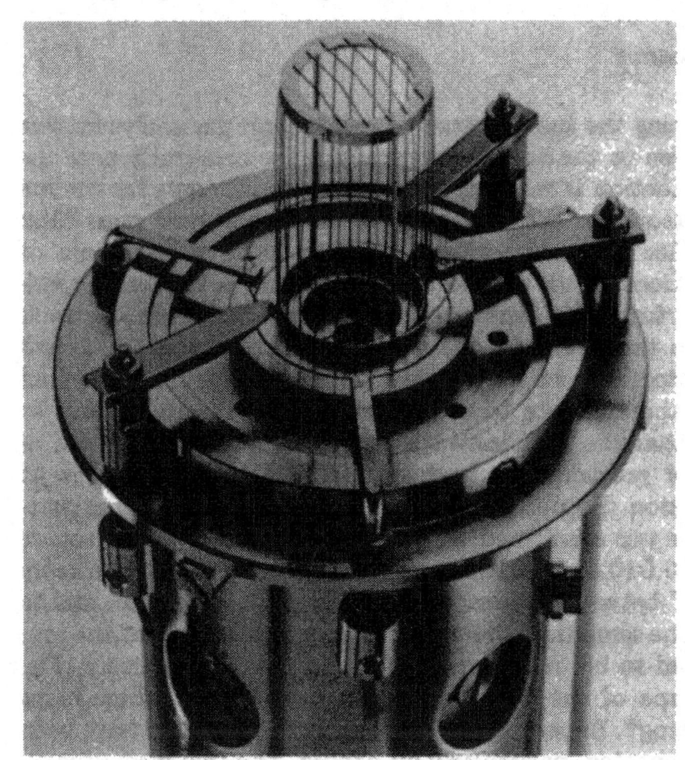

Fig. 6.10. Ion source as modified by Hofmann to give low gas desorption.

Adaptations of the basic design outlined above are often used in special applications. For example, in ultrahigh vacuum, the electrodes can be made easy to outgas either thermally or by electron bombardment. Hofmann [21] describes one version with a very open and light construction which, it is claimed, is easy to outgas by electron impact. This design follows the well established principles of the ultrahigh vacuum ionization gauge. It is shown in Fig. 6.10 which illustrates this point. This photograph also indicates the disadvantage pointed out by Hofmann; the flimsy construction makes the assembly vulnerable to mechanical damage and hence it is liable to loss of performance.

As an extreme example of the possible variation in source design, it is worth considering the work reported by Blum and Torney [22] who adapted the Redhead cold cathode ionization gauge [23]. Because of the wide spread in the energies of the ion source, full advantage of the high sensitivity could not be taken. In fact, the measured sensitivity of approximately 2.5×10^{-4} A/torr was of the same order as that achieved by quadrupoles with thermionic ion sources when operating with a similar resolution. This, coupled with the fact that (because it is a source of the Penning type) a magnetic field is required, must make the applications limited.

(2) *Alignment of the source*

Although, in discussing the manufacture of quadrupole gas analysers, due emphasis is usually given to the need for accuracy in the construction of the filter, very often no mention is made of the similar requirements for the ion source. One of the reasons for the popularity of the quadrupole mass filter over the magnetic deflection instruments is its insensitivity to variations of the ion entrance conditions, but it has, nevertheless, been the authors' experience that optimum performance can be achieved only if considerable care is taken to inject ions in the form of a beam of low divergence, close to and parallel with the principal axis of the analyser. It must be remembered that the fringing fields at the entrance to the analyser deform the ion beam so making it important that the ions leave the source in as narrow a beam as possible. This need for precision has been quantified by Arnold [5] in an experimental investigation. He noted a deterioration in three operating parameters when the source was misaligned. For example, when the whole source was moved off axis by 0.010 and 0.015 in., the resolution decreased markedly increasing the "valley" between masses 275 and 276 from 1% to 15% and to 30%, respectively. At the same time, the required injection energy of the ions V_z (see Section A) had to be increased to 11 and 17 eV, respectively (Fig. 6.11). Finally, the shape of the individual peaks deteriorated, there being significant "peak splitting". Experiences in the authors' laboratory have been consistent with those reported, particularly by Arnold. The two main effects, the deterioration in resolution and the required increase in injection energy have been observed. It has been consistently observed that this splitting phenomenon is an excellent indicator of any defect in the electron optics of the system either in the ion source or in the quadrupole field. This may, of course, be due either to a mechanical misalignment or to contamination.

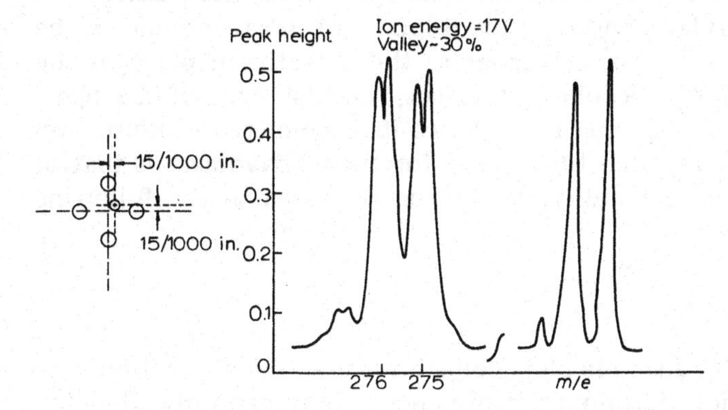

Fig. 6.11. Peak splittling observed by Arnold. The degradation in peak shape is caused by misaligning the source assembly 0.015 in. off the central axis.

As described in Section A and, in particular, by eqn. (6.5), a high resolution can be achieved only when ions are injected into the quadrupole with low energy. Thus the graphs in Fig. 6.12 are significant as they mean that any misalignment of the ion source must have a direct bearing upon the overall performance of the instrument. Results taken from Brubaker's [4] reported work and plotted in Fig. 6.12 show the typical relation between

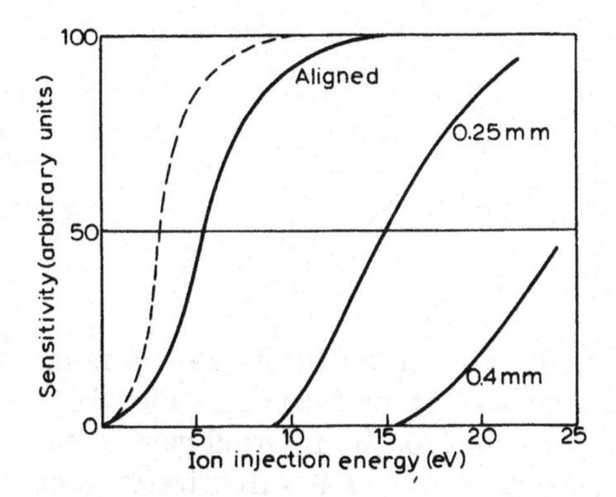

Fig. 6.12. The dependence of sensitivity on the ion injection energy when the source is either accurately aligned or displaced from the principal axis. The characteristic for the Brubaker ion source is shown by the broken line.

sensitivity and injection energy for a well aligned source. This reinforces the work published by the authors showing the optimum ion injection energy to be of the order of 5–10 eV.

D. ION DETECTION

A Faraday cup or plate used in conjunction with a sensitive electronic amplifier represents the simplest, cheapest, and most reliable means for detecting the ion beams emerging from the exit aperture of the quadrupole. The relatively high noise level and the slow response of this technique can be improved by using an electron multiplier as the detector in place of the Faraday cup. The multiplier becomes, in effect, the first stage of the signal amplifier where full advantage can be taken of its extremely low noise level and its fast response. The price to be paid for these advantages is that of sacrificing simplicity and reliability, as will be discussed in the following sections.

(1) *The Faraday cup*

When a Faraday cup or a simple plate collector is used in conjunction with an electrometer amplifier, optimum performance is obtained only when the connection between the two is made very short. Ideally, the first stages of the amplifier should be mounted in a separate "head" rigidly attached to the quadrupole housing with the lead from the collector no more than a few cm long. The electrostatic shielding must be near perfect, as must be the freedom from vibration. This is essential to reduce the spurious signal pick-up to a minimum. The short connection also keeps the input capacitance to a reasonably low value. A schematic diagram of this system is shown in Fig. 6.13.

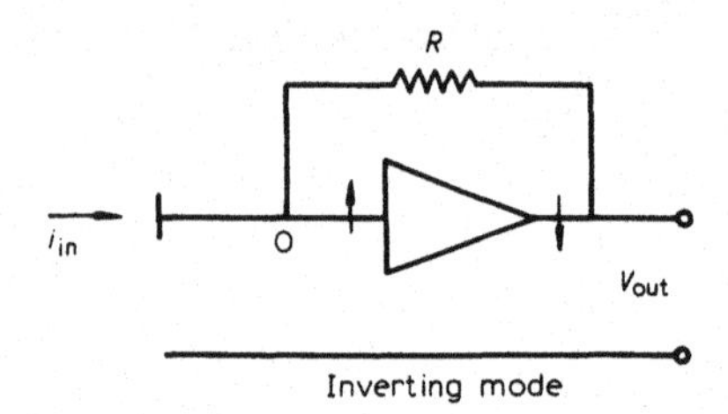

Fig. 6.13. Electrometer amplifier.

The all important measurement resistor, R, is usually chosen in the range 10^8 to $10^{12}\,\Omega$. Operated with negative feedback, as indicated in Fig. 6.13, the input line to the amplifier is a virtual earth. (As the internal gain of the amplifier is extremely large, the potential at point O is effectively fixed making the input impedance of the amplifier low and therefore the leakage paths to earth relatively unimportant. In addition, the effect of the input capacity is reduced.) It is practicable to use all solid state amplifiers with either junction FET or MOS FET transistors in the input stage. With good quality components, ion currents down to 10^{-15} A can be detected with a time constant of the order of 1 sec. Although it may be possible to improve upon this performance using very sophisticated amplifiers, the limit is largely set by the spurious signals injected at the collector. Thus no major improvement of more than an order of magnitude is likely without resort to extremely sophisticated systems. For a quadrupole operating over the low mass range, say up to 200 amu, relatively high sensitivities in the range 10^{-4} to 10^{-3} A/torr are practicable. Thus with these instruments, partial pressure detection limits can be as low as 10^{-11} or 10^{-12} torr using the simple electrometer amplifier. Naturally, a slow response must be accepted. The speed of response can be increased only at the expense of the minimum detection level. It is realistic, as a first approximation, to assume that the minimum detectable signal is inversely proportional to the time constant of the electrometer amplifier. This takes account, for example, of the practical necessity to reduce the resistor, R, progressively to increase the speed of the amplifier response. Table 6.2 summarizes the position for a typical application giving, of course, only order of magnitude values. Response time here is defined as the time interval between 10 and 90% for the total amplifier response to a step function.

TABLE 6.2

For calculation of the minimum detectable pressure, an instrument sensitivity of 10^{-4} A/torr has been assumed. The minimum detectable signal has been arbitrarily assumed equal to the "noise" level in the amplifier.

Response time (sec)	1	0.1	0.01	0.001
Minimum detectable signal (A)	10^{-15}	10^{-14}	10^{-13}	10^{-12}
Minimum detectable pressure (torr)	10^{-11}	10^{-10}	10^{-9}	10^{-8}

The pattern shown in Table 6.2 indicates the limitations of the simple electrometer amplifier for small signal measurement, especially where high accuracy and a fast response are required. In order to achieve a 1% accuracy of measurement with reasonable certainty, three times the instrument time constant must be allowed. Thus to measure a partial pressure of 10^{-8} torr (instrument sensitivity 10^{-4} A/torr) to about 1% accuracy, an amplifier with a time constant of 0.1 sec can be chosen and a dwell time of not less than 0.3 sec on the peak must be allowed. (This gives an inaccuracy of measurement and an uncertainty due to noise both rather below 1%.)

(2) *The electron multiplier*

Both the signal-to-noise ratio and the speed of response of the detector can be improved by replacing the Faraday cup with an electron multiplier. This, in effect, introduces a pre-amplifier with a current gain in the range 10^5 to 10^7. Its response time is effectively zero. Provided adequate precautions are taken to keep spurious input signals to a minimum, the signal-to-noise ratio is completely adequate for almost all quadrupole applications. In fact, with the multiplier fitted, the limitations of measurement are set by the statistical variations in the signal and not by the measuring circuits. For example, an ion current to the first dynode of the multiplier of 10^{-16} A (10^{-12} torr for an instrument sensitivity of 10^{-4} A/torr) corresponds to a flux of 600 ions/sec. To measure this signal to an accuracy of 10% with reasonable certainty requires a dwell time of about 0.1 sec. This "measurement time" is clearly inversely proportional to the signal level and to the acceptable error in measurement.

A quantitative comparison of performance obtainable with and without the multiplier can be made. A simple but important example is chosen in Table 6.3. It can be seen that the improvement with the multiplier is by rather more than three orders of magnitude, either in the minimum signal level with a given measurement time or in the measurement time with a given

TABLE 6.3

A comparison of the Faraday cup and electron multiplier detector systems. The measuring times are given for the determination of pressure to an accuracy of 10%. The quadrupole sensitivity (before electron multiplication) is assumed to be 10^{-4} A/torr. For the multiplier it is assumed that time must be allowed for 60 ions to reach the first dynode. For the Faraday cup, the assumption is that the amplifier time constant must be adequate to keep the signal-to-noise ratio > 10, a time interval twice the time constant being allowed for the measurement.

Pressure (torr)	10^{-8}	10^{-9}	10^{-10}	10^{-11}	10^{-12}	10^{-13}
Current (A)	10^{-12}	10^{-13}	10^{-14}	10^{-15}	10^{-16}	10^{-17}
Time (multiplier) (sec)	10^{-5}	10^{-4}	10^{-3}	10^{-2}	10^{-1}	1
Time (Faraday cup) (sec)	0.02	0.2	2			

signal level. Because the limitations of the two systems are different in nature, the comparison cannot be precise. Nevertheless, it is an adequate guide for practical operation. Because of the differences in individual electron multipliers and in electrometer amplifiers, a more detailed analysis would not be profitable in a general review. However, the factor of three orders of magnitude improvement in performance is, to a first approximation, independent of the quadrupole and its operating requirements.

Of particular importance is the ability to detect very low ion currents with the multiplier system. The noise level of the multiplier is usually significantly below an equivalent 10^{-17}A input current, provided the first dynodes are well shielded from spurious radiation. Direct photon pick up from a line of sight connection to the ion source is potentially a problem in the quadrupole analyser. This has been overcome in many applications by placing the multiplier at right angles to the central axis or set back in a side tube. The arrangement adopted by Goodings et al. [24] shown in Fig. 6.14 is typical. The ions are drawn to the first dynode by field penetration through the entrance aperture of the earthed shielding plate. The multiplier and mass filter enclosures were also at earth potential. For the particular experimental requirements described by Goodings et al. [24], the ion beam was modulated by impressing a square wave of 500 Hz on one of the pre-focussing electrode voltages. The a.c. voltage developed across the load resistor was coupled to a lock-in amplifier via a blocking capacitor. Potential requirements are favourable for the

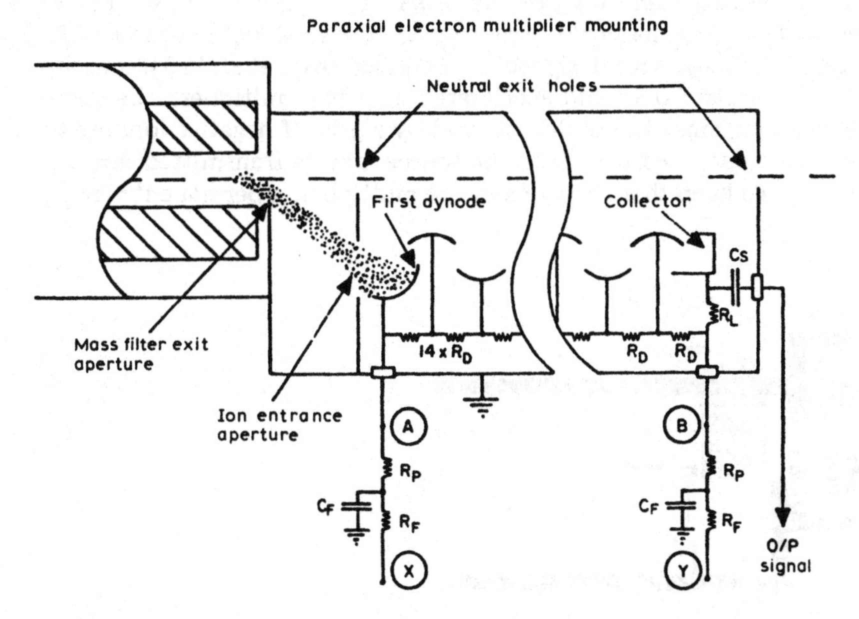

Fig. 6.14. The electron multiplier fitted "off axis" by Goodings et al. $R_D = R_L = 1\,M$; $R_p = R_F = 680\,K$; $C_s = 1000\,pF$; $C_F = 0.1\,\mu F$.

extraction of positive ions. It is normal for the low energy ion beam to leave the quadrupole field at earth potential and be accelerated through some 2000–5000 V to the first dynode. The secondary electrons are attracted back to the final collector at ground potential.

There are, unfortunately, inherent disadvantages to the electron multiplier used as a pre-amplifier for the measurement of positive ion currents in mass spectroscopy. The most important are, firstly, the relative instability of the gain and, secondly, the liability to a gain dependency on ion mass due to discrimination at the first dynode. Although multipliers have had a wide application in mass spectroscopy in the past two decades, unfortunately, neither of these factors is fully understood or predictable with certainty.

For negative ions, the position may be much simpler. Goodings et al. [24] find the multiplier sensitivity to be independent of ion impact energy over a wide range (300–3000 eV). They suggest that the primary electron comes from the impacting negative ion and not from the surface. In consequence, mass discrimination in the multiplier is not expected. Furthermore, only a small voltage on the first dynode is necessary for satisfactory negative-ion detection.

The channel electron multiplier, which consists of a glass tube, the inside of which is coated with a high resistance film, is suitable as signal detector in the quadrupole. It can be operated in a pulse-counting mode to give an absolute measurement of ion flux. If operated "off axis", the background spurious noise level can be extremely low, of the order of 1 count in 10 sec. Figure 6.15 shows that it is mechanically easy to place the small detector in an off-axis position. In this system, designed by Benninghoven et al. [25], ions are deflected by an electric field and accelerated into the multiplier. The quadrupole is shielded against the deflector field by a grid. If negative ions are to be detected, secondary electrons from the source may be transmitted through the quadrupole. To keep them away from the multiplier, a permanent magnet

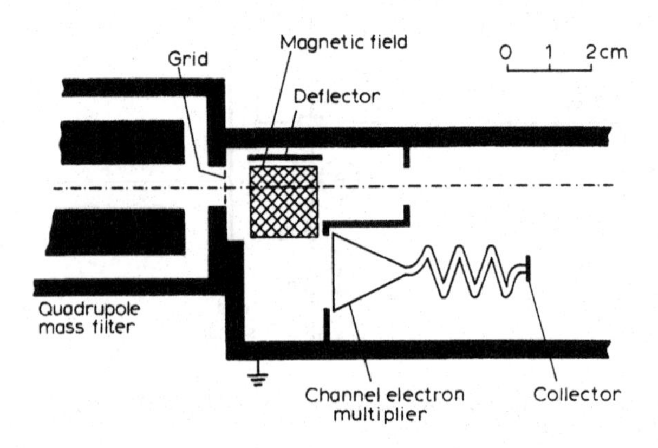

Fig. 6.15. The channel electron multiplier fitted "off-axis" by Benninghoven et al.

with a field of about 500 gauss can be installed as shown in Fig. 6.15. Ions are practically unaffected but electrons move in cycloidal paths away from the multiplier. This technique has been demonstrated by Benninghoven and his co-workers to improve the noise level by a factor of ten. (For further discussion of the application of electron multipliers to practical problems see Chapter XII, Section B(3)).

(3) *High pressure limit of operation*

The upper limit to the useful pressure range is determined by the collisions between ions and the gas molecules in the analyser. Experimental results show that linear operation can be expected up to pressures of the order of 10^{-4} torr [26]. This limit clearly depends upon the details of the instrument design, in particular upon the length of the quadrupole rods.

E. INSTRUMENT SENSITIVITY

(1) *Dependence on resolution*

As described in Chapter II, the amplitude of the ion oscillations in the filter depends both upon the entrance conditions (e.g. axial displacement, angular divergence, rf phase) and also on the ratio of the d.c. to a.c. voltages applied to the analyser rods ($V_{d.c.}/V_{a.c.}$). As this ratio is increased, a greater fraction of the ion flux is lost from the quadrupole by axial dispersion. For normal operation, where the entrance aperture to the analyser is relatively large, radius of the order of 50% of r_0, the effect is significant. As $V_{d.c.}/V_{a.c.}$ is increased to give increased resolution, so a greater fraction of the ions are lost and hence the sensitivity is reduced. The actual relationship between resolution and sensitivity is complex as it depends on the concentration and divergence of the ion beam leaving the source. It is complicated further by the defocussing action of the fringing fields between the ion source and the analyser. Because of the relatively long time they spend in these fields, low energy ions are more seriously defocussed and therefore transmitted less efficiently.

Figure 6.16 [20] shows as an example of a good, well constructed analyser the relationship between resolution and sensitivity for a particular quadrupole structure (rod 0.635 cm diam. × 14 cm long) at $m/e = 69$ and at $m/e = 502$.

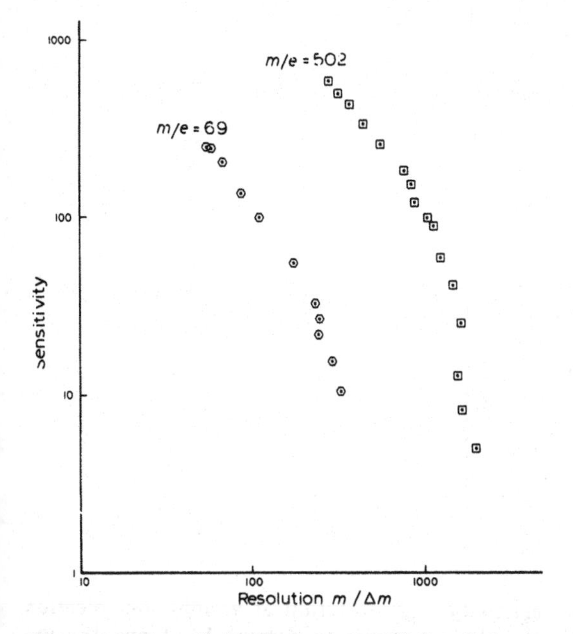

Fig. 6.16. Overall instrument sensitivity shown as a function of resolution for two mass fragments of PFTBA.

As can be seen, the sensitivity decreases at 1–1.5 times the rate of resolution increase until the resolution is twice to three times the mass. At this point, the sensitivity decreases 60 times faster than the resolution increases. That is, there is considerable decrease in sensitivity for very little change in resolution. It would not be practical to use this particular quadrupole at resolutions greater than 1000 at $m/e = 502$.

The general relationship between resolution and sensitivity as a function of injection energy is shown by the family of curves plotted in Fig. 6.17. These typical results were obtained by the authors for single charged argon ions using quadrupole electrodes 0.25 in. diam. × 6 in. long. For each injection energy, the resolving power increases asymptotically to a limiting value which, as was described earlier (Fig. 6.2) depends only upon the number of cycles the ions spend in the quadrupole field. Because of this factor, the lower energy ions have the higher limiting resolutions. However, the lower efficiency of the source and also the increased defocussing action in the fringing fields mean that the sensitivity for the low energy ions is relatively low. Consequently, to achieve the highest possible sensitivity at a given resolving power, the ion injection energy must be chosen carefully.

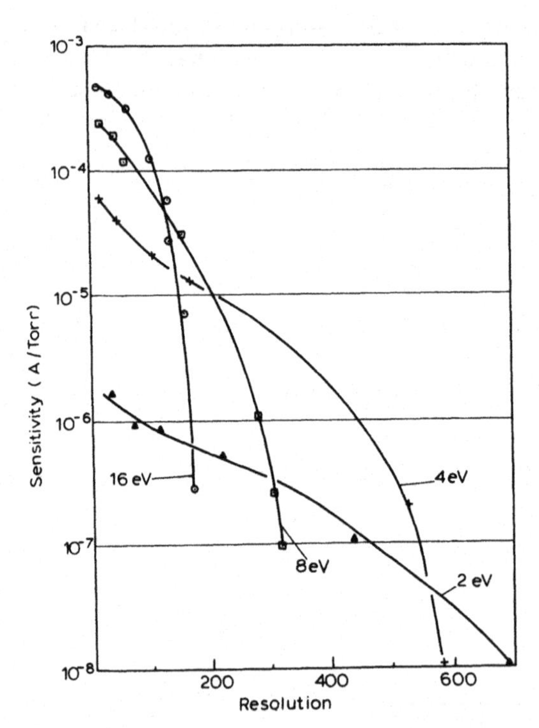

Fig. 6.17. The relationship between sensitivity and resolution at various ion injection energies. (The frequency was 4 MHz, the electron emission current 1 mA and the ion species Ar^+.)

(2) *Dependence on mass*

A desirable property of any quadrupole mass spectrometer is that the efficiency of transmission should not be dependent upon the charge-to-mass ratio of the ions, i.e. the sensitivity of the instrument should be the same for all ions there being no mass discrimination. Unfortunately, this is often difficult to achieve because, as pointed out above, the heavier the ion the longer the time spent in the fringing fields and, therefore, the greater the dispersion in the quadrupole field. Consequently, there is always a tendency for the heavier ions to be transmitted less efficiently so resulting in mass discrimination*. This effect is often observed when instruments are operated at the highest practical resolution. Up to a critical resolution there may be no discrimination throughout the mass range but beyond this point sensitivity falls very drastically at the high end of the mass scale as was shown in Fig. 6.16. The curves in Fig. 6.18 in which sensitivity is plotted, in effect, as a function of the $V_{d.c.}/V_{a.c.}$ ratio, emphasize this point. These curves for masses 129 (Xe^+) and 64.5 (Xe^{2+}) have been normalized on both axes to facilitate comparison. (The singly and doubly charged ions of an inert gas form convenient

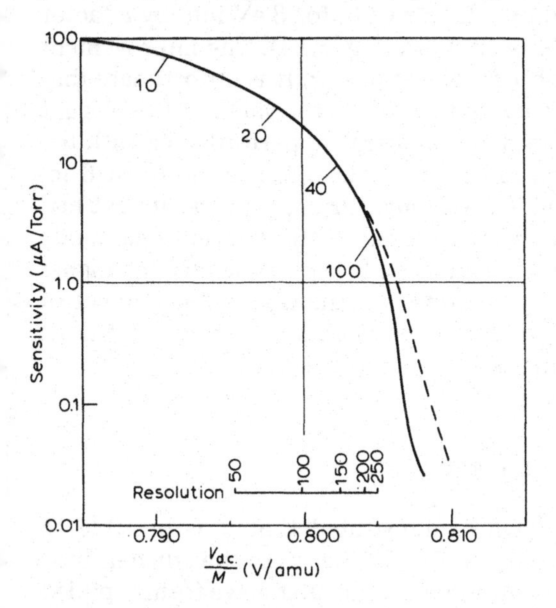

Fig. 6.18. The relationship between the normalized sensitivity and $V_{d.c.}/M$ for $^{129}Xe^+$ (solid curve) and $^{129}Xe^{2+}$ (broken curve). Values of resolution for the $^{129}Xe^+$ ions are indicated by the auxiliary scale. Also at particular points upon the $^{129}Xe^+$ curve are the changes in sensitivity caused by a 0.1% change in the d.c. voltage.

---

* See also Fig. 5.9 (p. 105).

monitors for this illustration.) At resolving powers up to about 150, the two curves are coincident indicating negligible mass discrimination. As $V_{d.c.}/V_{a.c.}$ is increased to give higher resolving powers the curves diverge, the lower mass having the higher sensitivity.

The magnitude of the discrimination and the resolving power at which it becomes significant are very much dependent on the design of the spectrometer. For best results, the ion source should be aligned accurately with the analyser and positioned as close as possible to the end of the rods in order to minimize fringing fields.

(3) *The delayed d.c. ramp*

The delayed d.c. ramp, proposed by Brubaker [4] as a means of eliminating the defocussing fields at the entrance to the analyser, consists of a set of four short auxiliary rods to which only the rf voltage, $V_{a.c.}$, is applied (see also Chapter V). In a detailed study of the delayed ramp, Brubaker reported a considerable improvement in sensitivity, particularly at low ion injection energies. He found that, typically, when the delayed ramp was used, the sensitivity for 15 eV ions was increased by a factor of 6, for 8 eV ions by a factor of 100 and for 4 eV ions by a factor of 200 (see Fig. 5.13). The improvement also depended to a small extent on the resolving power. It is unfortunate that there is little evidence from other workers to confirm the value of this elegant technique. In the authors' laboratory, improvements in performance with the ramp have been much smaller than reported by Brubaker, a factor of 6 instead of 200 for 4 eV ions, for example. Although Brubaker's instrument had rods of a larger diameter, the sensitivity achieved in the conventional mode was much less than that obtained by the authors. It is possible that the higher sensitivity can be attributed to more accurate construction and alignment of the ion source. If this is the case, then it appears that the ramp is of most value in correcting for non-uniformities in the source region.

F. OPERATION WITH NON-SINUSOIDAL FIELDS

Recently, Richards et al. [27] have demonstrated that it is possible to operate the quadrupole filter with electric fields other than the conventional sinusoidal. In particular, they have shown that rectangular wave forms applied between adjacent pairs of rods give a workable system. This, they point out, could have advantages from the point of view of a theoretical analysis (see Chapter IV, p. 83) and also in the design and construction of the electrical power supplies. The similarity of the new to the conventional system can be seen from the stability diagrams which are shown in Fig. 2.23. This diagram is drawn for various values of the duty cycle, δ, of the rectangular wave form. (δ is defined as the ratio of the time intervals for which the voltage is positive and negative during one complete cycle.) As δ is decreased from 0.5 (i.e. from a square wave) the diagram sinks towards the q axis, finally disappearing when $\delta = 0.390$. By operating with δ, such that only the tip of the diagram remains, it is possible to use the q axis as the mass scan line, the resolving power obviously being controlled by adjusting the duty cycle. With this system, no d.c. component is required and the difficulty of maintaining precisely the ratio $V_{d.c.}/V_{a.c.}$ is replaced by the problem of holding δ constant as the amplitude of the square wave is scanned.

Richards and his co-workers have deduced that, for satisfactory performance at resolving powers between 100 and 200, tolerances on the voltage amplitude, frequency and the length of the positive portion of the rectangular wave should all be of the order of 1 part in 10^4. This replaces the need to maintain $V_{d.c.}/V_{a.c.}$ to the order of 1 part in 10^4 when operating with a sinusoidal wave form.

A control unit has been constructed to generate a rectangular wave form of 50–100 V peak-to-peak at either 1 or 2 MHz. Due to capacitive loading, the waveforms were trapezoidal with rise and fall times of between 30 and 60 nsec. A resolving power of 230 (50% peak height definition) was obtained for krypton, this being considerably higher than the value of about 100 predicted by von Zahn's criterion [1] (eqn. (6.2)).

G. ELECTRICAL POWER SUPPLIES

The voltage applied across the two pairs of parallel rods consists of a continuously variable alternating component at a fixed radio frequency with a superimposed steady component such that the ratio $V_{d.c.}/V_{a.c.}$ is maintained at a pre-determined constant. For analytical instruments where optimum performance from the assembly is required, the stability and reproducibility of these supplies is important. There are, in fact, two most important parameters which are critically dependent upon the supply, (a) the ability to select and maintain the instrument tuned to a particular mass and (b) the ability to maintain a constant sensitivity.

The requirements for mass selection can be determined precisely from the basic equation relating ion mass, M, (for resonance), frequency, f, and voltage $V_{a.c.}$, which is

$$M = \frac{K V_{a.c.}}{f^2} \tag{6.11}$$

(see eqn. (6.1)). Thus

$$\frac{\Delta M}{M} = \frac{\Delta V_{a.c.}}{V_{a.c.}} - \frac{2\Delta f}{f} \tag{6.12}$$

The limiting factor in eqn. (6.12) is the voltage rather than the frequency stability. (It is relatively easy with conventional electronic oscillators to stabilize frequency to better than 1 part in 10^5 but an a.c. voltage constant to better than 1 part in 10^4 requires the highest quality electronic circuits.) Thus, for example, in a typical application, a mass stability of better than 0.1 amu over the operating range 0–200 amu can be achieved by specifying $\Delta f/f$ and $\Delta V/V$ to be below 2.5×10^{-4} and 5×10^{-4}, respectively.

It is not easy to be precise in defining requirements for a stable sensitivity. Here the important controlling parameter is the ratio $V_{d.c.}/V_{a.c.}$. The exact relationship depends upon a number of factors including the mechanical alignments of the quadrupole assembly and the resolution chosen. Figure 6.18 indicates for a particular instrument how sensitivity becomes more dependent upon the voltage ratio with increasing resolution. Even with operation at low resolution, the sensitivity change is of the order of 1% for a change in the ratio $V_{d.c.}/V_{a.c.}$ of 0.01%. For higher resolutions, the factor becomes more critical by almost an order of magnitude. Thus a design aim for the electrical power supplies should be a stability of both the a.c. and d.c. voltages to better than 1 part in 10^4 for analytical mass spectrometry. Virtually identical stability requirements have been specified for a commercial quadrupole with 0.5 in. diam. rods (U.T.i. type 100C). Bunyard [28] quotes "change of sensitivity" as a function of instrument resolution. He took the xenon 131 amu ion for his calibration. At a reasonable practical resolution, he reported a change in sensitivity of about 30% for a d.c. voltage change of 0.1%. He noted that the 131 peak fell from 0.50 to 0.34 when the d.c. voltage increased from 171.3 to 171.5 V.

It is not the intention of this section to review existing electronic circuits designed to generate the electric fields for quadrupole analysers but rather to provide useful information for circuit engineers and others who have occasion to design or manufacture such circuits. For most practical spectrometers, the frequency of operation lies within the range 500 kHz to 6 MHz, the upper limit of radio frequency voltage to be monitored is approximately 10^4 V, the lower limit is of the order of 1 V, and the dynamic range in any one instrument is equal to the ratio of M_{max} to M_{min} and is probably not greater than 200 to 1. These considerations lead to the conclusion that for all but the least demanding instruments the amplitude of the radio frequency voltage to the analyser rods can only be monitored to the required precision by the use of a vacuum diode preceded, if necessary, by a precision potential divider in a carefully designed circuit. The vacuum diode is preferred because of its better performance at radio frequencies and high voltages when compared with semiconductor diodes. The diode rectifier system, together with the radio frequency generator, a differential amplifier and a modulator are used to form a feedback loop designed to maintain the amplitude of the alternating component of the filter field constant to the required degree. The output from the rectifier system may also be used to control the amplitude of the direct voltage component of the filter field and so achieve and maintain the required $V_{d.c.}/V_{a.c.}$ ratio. A typical schematic diagram is shown in Fig. 6.19.

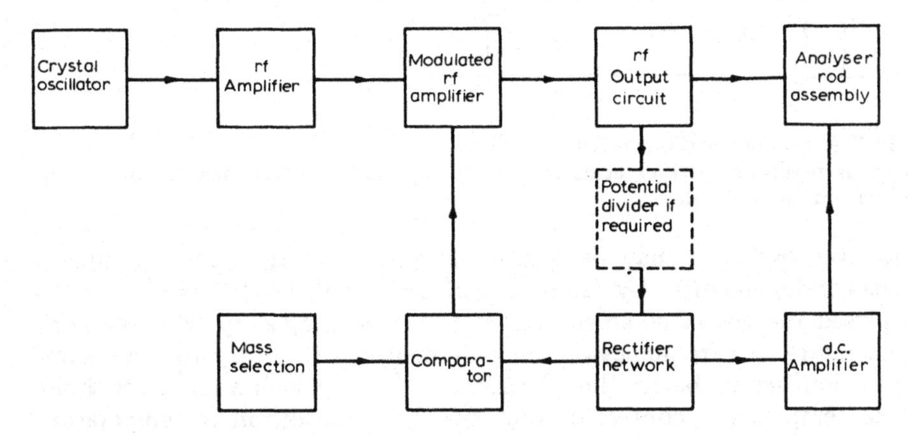

Fig. 6.19. Schematic diagram of a quadrupole power supply system.

The operation of cumulative diode rectifiers as shown in Fig. 6.20 has been analysed by Bell [29] and Scroggie [30, 31]. Amongst the important conclusions to be drawn from these authors are the following.
(1) The product $fCR \gg 5$.
(2) The value of C should be kept small by using as large a value of R as is practicable, since by so doing the peak current through the diode is minimized and the error due to the internal resistance of the diode is reduced.
(3) The effective input resistance of a shunt connected cumulative rectifier circuit is always less than $R/3$ and the actual value depends upon the form of any low pass filter which may be in circuit.
(4) The voltage developed across the load resistance by the diode splash current is very nearly proportional to both the logarithm of the resistance over a wide range of resistance values and to the supply voltage applied to the valve heater.

For use with radio frequency quadrupole spectrometers, the supply to the diode heaters will have to be sufficiently stable to enable the required $V_{d.c.}/V_{a.c.}$ ratio to be maintained. The degree of stability required will depend upon the actual type of diode in use but is typically 1 part in 10^4. Alternatively, a self-balancing circuit may be employed to achieve the required performance.

At the lower frequencies of operation, it is possible to use semiconductor diodes in conjunction with operational amplifiers as shown in Fig. 6.21 to produce precision rectifier systems. The maximum permissible input voltage to which these circuits may be subjected necessitates the use of an input voltage divider.

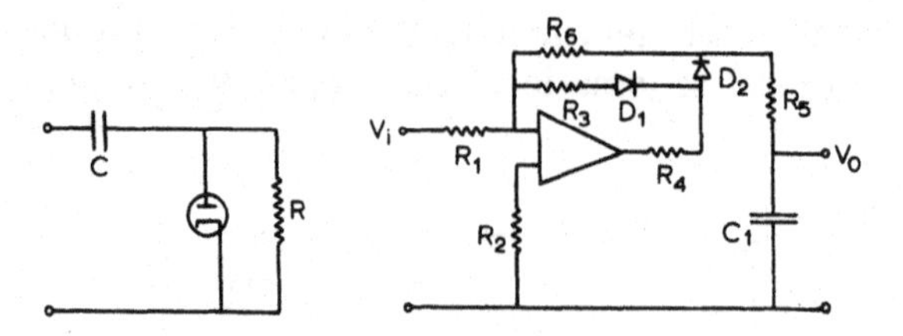

Fig. 6.20. The cumulative diode rectifier circuit.
Fig. 6.21. A precision rectifier using semiconductor diodes as feedback elements in an operational amplifier circuit.

A suitable oscillator may be readily designed using the circuit originally developed independently by Gouriet [32] and Clapp [33]. Baxendall [34] has analysed and described an equivalent oscillator using a crystal in its series resonant mode as a frequency-defining element and it is easy to obtain frequency stabilities of better than 1 part in 10^5 using such a circuit without resort to temperature control of the crystal. The addition of temperature control for the crystal enables stabilities greater than 1 in 10^7 to be readily obtained.

The basic circuit of the oscillator is shown in Fig. 6.22 and maintenance equation for oscillation is

$$R < \left[\frac{1}{\omega C_1}\right]\left[\frac{1}{\omega C_2}\right] g_m \tag{6.13}$$

where C_1 and C_2 are the feedback capacitors in the circuit, g_m is the mutual conductance of the active device and R is the equivalent series resistance of the resonant circuit or crystal. For a typical 4 MHz crystal at series resonance, R will be approximately 60 Ω and the g_m of a silicon transistor is 38.5 mA/V per mA of collector current so that the circuit values shown in the diagram are typical. Note that I_c is well defined by $-V/R_2$.

Since it is relatively easy to obtain a change of mass range in a quadrupole filter by changing the frequency of operation, it is advantageous to follow the oscillator by an untuned broadband amplifier which then drives the final modulated amplifier stage. Changing the mass range then only involves the substitution of the crystal and the final stage inductor followed by a single tuning adjustment.

The circuit diagram of a typical output circuit intended to resonate at 4 MHz and to supply a peak radio frequency voltage of up to 1000 V across rod pairs of an analyser assembly is shown in Fig. 6.23. Provision is made for monitoring the alternating voltage by a pair of rectifying diodes associated with an operational amplifier, for the balancing of the alternating voltage applied to the rod pairs and for the injection of the required steady voltages onto the rod pairs. In order to minimize the circuit losses due to the circulating radio frequency currents, the ratio of inductance to capacitance in the

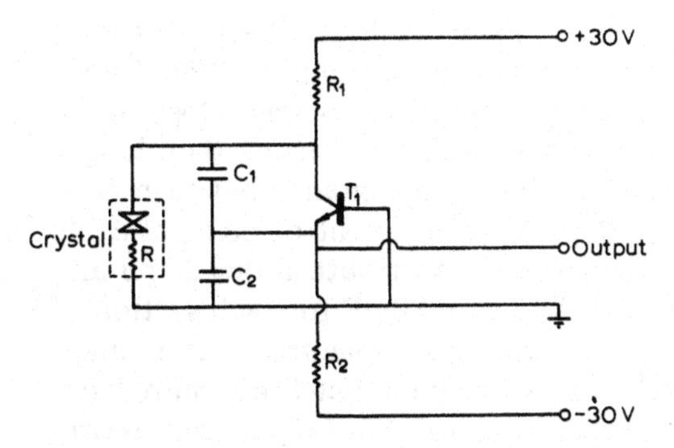

Fig. 6.22. The crystal oscillator circuit. C_1 = 680 pF; C_2 = 1200 pF; R_1 = 18 kΩ; R_2 = 27 kΩ; T_1 = ZTX 302.

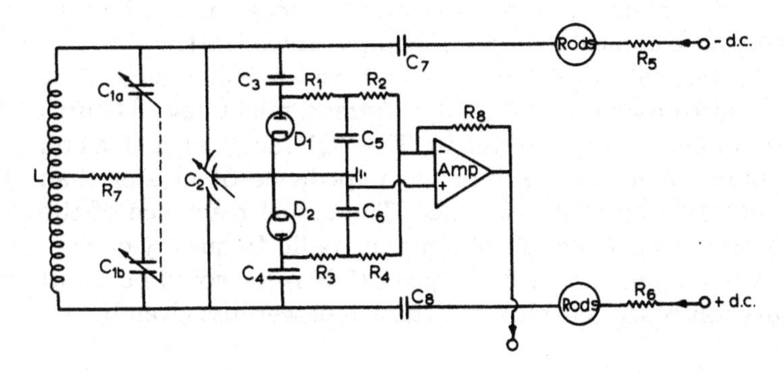

Fig. 6.23. The diagram of a typical radio frequency output circuit. C_{1a} = 5–43 pF; C_{1b} = 5–43 pF; C_2 = 3–10 pF; C_3 = 33 pF; C_4 = 33 pF; C_5 = 470 pF; C_6 = 470 pF; C_7 = 1000 pF; C_8 = 1000 pF; R_1 = 1 MΩ; R_2 = 1 MΩ; R_3 = 1 MΩ; R_4 = 1 MΩ; R_5 = 330 kΩ; R_6 = 330 kΩ; R_7 = 1 MΩ; R_8 = 25 kΩ; L = 20 μH.

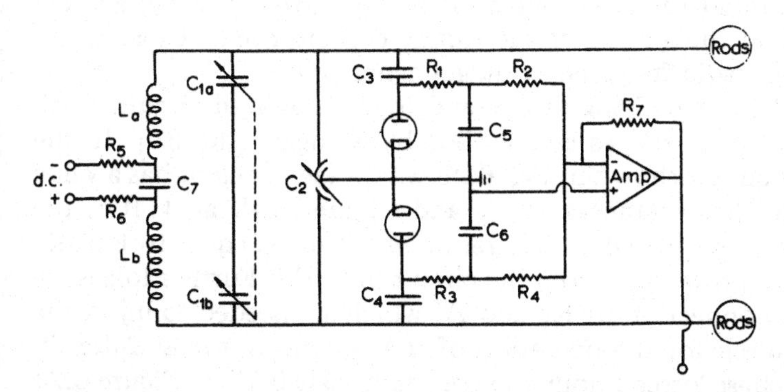

Fig. 6.24. Radio frequency output circuit modified for increased efficiency and high scan rates. C_{1a} = 5–43 pF; C_{1b} = 5–43 pF; C_2 = 3–10 pF; C_3 = 33 pF; C_4 = 33 pF; C_5 = 470 pF; C_6 = 470 pF; C_7 = 1000 pF; R_1 = 1 MΩ; R_2 = 1 MΩ; R_3 = 1 MΩ; R_4 = 1 MΩ; R_5 = 10 kΩ; R_6 = 10 kΩ; R_7 = 25 kΩ.

resonant circuit should be high. This requirement precludes the use of long leads between the inductor and the analyser and hence implies that the inductor, together with its associated circuitry, should be located close to, or integral with, the analyser. A typical analyser assembly, using 6 in. long × 0.25 in. diam. rods within an earthed tube, has a capacitance between each rod pair and earth of approximately 80 pF, whilst a foot of co-axial cable used to connect the analyser to the inductor will contribute about 25 pF and wiring strays a further small contribution. Allowing 30 pF for each section of the tuning capacitor will produce a total effective capacitance of perhaps 75 pF. A suitable coil to resonate with this capacitance at the required frequency may be designed from data published in ref. 35. Information is provided which enables a solenoid coil of the required inductance to be wound and a correction to be made for any decrease in the effective inductance due to the proximity of a conductive enclosure. The "Q" factor of the coil may be measured using a "Q" meter at the resonant frequency, or can be estimated from formulae and graphical data published by Howe [36].

Typically, a coil to resonate at 4 MHz would have an inductance L' of about 20 μH and, if constructed on a 2 in. diam. former, might have 34 turns of 20 s.w.g. wire wound over a length of 2.5 in. The "Q" factor of such a coil should exceed 500 and so produce a dynamic impedance of greater than 250 kΩ for the oscillatory circuit at resonance. The actual reactance of the coil would be about 500 Ω and hence the circulating radio frequency current carried by the coil when supporting 1000 V peak (700 rms) would be about 1.4 A rms. The circulation of this current will entail a power loss given by

$$P = \frac{\hat{V}^2}{2Q\omega L'} \tag{6.14}$$

and in the coil described this will amount to about 2 W. The tuning capacitors and any other capacitors which carry the circulating radio frequency current must, of course, be rated to handle the anticipated current without any excessive power loss and the air gap of the tuning capacitor must be adequate to withstand the high radio frequency voltages generated.

The operation of the rectifying diodes and their associated circuitry will absorb radio frequency power as is described by Scroggie [30, 31]. In the case of the rectifier circuit shown in Fig. 6.23 where each resistor has a value of 1 MΩ and where the reactances of C_5 and C_6 are small at 4 MHz, the equivalent load across each half of the resonant circuit is equal to 0.5 MΩ and results in a total power loss from the circuit of 0.5 W. Further losses are associated with the connection of R_5 and R_6 which are needed to inject the steady voltages required for the operation of the system as a mass filter. In the present circuit, these losses amount to approximately 0.75 W. Figure 6.24 shows an alternative arrangement of the circuit in which the losses associated with d.c. feed resistors are reduced considerably at the expense of a more complex coil assembly. This latter circuit also has advantages when it is desired to scan the mass range of the analyser at high rates since the current associated with the charging of the isolating capacitor C_7 flows through resistors R_5 and R_6 which may be of considerably lower value than the equivalent resistors R_5 and R_6 shown in Fig. 6.23. Hence the resulting voltage drop which tends to vary the $V_{d.c.}/V_{a.c.}$ ratio is reduced and both the relative sensitivity and mass resolution are less dependent upon the scan rate. Experience has shown that the circuits described are well able to achieve the electrical performance required for analytical use and that the limit of performance is principally set by the stability of resistance value displayed by resistors R_1, R_2, R_3, R_4, and R_8 in Fig. 6.23.

Sections A—F were written by A.E.H. and J.H.L. and Section G by W.E.A.

REFERENCES

1 U. von Zahn, Thesis, University of Bonn, 1956.
2 P. Kubicek, Czech, J. Phys., B19 (1969) 17.
3 A.E. Holme, W.J. Thatcher and J.H. Leck, J. Phys. E., 5 (1972) 429.
4 W.M. Brubaker, NASA Rep. NASW 1298 (1970).
5 W. Arnold, J. Vac. Sci. Technol., 7 (1970) 191.
6 M.S. Story, J. Vac. Sci. Technol., 4 (1967) 326.
7 D.F. Munro, Rev. Sci. Instrum., 38 (1967) 1532.
8 A.R. Fairburn, Rev. Sci. Instrum. 40 (1969) 380.
9 M.S. Story, paper presented at Pacific Conference on Chemistry and Spectroscopy October 6–9, 1969.
10 B. Scruton and B.M. Blott, J. Phys. E., 6 (1973) 472.
11 B. Goodwin and A.E. Holme, to be published.
12 P.H. Dawson and N.R. Whetten, J. Vac. Sci. Technol., 7 (1970) 440.
13 T. Hayashi and N. Sakudo, Proc. Int. Conf. Mass Spectrom., Kyoto, Japan, 1969.
14 G.F. Sauter, R.A. Gerber and H.J. Oskam, Rev. Sci. Instrum., 37 (1966) 572.
15 R.S. Narcisi and A.D. Bailey, in D.G. Kung-Hele, P. Muller and G. Righini (Eds.), Space Research, Vol. V, North Holland, Amsterdam, 1965.
16 H.J. Heinen, Ch. Hötzel and H.D. Beckey, Int. J. Mass Spectrom. Ion Phys., 13 (1974) 55.
17 A. Benninghoven, Surface Sci., 35 (1973) 427.
18 D.R. Sandstrom, J.H. Leck and E.E. Donaldson, J. Chem. Phys., 48 (1968) 5683.
19 W.M. Brubaker and J. Tuul, Rev. Sc. Instrum., 35 (1964) 1007.
20 M.S. Story, private communication.
21 G.A. Hofmann, Vacuum, 24 (1974) 65.
22 P. Blum and F.L. Torney, Rev. Sci. Instrum., 38 (1967) 1404.
23 P.A. Redhead, Can. J. Phys., 37 (1959) 1260.
24 J.M. Goodings, J.M. Jones and D.A. Parkes, Int. J. Mass Spectrom. Ion Phys., 9 (1972) 417.
25 A. Benninghoven, C. Plog and N. Treitz, Int. J. Mass Spectrom. Ion Phys., 13 (1974) 415.
26 J. Visser, J. Vac. Sci. Technol., 10 (1973) 464.
27 J.A. Richards, R.M. Huey and J. Hiller, Int. J. Mass Spectrom. Ion. Phys., 12 (1973) 317.
28 G.B. Bunyard, Quadrupole Mass Spectrom., 1 (4) (1974).

CHAPTER VII

THE MONOPOLE: DESIGN AND PERFORMANCE

R.F. Herzog

A. INTRODUCTION

The theory of the monopole and its historical development have already been described in earlier chapters (see especially pp. 37 and 107). This chapter* will cover primarily the practical aspects of construction and operation of a monopole and its performance evaluation [1, 2]. Nevertheless, a short summary of the trajectory formation will be helpful in understanding the experimental results. The accurate theory tends to conceal the real physical aspects of orbit formation behind the rather complex mathematics and usually requires computer analysis to obtain useful results. Fortunately, Brubaker [3] has described trajectory formation by an approximate method which provides an insight into the actions of the different forces on an ion which is moving along near the center z axis of the quadrupole field. This method is easily applied to monopoles.

B. ION OSCILLATIONS IN A MONOPOLE

Under the normal operating conditions of a monopole, only the oscillations in the y direction are of prime importance (see Fig. 2.3 and 2.25). The resonance in the x direction has little effect since the monopoles are usually operated with a much smaller a.c. voltage than quadrupoles, well inside the stability boundaries for the x direction. (Brubaker's method is even more appropriate here, Figs. 2.24 and 2.26.)

The equation of motion in the y direction for the monopole [see eqn. (2.24)] is

$$m\ddot{y} = F_{\mathrm{d.c.}} + F_{\mathrm{a.c.}} = y(e/r_0^2)(2U - 2V\cos\omega t) \tag{7.1}$$

The factor 2 of U and V results from the fact that the voltages of a monopole are measured with regard to the V electrode. $F_{\mathrm{d.c.}}$ and $F_{\mathrm{a.c.}}$ are the forces from the d.c. and a.c. fields, respectively.

It is convenient to introduce the system constant

* Based mainly on work carried out at GCA Corporation, Technology Division.

$$U^* = \frac{\omega^2 r_0^2 m_1}{e} \text{ (volt)} \tag{7.2}$$

where $m_1 = 1.66 \times 10^{-27}$ kg is the atomic mass unit.

The test monopole to be described later has $r_0 = 7.62 \times 10^{-3}$ m and $f = 1.8 \times 10^6$ Hz which results in $U^* = 77$ V. If M is the mass number of an isotope, one obtains

$$a = \frac{8(U/U^*)}{M} \tag{7.3}$$

$$q = \frac{4(V/U^*)}{M} \tag{7.4}$$

and eqn. (7.1) becomes

$$\ddot{y} = \frac{y(a - 2q \cos \omega t)\omega^2}{4} \tag{7.5}$$

(Note that the sign conventions used here for a and q are different from a_y and q_y in other chapters.)

The main obstacle to the integration of this equation is the variable factor y on the right side. However, in a first approximation, we can consider $y = \bar{y} =$ constant, thus replacing the variable y by its average value $\bar{y}$ during one a.c. cycle. We can also neglect a since it is always much smaller than $2q$. This leaves from eqn. (7.5) only

$$\ddot{y} = -\tfrac{1}{2}\bar{y} q \omega^2 \cos \omega t \tag{7.6}$$

which has the well known solution

$$y = \bar{y}(1 + \tfrac{1}{2} q \cos \omega t) \tag{7.7}$$

We know from Chapter II that q must be smaller than 0.706 to obtain stability in the x direction; for monopoles, the value of q can be much smaller. Equation (7.7) describes a "small" oscillation or ripple around the median distance $\bar{y}$ from the z axis with the normal a.c. frequency and an amplitude about $\bar{y}/3$ or smaller. These oscillations are the key to the focusing effect in the y direction; since the force is actually proportional to y, it is larger during the half cycle where $y > \bar{y}$ and smaller during the other half cycle. The average force of the a.c. field during one rf cycle is

$$\bar{F}_{\text{a.c.}} = -f \int_0^{1/f} y \frac{e}{r_0^2} \, 2V \cos \omega t \, dt \tag{7.8}$$

This integral can be easily calculated if we use the result of the first approximation, eqn. (7.7), for y. One obtains

$$\bar{F}_{\text{a.c.}} = -\bar{y} \frac{e}{r_0^2} \, 2Vf \int_0^{1/f} (1 + \tfrac{1}{2} q \cos \omega t) \cos \omega t \, dt$$

The integral has the value $q/4f$.

Using eqns. (7.2) and (7.4), one obtains

$$\bar{F}_{a.c} = \frac{-m\bar{y}\omega^2 q^2}{8} \tag{7.9}$$

This average force of the a.c. field is directed toward the z axis and is proportional to the distance $\bar{y}$ from the axis. This "strong" focusing force is partially reduced by the defocusing force of the d.c. field

$$\bar{F}_{d.c.} = \frac{+m\bar{y}\omega^2 a}{4} \tag{7.10}$$

which is directed away from the axis.

In the second appromation, the actual forces F are replaced by the average forces $\bar{F}$ and defining

$$\beta^2 = \tfrac{1}{2}q^2 - a \tag{7.11}$$

[see eqn. (3.16)] results in

$$\ddot{y} = -\bar{y}\left(\frac{\beta\omega}{2}\right)^2 \tag{7.12}$$

This equation describes a harmonic oscillator with the resonance frequency $\beta\omega/2$. The general solution is given by

$$\bar{y} = A \sin(\beta\omega/2)(t - t_0) \tag{7.13}$$

where the integration constants A and t_0 must satisfy the initial conditions.

If β^2 becomes negative, the solution of eqn. (7.12) is no longer periodic but increases exponentially. Therefore $\beta = 0$ describes the borderline between "stable" and "unstable" solutions. Considering the great simplifications made in the development of eqn. (7.11), it is comforting to see that the agreement with the accurate relation is within 5%. $\beta = 0$ describes the special case where the a.c. focusing force and the d.c. defocusing force tend to compensate each other and ion transmission becomes possible. If we call the corresponding d.c. voltage U_0, we obtain from eqns. (7.3), (7.4) and (7.11)

$$U_0 = U^*\left(\frac{U}{V}\right)^2 M \tag{7.14}$$

During a "normal" mass scan, the voltage ratio U/V is kept constant; therefore, one obtains a linear mass scale U_0 proportional to M. However if, instead of the ratio U/V, only the a.c. voltage V is kept constant, one obtains

$$U_0 = \frac{(V^2/U^*)}{M} \tag{7.15}$$

In this case, the d.c. voltage becomes inversely proportional to M. (See "extended" mass scan which will be described later in more detail.) For a monopole of limited length, eqns. (7.14) and (7.15) do not determine the accurate position of the top of a mass peak but rather the location of the steep slope at the low mass side of the peak as will be explained later. Nevertheless, the deviation is very small.

Assume an ion enters the field quite near the z axis ($y_0 = 0$) but with a slight radial velocity $\dot{y}_0$. In this case, the initial conditions for eqn. (7.13) are $t_0 = 0$ and $\dot{y}_0 = A\,(\beta\omega/2)$. Therefore

$$\bar{y} = \left[\frac{\dot{y}_0}{\beta\omega/2}\right] \sin\,(\beta\omega/2)\,t \tag{7.16}$$

The amplitude of the oscillation is proportional to $\dot{y}_0$ and inversely proportional to β. We will call this the "large" oscillation because β is usually very small. Superimposed over these "large" oscillations are the "small" oscillations described by eqn. (7.7). The "large" oscillations are primarily responsible for the performance of a monopole as a mass filter. One necessary condition for the monopole is that $\bar{y}$ must be always positive since otherwise the ion would impinge on the V-electrode. This means $(\beta\omega/2)\,t < \pi$. If the ion needs n a.c. cycles to pass the length L of the field

$$n = fL\left(\frac{m_1}{2e}\right)^{1/2}\left(\frac{M}{E_z}\right)^{1/2} \tag{7.17}$$

where E_z is the acceleration voltage of the ions [see eqn. (2.40)]. Then $t = n/f$ so that $\beta n < 1$ [see eqn. (2.52)]. Figure 7.1 shows the value of n for the test monopole and for different mass numbers M and acceleration voltages E_z. Except for hydrogen, n is always more than 10.

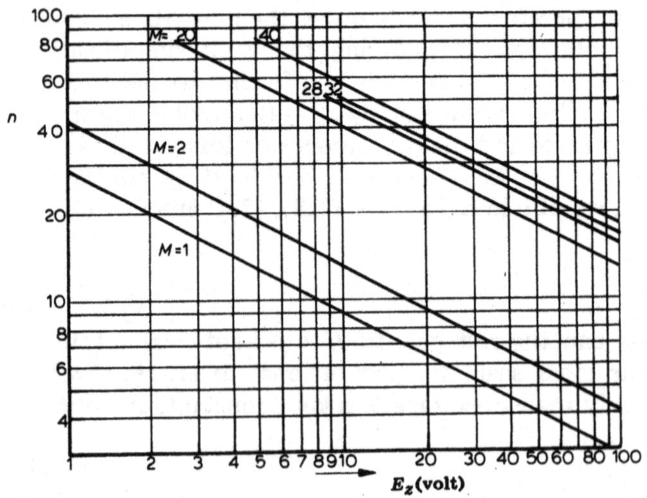

Fig. 7.1. The number, n, of the a.c. cycles required for an ion of mass number M and acceleration voltage E_z to pass the test monopole. Length (L) = 0.222 m; frequency (f) = 1.8×10^6 Hz.

In practice, most ions do not enter the field accurately at $y_0 = 0$ but as a beam with the center at about $y_0 = 0.1\,r_0$. These ions return earlier to the axis and require that βn is less than 1.

The "large" oscillations have the first maximum if $(\beta\omega/2)t = \pi/2$. Combined with the "small" oscillations, eqn. (7.7), we obtain the maximum amplitude

$$y_{max} = \dot{y}_0 t(2/\pi)(1 + \tfrac{1}{2}q) \simeq \dot{y}_0 t \qquad (7.18)$$

which is about equal to the position of the beam without the monopole field.

The geometry of the monopole (L, r_0, d_1, d_2) limits the maximum value of the radial velocity $\dot{y}_0$ of any ion which may pass. If $\dot{y}_0$ is negative, most ions are lost to the V electrode unless y_0 is relatively large.

Let us now consider the scanning of a mass spectrum by increasing U and V simultaneously and by keeping the ratio U/V constant and well below 0.168, corresponding to the tip of the stability diagram. For reasons of simplicity, we assume that the ion beam consists of only one mass and energy. For small voltages, we find from eqn. (7.11) that β^2 is first negative and the defocussing effect of the d.c. field is predominant over the focusing effect of the a.c. field. Therefore, most ions are lost to the rod; nevertheless a few ions can pass if they have very small values of y_0 and $\dot{y}_0$. The field is very weak near the z axis and has little effect on the ions. These ions produce a long peak tail of low intensity and contribute to a background on the lower part of the mass spectrum.

If the voltages are increased, one reaches the point where the d.c. and a.c. forces begin to compensate each other and β becomes zero. The number of ions which can pass a monopole of limited length L is now increased because the attraction of the rod is partially nullified and ions with larger values of y_0 and $\dot{y}_0$ can pass.

If the voltages are further increased, then β^2 becomes positive and the "large" oscillations become a reality. It was found experimentally that the top of the peak occurs if the first maximum of the "large" oscillations is near the beginning of the second half of the filter, or for

$$\beta n = 0.9 \qquad (7.19)$$

This figure depends slightly on the entrance and exit appertures.

If the voltages are further increased such that βn equals 1, then the "large" oscillations return to the z axis and the ions are removed by the V electrode (see Fig 2.26). The cut-off point is not sharp since it depends on y_0 and $\dot{y}_0$ within limits given by the apertures of the ion source. One can see from the above that the main part of the peak is essentially confined between $0 < \beta n < 1$. The boundary, $\beta = 0$ corresponds to the d.c. voltage U_0 [see eqn. (7.14)]. If the d.c. voltage is increased to $U_0 + \Delta U_0$ keeping the ratio U/V constant, one obtains from eqn. (7.11)

$$\beta^2 = 8\left(\frac{U}{V}\right)^2 \frac{\Delta U}{U_0} \qquad (7.20)$$

References p. 180

Using eqns. (7.2) and (2.40) one obtains

$$\beta^2 n^2 = \left(\frac{L}{\pi r_0}\right)^2 \frac{\Delta U}{E_z} \tag{7.21}$$

and finally from eqn. (7.19)

$$\Delta U = E_z 0.8 \left(\frac{\pi r_0}{L}\right)^2 \tag{7.22}$$

This equation indicates that, compared with eqn. (7.14), all mass peaks are shifted slightly toward higher masses by ΔU. Since ΔU is independent of M, the mass scale remains linear and only the zero point is shifted. This small shift is proportional to the acceleration voltage E_z of the ions, and could be verified experimentally.

The peak width at half height naturally depends on the size of the entrance and exit aperture. A compromise has to be made between resolution and sensitivity. It has been found experimentally (in agreement with theoretical considerations) that the peak width at half height corresponds to $\beta^2 n^2 = 0.4$. Therefore from eqn. (7.21)

$$\Delta U_{1/2} = E_z 0.4 \left(\frac{\pi r_0}{L}\right)^2 \tag{7.23}$$

The peak width is independent of the mass and therefore the resolution $R = M/\Delta M$ is proportional to the mass. High resolution requires a small acceleration voltage E_z which reduces the sensitivity; here too, a compromise has to be made. The resolution R is given by

$$R = \left(\frac{U_0}{E_z}\right) 2.5 \left(\frac{L}{\pi r_0}\right)^2 \tag{7.24}$$

High resolution can also be obtained if $L \gg r_0$ which again reduces the sensitivity.

C. ION ENTRANCE AND EXIT

The influence of fringing fields has been discussed in Chapter V. Unfortunately, there is limited experimental evidence concerning these fields. With very short or zero fringing fields, the ions entering the monopole receive transverse impulses depending on the initial phases of the field [see Fig. 2.27(a)]. The phases (ξ_0) near 0 and $\pi/2$ will be the main contributors to ion transmission. As described in Chapter V, these effects may be modified by gradual entry into the field.

The fringing field also accelerates or decelerates the ions at the entry or exit. The potential inside the monopole is given by $(y/r_0)^2 V \cos \omega t$ (again neglecting U). This potential has the maximum value $\pm(y/r_0)^2 V$ if the rf phase of the field, ωt_0, is either 0 or π. In practice, the test monopole required $V = 474$ V to be tuned at mass 40. Assuming an entrance aperture $y = d_1 = 0.1 r_0$, the maximum energy gain or loss will be about 5 V, which is still tolerable since the original ion energy was usually about 20 eV. However, if $y > 0.2 r_0$, some of the ions will be prevented from crossing the fringing field resulting in a loss of transmission. A similar process occurs at the exit aperture, which is usually made larger than the entrance aperture in order to obtain high sensitivity. It is obvious that larger apertures than the ones mentioned above would improve the sensitivity only slightly but reduce the resolution significantly. Note that a monoenergetic beam from the ion source becomes heteroenergetic in the monopole field and n depends not only on E_z, M, L, and f but also on y_0, the phase, and V. As a result, the large oscillations are more spread out and the focussing properties described by Lever (see p. 57) are less pronounced. This is an advantage since it reduces the sensitivity variations, which will be described later.

D. MECHANICAL CONSTRUCTION OF THE MONOPOLE

A laboratory model of a monopole mass spectrometer was designed and built with the basic parameters of rod length $L = 0.222$ m, field radius $r_0 = 7.62 \times 10^{-3}$ m, and rf frequency $f = 1.8 \times 10^6$ Hz.

The following design features were incorporated. The entrance and exit apertures were triangular in shape, two sides coinciding with the V electrode and the third side adjustable from outside, thus varying the heights d_1 and d_2 of the triangles. The apertures were made of thin sheets of gold to minimize charging effects. Later, the entrance aperture was replaced by a round hole of 1.2×10^{-3} m diameter offset by 0.85×10^{-3} m from the apex of the V electrode in order to prevent obstruction of the beam. The entrance aperture was insulated from the V electrode to permit retardation of the ions after the injection, which was known to be beneficial for quadrupoles but, as it turned out, made no improvement for the monopole.

The monopole rod and the V electrode were made of 304 stainless steel and gold plated. The V electrode consisted of two rectangular blocks screwed together, since machining is easier and more accurate than making the whole V electrode in one piece. Figure (7.2) shows some details of mounting the rod and V electrode in relation to the entrance aperture. A standard Nier type ion source with an iridium filament was aligned with the entrance aperture. Solid gold of high purity was used for the electrodes of the ion source because most work was done with atomic oxygen. A high gain 16-stage Cu–Be electron multiplier was incorporated to permit counting of single ions.

The instrument was in continuous use for over three years and showed no degradation of performance. The spectra obtained were well reproducible.

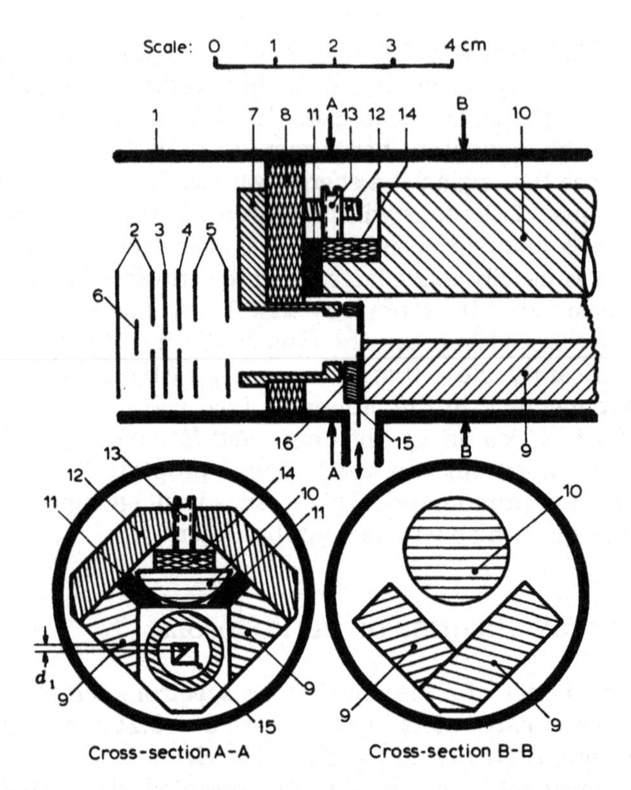

Fig. 7.2. Simplified sketch of the monopole mounting and the entrance aperture. 1, Housing, precision reamed after welding; 2–6, Nier-type ion source consisting of 2, ionization chamber (3×10^{-3} m hole, + 26 V); 3, draw out electrode (1.6×10^{-3} m hole, − 155 V); 4, acceleration electrode (3×10^{-3} m hole, − 345 V); 5, focusing electrode (6×10^{-3} m hole, − 3.8 V); 6, electron beam, energy (88 eV); 7, shield to prevent ions from striking the insulators (usually on ground potential); 8, main mounting disk of the whole monopole structure. This disk fits tightly into the housing 1 and permits a higher pressure in the ion source than in the monopole proper. It is made of machinable ceramic (Aremcolox 502-600) which is also used for the other insulators 11 and 14; 9, V-electrode, made of two rectangular bars, bolted together and to the disk 8. A similar mounting arrangement is at the other end of the V-electrode. This electrode is insulated from ground, but was usually grounded; 10, rod, positioned precisely against V-electrode by the spacers 11; 11, insulator which assures proper distance and alignment of rod and V-electrode; 12, bracket, bolted on the V-electrode; 13, setscrew to press the rod against the insulator and V-electrode; 14, ceramic block to insulate the rod from the setscrew; 15, entrance slit, which can be moved up and down to adjust the height, d_1, of the entrance aperture. This slit is insulated with mica from the V-electrode and from ground. Except for special tests, this slit was kept grounded; 16, holder for the slit. Bolted to and electrically connected with the V-electrode but insulated from the shield.

Ceramic insulators were used for the rod and the ion source to permit bake-out at 450°C. No electrical breakdown occured.

Great care was used for the alignment of the rod against the V electrode. However, it appears from theoretical considerations that the alignment of the monopole is much less critical than for a quadrupole. This is based on the fact that the ions stay in a monopole only for less than half of one large oscillation, but for many large oscillations in a quadrupole. Unfortunately, no experimental data are available yet to confirm this conclusion.

E. THE POWER SUPPLY*

The power supply for this monopole has been built in such a way that it permits testing of the instrument under a wide variety of situations. Any value of the voltage ratio U/V can be selected. A normal scan is performed by keeping this voltage ratio constant. The scanning speed can be adjusted over a very wide range and the upper and lower limits of the scan can be adjusted separately. It is possible to use either single scan or automatically repeated scans. In addition, manual selection of any peak is possible. For the extended mass range to be described later, the a.c. voltage is kept constant and the d.c. voltage is reduced. The scanning speed of the extended mass range can be adjusted independently from the scanning of the regular mass range. The extended mass range scan can be started either manually or automatically. Finally, it is possible to reverse the polarity of the d.c. voltage to permit analysis of negative ions. This power supply will be described below in greater detail.

General description. The a.c. power supply is used to generate a regulated 1.8 MHz rf signal, amplitude modulated from 2 V peak-to-peak to 1800 V peak-to-peak, and a superimposed d.c. signal proportional to this rf amplitude. The power supply consists of two basic parts: the sweep chassis and the rf unit.

Circuit description. See simplified schematic diagram Fig. (7.3).

(1) *Sweep unit*

A basic saw-tooth sweep voltage is generated by a high gain operational amplifier, 1, in an integrator circuit. The output is linear with time and the slope is dependent on RC and the applied voltage. The capacitor values are selectable by the front panel coarse speed control which provides the basic range of 1, 10, and 100 sec. The fine speed control permits the selection of intermediate values by changing the input current to the integrator. The

* Designed by John D'Angelo.

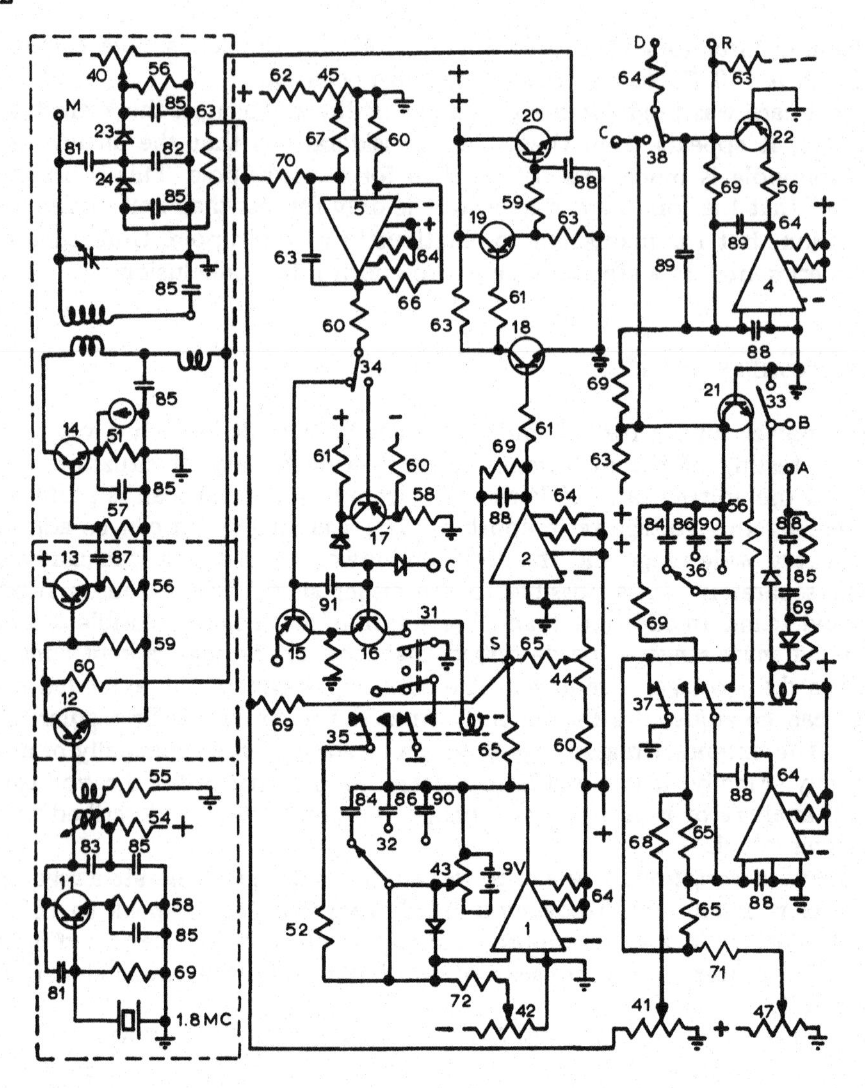

Fig. 7.3. Simplified schematic of the power supply. All points marked + are connected to + 15 V against ground. All points marked − are connected to − 15 V against ground. All points marked ++ are connected to + 100 V against ground. All points marked −− are connected to − 100 V against ground. Note that the following connections are not shown in the schematic diagram: A to A, B to B, C to C, and D to D. Terminal M is connected with the monopole rod and terminal R with the recorder. The second pole of the recorder and the V-electrode are grounded.

Parts list: 1–5 Operational amplifiers, Nexus SD1. Transistors and diodes: 11, 2N 3819; 12, 13, 15, 17, 2N 697; 14, TRS 4405 S; 16, 2N 404; 18, 19, 21 2N 699; 20, 2N 3583; 22, 2N 398; 23, 24 FD 3309; all other diodes, IN 645. Switches and relays: 31, DPDT, sweep on–off, shown in on position; 32, normal scan, speed adjustment coarse; 33, SP, extended mass range, on–off, shown in off position; 34, SP, automatic scan of extended mass range after normal scan, on–off, shown in on position; 35, extended mass range,

scan speed coarse; 36, relay for normal scan; 37, relay for extended mass scan; 38, SP, d.c. output polarity, shown in the minus position. Potentiometers and controls: 41, 500K 10T, ratio *U/V* adjustment, normal scan; 42, 50K, normal scanning, speed fine adjustment; 43, 100K 10T, upper end adjustment of normal scan; 44, 10K 10T, lower end adjustment of normal scan, also manual adjustment of a peak without scan; 45, 100K, trigger level adjustment for automatic extended mass scale scan; 46, 50K, compensation for non-linearity of the diodes for very small voltages; 47, 50K, extended mass range, scan speed fine adjustment. Resistors: 51, 10 Ω; 52, 12 Ω; 53, 47 Ω; 54, 100 Ω; 55, 330 Ω; 56, 1 kΩ; 57, 1.2 kΩ; 58, 2.2 kΩ; 59, 2.7 kΩ; 60, 4.7 kΩ; 61, 10 kΩ; 62, 18 kΩ; 63, 22 kΩ; 64, 47 kΩ; 65, 100 kΩ; 66, 150 kΩ; 67, 220 kΩ; 68, 470 kΩ; 69, 1 MΩ; 70, 2.2 MΩ; 71, 3.9 MΩ; 72, 10 MΩ. Capacitors: 81, 3 pF; 82, 47 pF; 83, 68 pF; 84, 0.02 μF; 85, 0.1 μF; 86, 0.2 μF; 87, 0.47 μF; 88, 0.01 μF; 89, 0.002 μF; 90, 2 μF; 91, 10 μF

slowest possible speed is approximately 500 sec and the highest speed approximately 1 sec.

The maximum sweep voltage can be limited to any value within the sweep range by adjusting the high level potentiometer, 43. The generated sweep voltage is applied to the summing junction S of a d.c. amplifier which drives the rf amplifier. The rf amplifier produces a 1.8 MHz saw tooth modulated signal which varies between 2 and 1800 V peak-to-peak. A diode rectifier produces a d.c. voltage proportional to the a.c. signal with a maximum value of 100 V. This d.c. voltage is used as a feedback signal for 2, thus reducing any modulator non-linearity to a very low value and assuring an rf voltage strictly proportional to the generated saw tooth voltage.

An additional current can be applied to the summing junction S of 2 by means of the low level adj. potentiometer, 44. This control can be used for the manual sweep of the rf voltage or for adjusting the starting point of the sweep. The low level adj., in conjunction with the high level adj. is normally used for the limited scan of any portion of the spectrum.

A d.c. voltage proportional to the rf signal is also applied to 3 by means of the d.c./a.c. ratio control, 41. When relay 37 is de-energized, 3 is used as a precision d.c. amplifier. With 37 energized, amplifier 3 becomes an integrator which produces a linear down scan of the d.c. voltage. The down scan speed is adjustable over essentially the same range as the up scan speed. Amplifier 4 is a precision d.c. inverter. Either a positive or negative d.c. voltage proportional to the a.c. voltage can be selected by the d.c. pol switch 38 and be applied to the low end of the rf coil for the analysis of negative or positive ions in the mass spectrometer. The maximum value of the d.c. voltage is about 100 V.

Amplifier 5 is a precision level detector adjustable to any point of the sweep range by means of the trig level control, 45. It is used to trigger either the sweep generator reset (thus automatically starting a new cycle), or the d.c. down scan (which in turn resets the sweep at the end of the cycle to maintain automatic operation).

References p. 180

(2) *Rf unit*

The 1.8 MHz oscillator uses a FET transistor 11 and a parallel resonant quartz crystal. The oscillator output is applied to 12 which is connected as a modulated amplifier. The emitter follower 13 provides a low impedance drive for the power amplifier 14. The low impedance of the emitter follower 13 and modulation of both the voltage amplifier 12 and power amplifier 14 permits stable operation over the very wide dynamic range (2–1800 V peak-to-peak) without the need for neutralization.

For the proper operation of the monopole filter, it is of great importance that the power supply produces an undistorted sine wave a.c. voltage. Harmonic distortions not only change the peak heights and shapes but also the mass scale. These effects depend on the relative phase angles and are, therefore, strongly dependent on the position of the tuning capacitor. An adjustment for minimum emitter current results in better mass spectra than an adjustment for maximum a.c. voltage. Besides the intended slow change of amplitude during a mass scan, there should be no other modulation of amplitude or frequency since this would have a strong effect on the peak shape. These clean output conditions must be maintained over the whole wide range of output voltages. Occasionally, wild oscillations have been observed over a limited voltage range which resulted in complete suppression of the corresponding mass peaks. Careful layout of the wiring and shielding between the rf stages have eliminated these difficulties.

F. COMPARISON OF THE EXPERIMENTAL OBSERVATIONS WITH THEORETICAL EXPECTATIONS

(1) *Mass scale and dispersion*

The normal mass scan of a monopole mass spectrometer is performed by increasing both the d.c. voltage U and the a.c. voltage V in proportion to each other, thus keeping the voltage ratio U/V constant. If the ion accelerating voltage is kept below 10 V, then all ions except hydrogen require at least 20 rf cycles to pass the field. In this case, the mass scale is essentially linear and is given by eqn. (7.14).

The constant U^* can be calculated either theoretically from eqn. (7.2) by use of the basic system parameters ω and r_0 or from experimental results by measuring U_0 and V_0 for a known mass number M and by using eqn. (7.14). The argon peak $M = 40$ was observed at $U_0 = 64$ V and $V_0 = 474$ V which results in $U^*_{\text{experimental}} = 88$ V as opposed to $U^*_{\text{theoretical}} = 77$ V. Considering the difficulty of accurately measuring V, these values are in fair agreement. For technical reasons it was not possible to measure V directly on the rod but only on the coil. A voltage drop of 7% along the transmission line would explain the discrepancy.

The dispersion of the monopole mass spectrometer is the voltage difference for adjacent masses and given by $U^* (U/V)^2$. The maximum value of U/V is 0.168 because of the stability limitation in the x direction. Therefore, the maximum dispersion depends only on U^* which is proportional to f^2 and r_0^2. The dispersion quadruples if either the frequency or the field radius is doubled. Unfortunately, in order to cover the same mass range, it is then necessary to provide voltages four times larger.

If the ion energy is increased, all peaks are shifted from the position U_0 to $(U_0 + \Delta U)$ where ΔU can be calculated from eqn. (7.22). (See also Chapter II, p. 000.) We have previously seen that a peak occurs for $(\beta n) = 0.9$ if $n \geqslant 10$ and if a medium size exit aperture is used. For the laboratory model we obtain $(\beta n)^2 (\pi r_0 /L)^2 \simeq 0.01$, so that $\Delta U = 0.01\, E_z$. This shift is the same for all masses and, therefore, does not disturb the linearity of the mass scale. For instance, the $M = 40$ peak, observed at $U_0 = 64$ V for a very small energy, would be shifted by 0.8 V to the new position, 64.8 V, if the ion energy were increased to 80 V. This means that the zero point of the whole mass scale is now offset by 0.8 V. However, at the low end of the mass scale, this would no longer be true. For $M = 1$, n is smaller than 10 and the shift of the atomic hydrogen peak is less than for the molecular hydrogen peak and the shift of the latter is less than the shift for the heavier mass peaks. This results in a slightly larger dispersion between H_1 and H_2 than between adjacent heavier masses. Under normal operating conditions, for $E_z = 25$ V and $U/V = 0.135$, the effect is still quite small. The dispersion between the hydrogen peaks is only about 4% larger than between heavier adjacent masses.

The mass scale for very low ion energies should be practically linear. The observed deviations from linearity were much greater than theoretically expected and could be traced to a non-linear characteristic of the diodes and transistors used in the power supply. Special selections of the diodes and the application of compensating voltages have considerably improved the linearity. However, it is felt that more improvements in this regard are still possible and desirable. Nevertheless, the mass scale is sufficiently linear to permit easy determination of the mass numbers. This feature has been found to be extremely convenient in comparison with magnetic mass spectrometers. The lack of any hysteresis effect and the accurate reproducibility of the peaks are very valuable.

(2) *Scanning methods and extended mass range*

Several methods of scanning the mass scale are possible. In the first prototype instrument as described by von Zahn [4], the d.c. and a.c. voltages were kept constant and the frequency was changed. In this case, the mass is inversely proportional to the square of the frequency. A wide mass range, e.g. mass 1–400, can be covered with a much smaller frequency range, e.g. 20–1. However, the resolution $(M/\Delta M)$ of such an instrument is constant, which is a disadvantage for fast recording, since peaks at the low mass end are too sharp to prevent clipping and peaks at the high mass end are not sufficiently resolved.

Another method of scanning has been described by Grande et al. [5]. In this case, the frequency is kept constant and the d.c. and a.c. voltages are increased proportionally; the mass is essentially a linear function of the applied voltages. The resolution of a monopole of finite length is proportional to the mass if the ion energy is kept constant. This means that the peak width stays essentially constant over the whole mass range and unit resolution can be obtained simultaneously for all masses. The power supply for such an instrument is much simpler since the frequency is kept constant and tuning has to be performed only once. Unfortunately, rather high voltages are still required if very high masses are to be detected. It is the unique advantage of the monopole over the quadrupole filter that this drawback can be partially removed by drastically reducing the d.c./a.c. voltage ratio. This involves either a reduction of the resolution over the whole mass range, if the ion energy is kept constant, or a reduction of sensitivity, if the ion energy is reduced. In both cases, the mass range is still limited by the maximum a.c. voltage delivered by the power supply.

In order to avoid this limitation of the mass range and to maintain the highest sensitivity and resolution possible with a given power supply, a new type of mass scan has been tested, consisting of two parts.

(a) The low mass range is swept in the conventional manner by increasing the a.c. voltage and keeping the d.c./a.c. voltage ratio constant. This ratio should be rather high and close to the stability limit in order to obtain good resolution even for the quite high ion energies which are desirable to obtain good sensitivity. The highest mass which can be measured in this way depends primarily on the maximum a.c. voltage produced by the power supply.

(b) The adjacent higher mass range is then swept by reducing the d.c. voltage and keeping the a.c. voltage constant at its maximum value. The mass scale is given by eqn. (7.15). The mass is now inversely proportional to the d.c. voltage U_0.

In this regard the mass scale is identical to the one obtained in a conventional magnetic mass spectrometer if the magnetic field strength is kept constant and the scanning is performed with the acceleration voltage E_z. Figure 7.4 demonstrates the similarity with the mass spectrum of perfluorodimethylcyclohexane (C_8F_{16}). The upper spectrum has been obtained with a monopole mass spectrometer by use of the novel scanning method. Below is the same spectrum obtained with a small magnetic vacuum analyzer. It is obvious that the monopole mass spectrometer provides much better transmission and sensitivity for high masses. In this regard it is similar to a magnetic mass spectrometer with magnetic sweep, since the acceleration voltage E_z is kept constant in both instruments. The experimental mass scale is in close agreement with eqn. (7.15) as can be seen from Fig. 7.5.

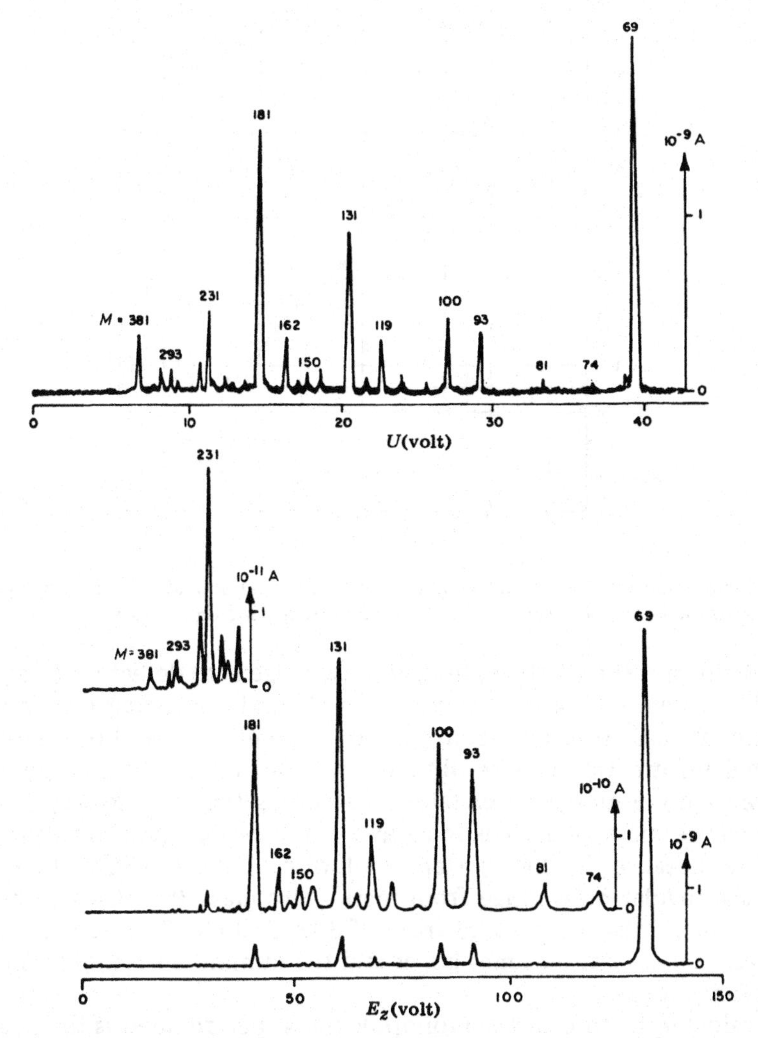

Fig. 7.4. Mass spectra of C_8F_{16}. The upper spectrum was obtained with a monopole by keeping the a.c. voltage constant and reducing the d.c. voltage, U_0. The lower spectra have been obtained with a magnetic mass spectrometer by reducing the accelerating voltage, E_z. Notice that mass scale and resolution are about the same, but that the monopole has higher sensitivity for heavier masses.

The mass resolution $M/\Delta M$ increases proportionally with the mass number M during the regular scan over the low mass range and decreases inversely proportionally to M during the new scan over the extended mass range. This results in a constant peak width ΔU for all masses, which permits the use of the fastest constant sweep rate compatible with the electrometer–recorder system. No sudden change of the resolution occurs at the transition point between the low and the extended mass range.

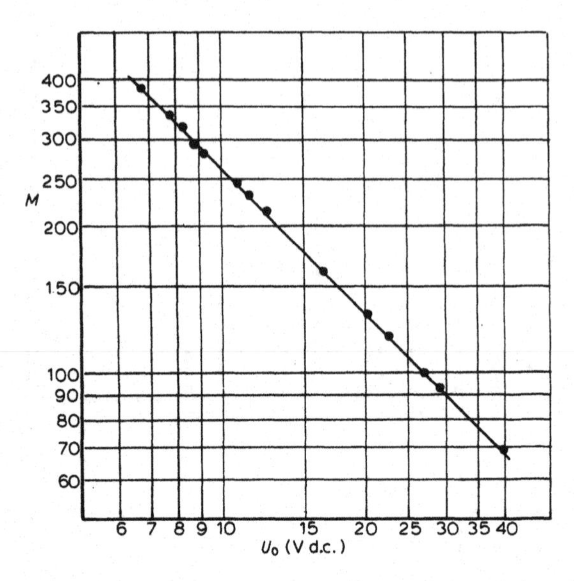

Fig. 7.5. Mass scale of the extended mass range. Comparison of the observed peak positions (points) with the theoretical straight line representing eqn. (7.15).

The main purpose of the extended mass range is to detect with high sensitivity the presence of any ions which are beyond the regular mass range and to permit at least a rough determination of their masses. Fortunately, this additional information can be obtained without increase of the power consumption. Only a moderate increase of scanning time is necessary. However, if an accurate mass analysis is mandatory, the regular mass range would be preferable because of the better resolution. Unfortunately, in order to extend the regular mass range from mass 48 to mass 381, it would be necessary to increase the a.c. voltage from 478 to 3800 V. The power consumption would be at least 64 times larger and the minimum scanning time would be four times longer.

One valuable feature of the monopole mass spectrometer is the possibility to trade off sensitivity for resolution simply by reducing the energy of the ions. For applications where resolution is more important than sensitivity, it is possible to increase the resolution during the extended mass range by simultaneously reducing the ion energy. Scanning the ion energy is not critical since it has very little effect on the position of the peaks. If the ion energy E_z is reduced proportionally to U, the resolution stays constant over the extended mass range and the performance of the monopole mass spectrometer becomes quite similar to a magnetic spectrometer with electric scan. Even in this case, no increase of the power consumption is required, but the maximum scanning speed has to be reduced in proportion to the smaller width of the peaks.

Equation (7.15) describes the extended mass scale for a very long monopole where n is close to infinity. For real monopoles of limited length, a slight correction is necessary. The end point of the extended mass range is characterized by $U = 0$ and $a = 0$*. This does not prevent the existence of the "large" oscillations, which results here in a mass spectrum of very low resolution. Experimentally, the argon-40 peak has been observed at $V = 56$ V if the ions were accelerated by $E_z = 88$ V. The mass spectrum obtained can be described by

$$V = 0.95\,(E_z M)^{1/2}$$

The mass number becomes proportional to V^2. This mass scale has been well confirmed experimentally; it is compressed for higher mass numbers and therefore the resolution becomes very small. Peak 20 is only a small hump on the flank of peak 40. However, peak 4 is well separated. This represents the practical limit of the extended mass range. Fortunately, this limit is high, for example $M = 18{,}000$ for $E_z = 40$ V and $V = 800$ V.

(3) *Peak width and peak shape*

A large number of measurements have been performed to study the peak width at half height $\Delta U_{1/2}$ as a function of the operating parameters. According to eqn. (7.23), $\Delta U_{1/2}$ should be proportional to E_z and independent of M and U/V. These facts were verified experimentally for mass 40 and mass 20 and for $E_z = 15$, 25, and 90 V and for $U/V = 0.042$, 0.080, 0.116, and 0.158. In all cases, it was found that $\Delta U_{1/2} = 0.048\, E_z$. The dependence upon r_0/L could not be checked since this would have required testing many monopole structures of different lengths and diameters. The factor 0.4 in eqn. (7.23) depends only slightly on the size of the entrance and exit apertures and was obtained with a round entrance aperture of 1.2×10^{-3} m diameter, sufficiently offset from the apex of the V electrode in order not to be obstructed, and a triangular exit aperture of 2×10^{-3} m height. The peak width at half height is very little effected by the size of these apertures. However, the peak tails are increased if the apertures are increased above the values mentioned. On the other hand, a reduction of the size of these apertures results in a severe loss of sensitivity. The dimensions given represent a good compromise for good resolution and sensitivity and have been used for most of the measurements.

The normal peak shape is triangular with a sharp top, the upper part with practically straight slopes. The tails of the peaks extend over several mass numbers and the one on the side of higher mass numbers is more intense. For this investigation, argon was admitted to the instrument to a pressure of

* This case corresponds to the operation of a quadrupole where the d.c. voltage is zero and the output signal is the integral of the mass spectrum from M to infinity.

TABLE 7.1

Experimental peak shape
Argon peak at mass 40.0

Height	M	ΔM
0.5	39.95–40.1	0.15
0.1	39.9–40.5	0.6
0.05	39.9–40.7	0.8
10^{-2}	39.9–41	1.1
10^{-3}	39.5–42	2.5
10^{-4}	39–44	5
10^{-5}	37–46	9

3×10^{-4} torr. Table 7.1 shows the width of the 40 peak at different relative heights.

The interference with the adjacent mass 39 is 10^{-4} and with 41 is 10^{-2}. Outside masses 36–48, interference becomes negligible which should permit there the detection of trace impurities in the parts per million range. The higher tail at the high mass side is in contradiction with the theoretical expectations. A steeper slope at the high mass side should be expected and was actually observed in the upper half of the peaks. The tail must come from secondary effects, not considered so far. The first observation of the higher tail was made by von Zahn [4]. Later, Hamilton [6] reported severe distortions of the peak shape, especially if the monopole is operated with high ion currents. He proposed that this effect is caused either by reflection of ions incident on the V electrode at grazing incidence or by release of secondary ions from absorbed gas layers. It is a fact that the intensity of the so-called ghost peaks is highly dependent on the previous use of the monopole. They become a real problem if the monopole has to be used at higher pressures ($\simeq 10^{-4}$ torr) and without frequent bake-out.

The experiments with the test monopole were made at high and ultrahigh vacuum and at least at the beginning of the experiments the spectra were perfectly clean. However, after about 2 years of operation, small ghost peaks did appear. Their height was usually less than 1% of the adjacent main peak. They were always at the high mass side of the main peak, sometimes only a tiny hump in the slope, sometimes really separated. The distance between the main peak and the ghost depends on the U/V ratio: if the ratio is larger than 0.1, the distance is about 1/2 atomic mass unit. If the ratio is 0.04, the distance becomes about three atomic mass units and is finally inversely proportional to the voltage ratio. The ghosts are a real nuisance if one is interested in trace impurities because misinterpretation is easily possible. However, real peaks do not change their position relative to the other peaks if the voltage ratio is changed. The ghost peaks can be ignored if one is only interested in the main constituents. Nevertheless, more research would be valuable to determine the origin of the ghosts and to eliminate them.

(4) *Resolution and peak separation*

The resolution R is defined as the ratio $M/\Delta M$ where ΔM is the peak width measured in mass units at a given fraction of the maximum peak height. For this fraction, the values 50 and 5% are commonly used. We will use 50%, or half height resolution, to permit comparison with published data. One obtains from eqn. (7.20) and from $(\beta n)^2 = 0.4$

$$R = \frac{M}{\Delta M} = \frac{U_0}{\Delta U} = 20 \left(\frac{U}{V}\right)^2 n^2 \qquad (7.25)$$

It can be seen that the resolution depends only on the voltage ratio and on the number, n, of rf cycles needed to pass the field.

$$n = \frac{0.22 R^{1/2}}{U/V} \qquad (7.26)$$

For $U/V = 0.168$, corresponding to the top of the stability diagram, the factor 1.32 compares favorably with von Zahn's [4] value of 1.5 and Dawson and Whetten's [7] value of 1.2. At lower U/V ratios, the resolution rapidly decreases. (See also Chapter II, p. 000 for a further discussion.)

If all ions are accelerated by the same voltage E_z, then n is proportional to $M^{1/2}$ and $M/\Delta M$ is proportional to M. Therefore, the peak separation stays constant over the whole mass range and is a more useful parameter in describing performance than the resolution. The peak width at half height expressed in atomic mass units can be obtained from eqns. (7.23) and (7.14).

$$\Delta M_{1/2} = \frac{0.4(\pi r_0/L)^2 E_z}{[U^*(U/V)^2]} \qquad (7.27)$$

The peak width at 5% height is about 5 times larger. Two adjacent peaks of equal height are, for most applications, adequately resolved if they are separated by a 10% valley, corresponding to $\Delta M_{5\%} = 1$. The maximum permissible accelerating voltage for obtaining a 10% valley is

$$E_{z\,\max} = 0.5 \left(\frac{U}{V}\right)^2 U^* \left(\frac{L}{\pi r_0}\right)^2 \qquad (7.28)$$

which is in good agreement with the experimental results. It should be noticed that $E_{z\,\max}$ does not depend on the field radius r_0 or on the mass number M. The ion acceleration voltage E_z should be kept above about 10 V, preferably closer to about 100 V in order to minimize charge up effects. This limits the product of the minimum length and frequency which can be used.

(5) *Sensitivity*

Absolute sensitivity measurements are difficult to perform since they depend on the multiplier gain. However, numerous relative sensitivity measurements have been made to study the sensitivity dependence on the size of the

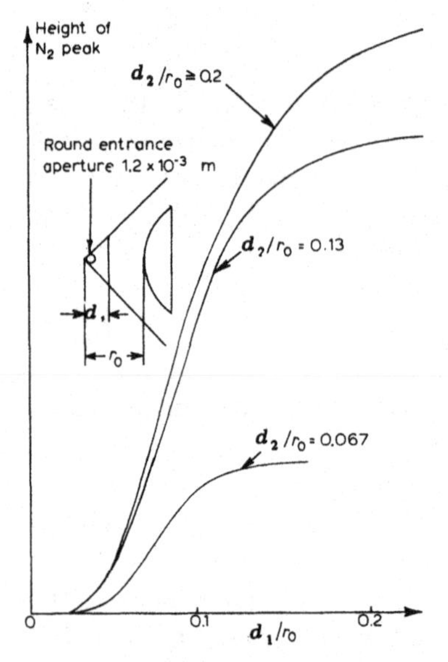

Fig. 7.6. Sensivity as a function of the entrance aperture, d_1, and the exit aperture d_2.

entrance and exit apertures, the ion energy, and the voltage ratio. The following results have been found experimentally.

(a) *Entrance and exit apertures*

If the triangular height of the entrance aperture, d_1, and of the exit aperture, d_2, are very small compared with r_0, then the sensitivity is proportional to $d_1^2 \times d_2^2$. For larger apertures, the increase is slower (see Fig. 7.6) and above $d_1/r_0 = 0.2$ and $d_2/r_0 = 0.2$ there is practically no further increase of sensitivity. This indicates that the monopole rod now limits the beam acceptance. Further increase of the sensitivity requires a larger instrument where d_1, d_2, and L are increased in proportion to r_0. In this case, the sensitivity is proportional to r_0^2 as long as the ion beam completely fills the entrance aperture.

(b) *Voltage ratio*

It has been found experimentally that the sensitivity is roughly proportional to $[0.168-(U/V)]^{1/2}$. If $U/V = 0.168$, corresponding to the top of the stability diagram, the sensitivity becomes zero. For smaller values of U/V, the sensitivity increases first rapidly and later more slowly. About half of the maximum sensitivity is reached for $U/V = 0.126$.

(c) *Ion energy*

The effects of the ion energy are more difficult to study because they are partly caused by the ion source and partly by the monopole. Increase of the energy reduces the angular spread of the beam and increases the transmission of the monopole as long as the beamspread is larger than the acceptance angle of the monopole. This results in an increase in sensitivity with an increase of the acceleration voltage up to about 20 V and constant sensitivity for higher voltages. The transition point depends on the geometry of the ion source and the monopole. In addition, ions with smaller energies are more influenced by charge up effects which can never be completely eliminated.

Let us assume that separation of adjacent peaks is required and that eqn. (7.28) has to be fulfilled. In this case, E_z and U/V are interrelated; small values of U/V require very small values of E_z which reduce the sensitivity. In between, there is a broad sensitivity maximum between $U/V = 0.1$–0.15 and $E_z = 40$–100 V. These ion energies are much larger than those usually used with quadrupole instruments. Therefore, a wider energy spread can be tolerated which is an important advantage of the monopole.

(d) *Focusing effects*

In order to check the capability of the monopole mass spectrometer for quantitative measurements, the heights of peaks 32 (O_2) and 28 (N_2) were measured in several air spectra and the ratio was computed and tabulated. The peak height is not a monotonic function of the ion energy. It contains maxima and minima which have to be expected from Lever's work on focusing [8] (see also Chapter II, p. 57). These maxima and minima are very reproducible. Dependent on the value of n, some of the ions will either pass through the monopole or be removed by one of the electrodes. Therefore, the transmission is a semi-periodic function of E_z. Since the maximum of the 28 peak occurs at an ion accelerating voltage, E_z of 130 V and that of the 32 peak at 140 V, large variations of the peak height ratio do occur and typical values are tabulated in Table 7.2. Columns 2 and 3 contain the results from abnormal operating conditions which have been purposely selected since here the focusing effect is most pronounced. In this worst case, the maximum deviation of the 32/28 peak height ratio for all ion energies is 55%. Columns 4–6 demonstrate that much better results can be obtained when the size of the exit aperture is increased and the d.c./a.c. voltage ratio is reduced. Even further improvements are possible with individual calibration for each mass if the ion energy is kept constant thereafter. This has to be expected since the reproducibility of the spectra is good and the average deviation between columns 2 and 3 for the same ion energies is only 6%.

The sensitivity becomes independent of the mass if the accelerating voltage is increased proportionally to the mass since, in this case, n becomes a constant. Unfortunately, this results in constant resolution $M/\Delta M$, which means that the peak width increases proportionally to the mass. Under these conditions, it is difficult to achieve unit mass resolution for high masses. In addition, the peaks at low mass numbers are unnecessarily sharp which limits the maximum scanning speed.

TABLE 7.2

Ratio of peak heights for O_2/N_2

Ion accelerating voltage	Exit slit 0.7 x 10^{-3} m voltage ratio 92% of cut-off value		Exit slit 2.2 x 10^{-3} m voltage ratio 92% of cut-off value	Exit slit 2.2 x 10^{-3} m voltage ratio 86% of cut-off value	Exit slit 2.6 x 10^{-3} m voltage ratio 25% of cut-off value
170	0.18	0.20	0.18		
160	0.16	0.17	0.15		
150	0.14	0.17	0.15		
140	0.31	0.30	0.18		
130	0.11	0.11	0.12		
120	0.10	0.11	0.11	0.12	
110	0.12	0.11	0.13	0.16	
100	0.17	0.18	0.16	0.15	
90	0.24	0.23	0.16	0.15	
80	0.13	0.17	0.14	0.15	0.12
70	0.10	0.09	0.12	0.14	0.13
60	0.24	0.23	0.13	0.15	0.16
50	0.12	0.09	0.11	0.13	0.13
40	0.15	0.16	0.08	0.17	0.15
30	0.10	0.13	0.13	0.12	0.13
20	0.13	0.17	0.14	0.17	0.13
Maximum deviation (%)	0.20 ± 0.11 55		0.13 ± 0.05 38	0.145 ± 0.025 17	0.14 ± 0.02 14
Average value standard deviation (%)	0.16 ± 0.06 37		0.14 ± 0.03 21	0.15 ± 0.02 13	0.14 ± 0.14 10

Since the sensitivity variations are caused by the exit aperture, they can be reduced to practically zero if the exit aperture is opened to the full width of the rod distance, r_0. Unfortunately, in this case, the peak tail becomes rather large. A compromise is therefore necessary. Entrance and exit apertures should be chosen to be as large as is compatible with adequate resolution and sufficiently low background. By this choice, and by operating the instrument with a lower voltage ratio, it is possible to keep the sensitivity variations within tolerable limits.

(e) *Sensitivity gain by a magnetic field*

We have seen previously that for $\beta n = 1$ a well focused image of the entrance aperture would occur if the beam did not strike the V electrode. The effect of the magnetic field is to shift this image from negative to positive y values where it really can be used. Accurate calculation of the transmission under this condition becomes very complicated. Because the deflection of

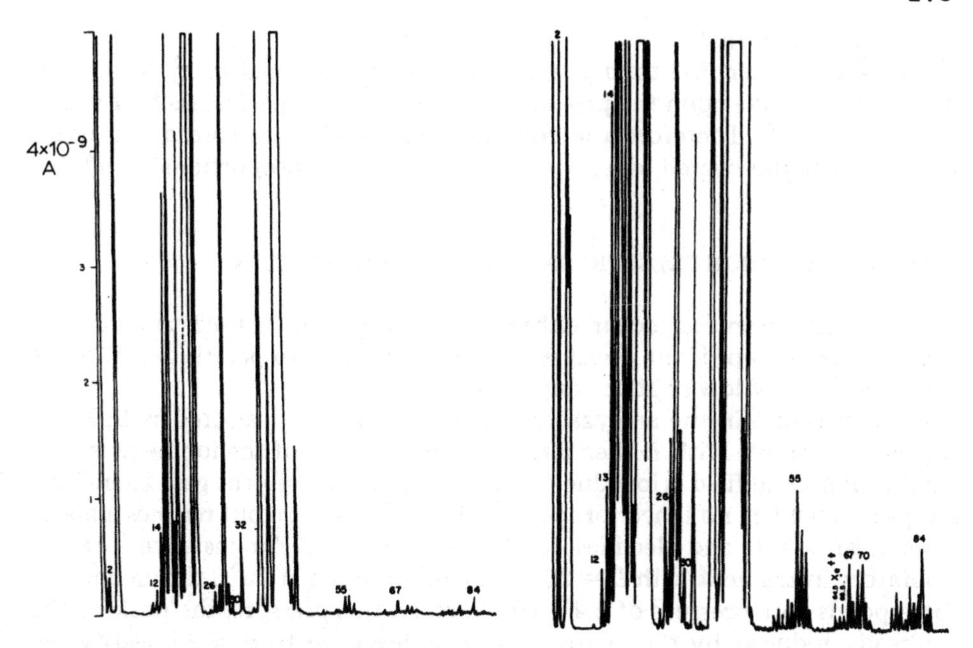

Fig. 7.7. Effect of the magnetic field. Both spectra have been obtained under identical conditions except that in the right spectrum a small permanent magnet deflected the beam away from the V-electrode.

the beam is gradual, one must expect that almost one half of the beam is lost at the V electrode near the entrance aperture. Therefore the sensitivity gain should be quite moderate. However, a substantial gain was observed as shown in Fig. 7.7 which gives spectra obtained with and without a magnet. The imposition of the magnetic field increases peak heights by about a factor of ten.

The improvement can be understood if one also considers the focusing effect in the x direction. Since the magnetic field shifts the image away from the V electrode, the acceptance angle in the x direction may be greatly increased. The peak heights decreased by an order of magnitude if the magnetic field was reversed.

The magnetic field was produced with a 5 in. bar magnet located near the entrance aperture at a distance of about 0.11 m from the axis. The field strength at the aperture was approximately 7.5 gauss and diminished gradually towards the exit. With an ion energy of 25 V, the resulting beam deflection would be quite small. This deflection has to be gradual and distributed over the beam. It cannot be replaced by electrical deflection before the ion entrance. However, it is better to have a larger field near the entrance aperture than a homogeneous field along the whole monopole. The optimum magnet location is very slightly mass dependent but the positioning is not critical. The magnetic field has no effect on the mass scale or on the peak shape.

(f) *Geometrical effects*

As stated earlier, the total aperture area is proportional to r_0^2. The sensitivity also depends upon the product of the acceptance angles, each of which varies with r_0/L. Therefore the combined effect of beam spread and beam width is that the overall sensitivity is approximately proportional to r_0^4/L^2.

G. EFFECT OF A HIGH PRESSURE IN THE MONOPOLE SECTION

Such a situation can occur either if the instrument is used to analyze a mediocre vacuum or if the pressure is purposely raised to permit detection of impurities at very low relative concentrations.

If the pressure in the analyzer is increased until the mean free path, λ, becomes comparable with L then a fraction $\exp(-L/\lambda)$ of the ion beam entering the monopole field is lost due to scattering on the ambient gas. Therefore, the peak height is no longer proportional to the pressure but reaches a maximum near $\lambda = L$ and declines for higher pressures. For instance, for the laboratory instrument with $L = 22.2$ cm and for argon at 25°C this maximum corresponds to a pressure of 2.4×10^{-4} torr. At this pressure, the peak height is already reduced by the factor 0.37. The deviation from the linearity becomes noticeable at a pressure of 2×10^{-5} torr. Nevertheless, it is possible to operate the monopole at a much higher pressure which is only limited by a glow discharge in the monopole or in the multiplier. The instrument has been successfully operated with argon at a pressure of 2×10^{-3} torr. The peak width is not increased at this high pressure which is an important advantage over magnetic vacuum analyzers.

A small fraction of the scattered ions can reach the multiplier and cause a background current which is independent of the mass. This background current is proportional to the square of the pressure because it depends on both the ion current and the scattering probability. At an argon pressure of 2×10^{-4} torr, the background was 140 ppm of the 40 peak, but since it was quite constant, it was still possible to see superimposed peaks of about 10 ppm. Nevertheless, the background interferes with the detection of trace impurities since it prohibits the use of high electrometer sensitivity or of the more sensitive pulse counting technique. It is expected that better shielding of the multiplier entrance aperture and differential pumping of the analyzer could significantly reduce this background and improve the capability to detect trace impurities in the part per million range.

H. NEGATIVE IONS

The monopole mass spectrometer should also be useful for the detection of negative ions if a positive d.c. voltage is applied to the rod. For the detection of negative ions, it is necessary to keep the first dynode at about + 500 V which results in an anode potential of about + 3500 V above ground.

The detection of small currents on an electrode which is at such a high potential above ground causes considerable experimental difficulties which can be solved in various ways. The method successfully adopted here was to use a capacitor to separate the high voltage of the anode from the low voltage of the pulse amplifier input. (An isolation transformer can also be used for this purpose.) The multiplier power supply must be perfectly free of any ripple and transients. The coupling condenser must be of highest quality and the whole circuit must be very well shielded in order to permit the use of the full gain of the pulse amplifier necessary for the detection of single ions. Pulses which are above an adjustable background level are used to trigger secondary pulses of uniform length and height. These pulses are integrated and measured with a logarithmic electrometer. This method of ion detection is preferable to the straight electrometer type measurement since it actually measures the count rate which is essentially independent of the multiplier voltage and gain.

A gas mixture consisting of sulphur hexafluoride and water vapor was introduced into the mass spectrometer and ionized in the conventional way by electron impact. Figure 7.8 shows the mass spectrum thus obtained. The

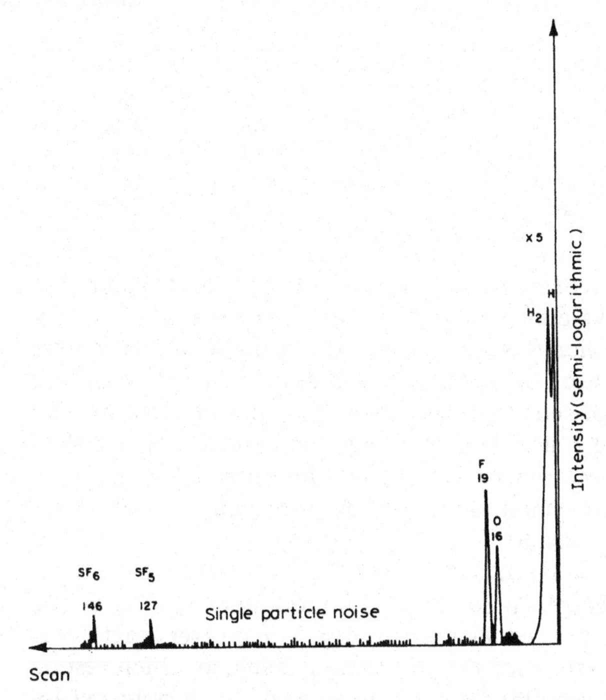

Fig. 7.8. Negative ion spectrum of a mixture of SF_6 and H_2O.

References p. 180

following negative ions have been observed: large peaks of atomic and molecular hydrogen and smaller peaks of atomic oxygen, fluorine, SF_5, and SF_6. Occasionally, a very small peak at mass 17, probably OH, has been observed.

I. POWER CONSUMPTION

Assuming that the main power losses of a monopole are caused by the resistance of the coil, one can show that the power consumption, P, is proportional to $U^{*2} E_z MR$. The power consumption does not depend on the voltage ratio U/V. But, most important, U^* should be small, which is also desirable to avoid high voltage breakdown of the insulators. Since U^* is proportional to $f^2 r_0^2$, P becomes proportional to $f^4 r_0^4$.

Table 7.3 compares five typical cases, all with the same peak separation $\Delta M_{5\%} = 1$ which requires that f is inversely proportional to L. It was previously explained that the sensitivity, S, is proportional to r_0^4/L^2.

TABLE 7.3

Design trade-offs

Case	Field radius	Field length	Frequency	U^*	Power	Sensitivity
1	r_0	L	f	U^*	P	S
2	$2r_0$	$2L$	$f/2$	U^*	P	$4S$
3	$2r_0$	L	f	$4U^*$	$16P$	$16S$
4	r_0	$L/2$	$2f$	$4U^*$	$16P$	$4S$
5	$2r_0$	$4L$	$f/4$	$U^*/4$	$P/16$	S
6	$r_0/2$	$L/4$	$4f$	$4U^*$	$16P$	S

The table shows the effect of r_0 and L on P and S. If the instrument is enlarged in proportion, P stays constant and S is increased (case 2). If only r_0 is increased or L reduced, then S is also increased but the increased power requirement makes this increase of the sensitivity rather costly (cases 3 and 4). One obtains a better increase of sensitivity for the same amount of additional power if r_0 is increased than if L is reduced. A larger and much longer instrument (case 5) provides the same sensitivity but can be operated with much less power. On the other hand, a small instrument requires much more power to obtain the same sensitivity (case 6).

The following conclusions can be drawn.

(1) It is, in general, an advantage to make the monopole as long as the available space allows despite the reduced sensitivity. The greater length permits operation with lower frequencies for the same resolution which results in smaller voltages and much smaller power consumption. The only excep-

tions to this rule are applications in which the instrument has to be operated at a relatively high pressure, where the mean free path limits the useful length of the monopole.

(2) The field radius should be made as large as possible with the available power. This limitation is quite strong since a small increase of the field radius requires a very large increase of the power (case 3). The sensitivity obtained is proportional to the available power.

One can see from the above that for all the desirable features, high mass range, M, high resolution, $M/\Delta M$, high sensitivity, S, and high ion accelerating voltage, E_z, one has to pay either with the larger size of the instrument or with much higher power consumption. The choice between these alternatives depends on the particular application.

J. CONCLUSIONS

The monopole mass spectrometer is a very useful and reliable instrument as long as it can be kept perfectly clean and is used only at high vacuum. Compared with the quadrupole it has the following advantages.

(1) The mechanical construction is simpler and the alignment of the rod is less critical.

(2) The power supply is much simpler since only one single output is required and the d.c./a.c. voltage ratio is not critical.

(3) The power consumption is much lower and lower voltages are necessary.

(4) The whole mass range can be covered with one single scan and practically uniform peak separation.

(5) The resolution is better for the same acceleration voltage or the acceleration voltage can be increased for the same resolution.

The monopole suffers from the following disadvantages.

(1) Theoretically, the sensitivity should be lower. Practically, this can be compensated by a higher accelerating voltage and, especially, by a magnetic field near the entrance aperture.

(2) Ion source requirements are more stringent. Therefore, contamination of the electrodes and the resulting charge up have a greater effect.

(3) So far, the peak tops are sharp which is a disadvantage for accurate peak height measurements.

(4) Sensitivity variations impose difficulties for an accurate quantitative analysis. They can be neglected for a semi-quantitative analysis.

(5) The detection of trace impurities near the main peaks is difficult because of interference with ghost peaks.

Further research is desirable to eliminate the ghost peaks and the general background and to study experimentally the effect of the fringing fields.

References p. 180

REFERENCES

1 R.F. Herzog, NASA Contract NASW-1314, final report of phase 1, 1971.
2 R.F. Herzog, NASA Contract NASW-1314, final report of phase 2, 1968.
3 W.M. Brubaker, Congress International des Techniques et Applications du Vide, Paris, 1961.
4 U. von Zahn, Rev. Sci. Instrum., 34 (1963) 1.
5 R.E. Grande, R.L. Watters and J.B. Hudson, J. Vac. Sci. Technol., 3 (1966) 329.
6 N.E. Hamilton, 14th Ann. Conf. Mass Spectrosc. Allied Topics, Dallas, 1966, p. 762.
7 P.H. Dawson and N.R. Whetten, Rev. Sci. Instrum., 39 (1968) 1417.
8 R.F. Lever, IBM J. Res. Develop., (1966) 26.

CHAPTER VIII

QUADRUPOLE ION TRAPS

J.F.J. Todd, G. Lawson and R.F. Bonner

A. INTRODUCTION

In this chapter we consider the application of three-dimensional quadrupole fields to the trapping of ions for extended periods. Two basic types of ion trap have been described in the literature, one comprising a three-electrode structure [1, 2], the other a six-electrode arrangement [3]. In addition, ions have been contained in a field generated by a set of conventional quadrupole rods bent in the form of a circle [4] or race-track [5]. The three-electrode trap has received by far the greatest attention to-date, and is considered in most detail here. In the absence of a generic name, reports describing the device have referred to it as the "three-dimensional quadrupole ion trap" [6], the "3DQ" [7] and the 'quistor' (quadrupole ion store) [8].

The equations describing the motion of ions within the trap were derived from the expression for the basic form of the potential in a quadrupole field in Chapter II [eqns. (2.26)–(2.31)]. Because of the requirement to satisfy the Laplace condition $\nabla^2\phi = 0$ (in the absence of stored ions), an asymmetry in the potential gradient must exist. Consequently, the three-electrode trap is not simply a solid of revolution generated by rotating the cross-section of an array of mass filter rods about a vertical axis, but an inequality between the r_0 and z_0 dimensions ($r_0^2 = 2z_0^2$) has been introduced (see Fig. 8.1). The required fields can be formed simply by coupling the rf signal between the ring electrode (a single-sheet hyperboloid) and the two end-cap electrodes (a two-sheet hyperboloid), rather than having to apply a complex form of differential biassing which a device with symmetrical internal dimensions would require. This contrasts with the six-electrode trap (see section C.(1), below) where cubic symmetry is retained but the field is generated from a three-phase rf supply.

The asymmetric fields within the trap lead to the Mathieu stability diagram which has already been examined in Chapter II [Fig. 2.30(b)]. The ion trap can be operated in a "total pressure mode" along the $a = 0$ line, with zero d.c. bias, or in the "mass spectrometric mode" with the pair of end-cap electrodes biassed negatively with respect to the ring to give a working point close to the lowest apex. In both these forms of operation, the behaviour of the trap is

References pp. 222–224

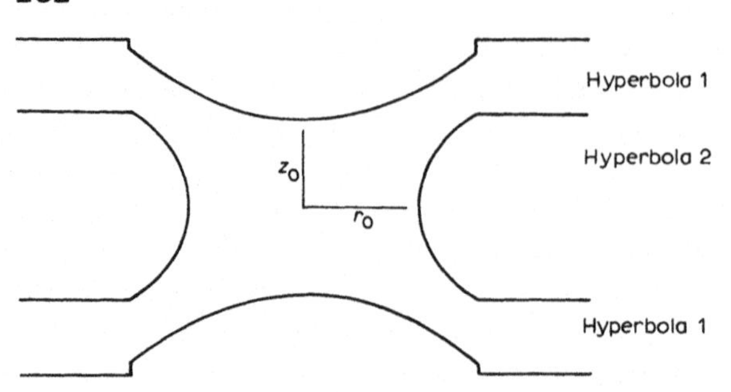

Fig. 8.1. Section through a three-electrode ion trap showing the hyperbolic surfaces employed [see also eqn. (8.1).] The dimensions r_0 and z_0 are related by $r_0^2 = 2z_0^2$.

exactly analogous to that of the mass filter: where the difference occurs is that for the trap, ions must be created and, in principle, be detected within the device. This latter aspect has already been considered theoretically in Chapter II (p. 49) and will be examined in more practical terms below.

B. THE THREE-ELECTRODE ION TRAP. CONSTRUCTION AND INSTRUMENTATION

(1) *The electrodes*

(a) *Geometry*

As with the mass filter and the monopole, the conducting surfaces in the three-electrode trap should ideally be hyperbolic in section. Because the manufacture of the appropriate shapes is generally a specialist machinist's job [there are no readily available pre-machined forms, unlike the mass filter where round rods are conveniently substituted for hyperbolic surfaces, see Chapter VI], most of the devices which have been reported have in fact retained this ideal geometry, although the trap constructed by Rettinghaus [18] incorporated spherical surfaces. The cross-sections of the two sets of hyperbolic surfaces should be complementary and follow the equations

$$\text{Hyperbola (1): } \frac{r^2}{r_0^2} - \frac{z^2}{z_0^2} = 1$$
$$\text{Hyperbola (2): } \frac{r^2}{r_0^2} - \frac{z^2}{z_0^2} = -1 \tag{8.1}$$

Since, to satisfy the Laplace condition, we have $r_0^2 = 2z_0^2$, then

$$r^2 - 2z^2 - r_0^2 = 0$$
$$r^2 - 2z^2 + r_0^2 = 0 \tag{8.2}$$

and from eqn. (8.2) we can specify the lathe settings for r and z which allow the surfaces to be turned out. In practice, this must result in the production of a stepped surface which is then finished by fine polishing. The choice of a value for r_0 depends upon the amplitude and frequency of the rf power supply available and the desired mass range. However, in contrast to the quadrupole and the monopole, a higher ratio of a/q (ca. 0.524) is required to achieve a comparable resolution, for example along the "mass scan line" in Fig. 2.30(b), in the ion trap.

(b) *Materials and methods of mounting*

The foregoing discussion clearly implies that the electrodes are fabricated from solid metal, and in most reports the material has been stainless steel. However, in a number of instances, the solid surfaces have been replaced by wire mesh made of stainless steel [10], copper [11] or tantalum [12]. The fields within these latter, rather more crude, devices are clearly less than ideal, but the construction is simpler since the mesh electrodes can readily be attached to supporting wires (Fig. 8.2). The heavier solid electrodes require to be mounted more rigidly: one such method, which employs four equally spaced lengths of stainless steel studding sleeved with accurately machined pyrophyllite insulators, is shown in Fig. 8.3. An early version described by Dawson and Whetten [13] incorporated ceramic rings to provide the insulation between the end-caps and the ring electrode but was discarded because it was suspected that possible charging up of the ceramic surfaces would adversely affect the ion trajectories.

(2) *Ion creation*

In principle, ions may be either formed within the trap, for example by electron bombardment or photoionization, or be injected from outside. Although means of implementing this second method have been considered in some detail [14, 15], in practice, it is likely that problems would be encountered since except for these ions entering along the nodal surfaces, exposure to the rf fields would generally lead to large-amplitude trajectories because of the large initial displacements of the ions.

In practice, the method most commonly employed has been to create ions within the trap by injecting an electron beam. This can be done equatorially through holes in the ring electrode as in Fig. 8.3, or axially through holes in one of the end-caps. The filament itself can be of a conventional design, the electron energy being determined by the potential difference between the filament and the adjacent electrode of the ion trap. However, an essential feature of any application where it is intended to create a discrete ion packet is that a gating electrode be incorporated so that the electron beam may be deflected and prevented from entering the trap. In such a system it is important to ensure there is adequate shielding to prevent stray electrons from

Fig. 8.2. A three-electrode ion trap fabricated from 38-mesh copper gauze showing the method of support.

reaching the trap inadvertently, a possibility which is especially likely when the electrodes are fabricated from wire mesh.

(3) *The rf power supply*

The basic unit of all the types of electronic circuit which have been associated with the operation of the ion trap is the d.c./rf power supply. The performance characteristics demanded of this unit are basically the same as those required for successful operation of the mass filter (see Chapter VI). However, a notable practical difference between the two systems which has been incorporated since the earliest reported experiments on ion traps [9, 16, 17] has been the use of a unipolar rf supply connected to the ring electrode only,

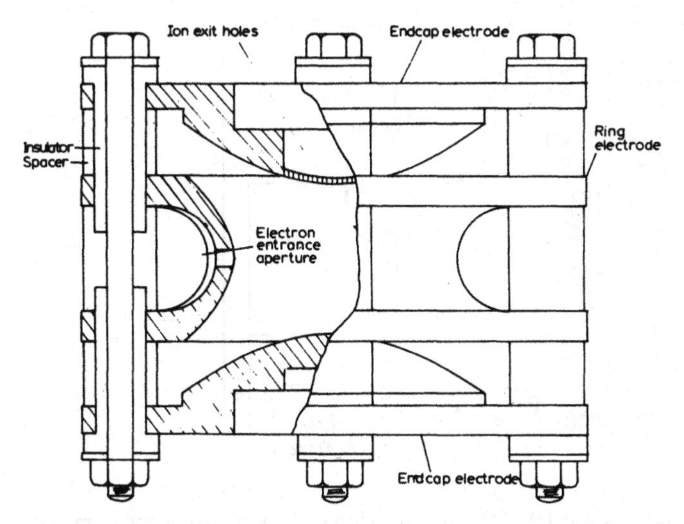

Fig. 8.3. A section through a three-electrode ion trap with solid surfaces showing the use of insulators and spacers to achieve rigid support of the electrodes.

rather than a bipolar rf supply connected between the end-caps and the ring. Whilst there is no fundamental objection to this latter method, which has indeed been employed in at least one instance [13], the former approach avoids the otherwise severe problems of isolating the rf potentials from the circuitry associated with ion detection. The major disadvantage is that for a given available rf amplitude, the mass range is immediately divided by two since the value of V_m in eqn. (6.1) is now the maximum zero-to-peak amplitude, not that given by the peak-to-peak difference. Furthermore, the potential at the centre of the trap is oscillating through $\pm V_0/2$ relative to ground, and whilst this does not affect the motion of ions within the device, adverse focussing conditions could exist when ions are ejected for subsequent detection.

(4) *Ion detection*

The theoretical aspects and limitations of the several methods available for ion detection have already been discussed in Chapter II (p. 49) and the results obtained with these different systems are examined below [Section D.(1)]. We consider here how the various techniques are implemented.

The two earliest means of ion detection were *mass selective* in that they relied upon the interaction between the motions of ions whose a and q values corresponded to a particular working point on the stability diagram and the characteristics of a sensing circuit coupled between the two end-cap electrodes. The circuit employed by the Bonn group and described in detail by Fischer [9] is shown in Fig. 8.4. Ions were created continuously by the passage of a beam of energetic electrons through a hole in one of the end-caps and were

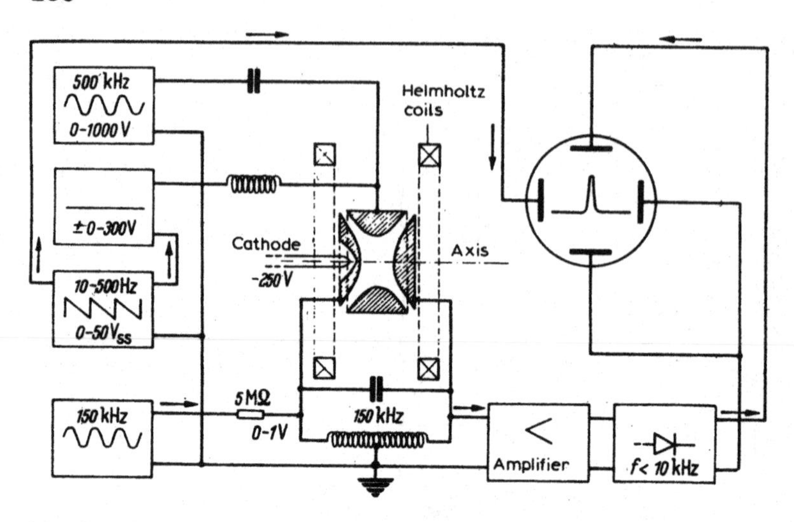

Fig. 8.4. Circuit employed by Fischer [9] for the detection of ions by using the damping of an auxiliary rf circuit tuned to the fundamental frequency of ion motion.

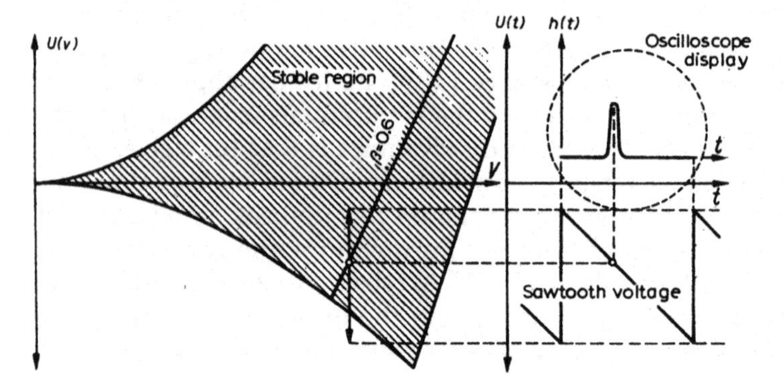

Fig. 8.5. Illustration of the means by which Fischer [9] employed the circuit of Fig. 8.4 to generate a mass spectrum.

maintained in stable trajectories with an oscillating potential having a maximum amplitude of 1000 V at 500 kHz. The secular motion of ions back and forth along the rotational axis of the trap was then detected by resonant absorption of power from an auxiliary generator oscillating at 150 kHz, the output from which was applied across a 5 MΩ resistor and across half the pure resistance of the resonator. The voltage developed across the resonator was then proportional to its resistance and inversely proportional to the attenuation. This auxiliary frequency corresponds to working along the line $\beta_z = 0.6$, and successive ions could be brought into resonance by applying a sawtooth signal as part of the d.c. voltage on the ring electrode (see p. 51). This is

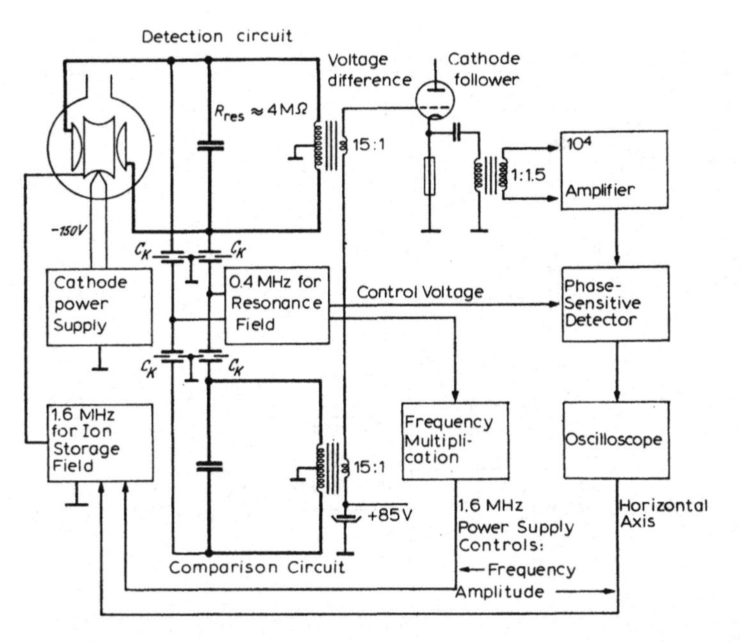

Fig. 8.6. Circuit employed by Rettinghaus [18] for the resonant detection of stored ions

shown in detail in Fig. 8.5. A mass scan corresponds to moving vertically through the operating point (Fig. 8.5, open circle), and if the saw-tooth potential is used simultaneously to drive the x axis of an oscilloscope, then intersection of the locus of the value of the applied d.c. potential with the $\beta_z = 0.6$ line leads to a signal output from the demodulation circuit proportional to the number of stored ions of that mass.

The alternative resonance detection system employed by Rettinghaus [18] is shown schematically in Fig. 8.6. A frequency-tuned detection circuit coupled between the end-caps was balanced without any ions being present in the trap. When ions were created at low pressure (ca. 2×10^{-9}torr) by bombardment with an equatorially introduced electron beam for 5 sec and then stored, their motion could be detected as an induced alternating potential provided that the frequency of the secular motion was equal to that of the tuned circuit. In the system used, this was achieved by operating along the $a = 0$ line (zero applied d.c.) at a value of $\beta_z = 0.5$. An initial value of the rf amplitude was chosen to correspond to the lowest mass in the range being detected and then slowly swept to bring successively higher masses into resonance. This method suffers from broadly the same objections as that of Fischer (see Chapter II), particularly in that at resonance for a particular species there are other ions present in the trap; as the higher masses are being monitored, the lower mass ions are totally rejected from the trap so the environment within the trap changes during the scan. As a detector for very low pressure residual gases,

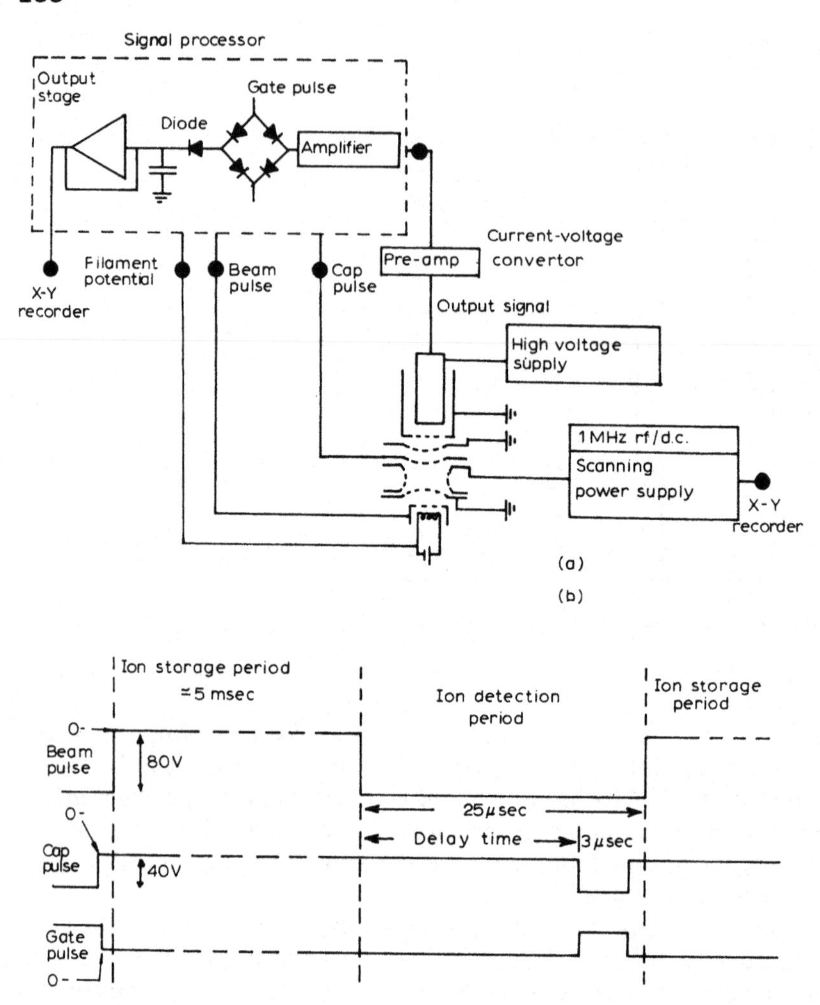

Fig. 8.7. (a) Circuit employed by Dawson and Whetten [52] for the detection of ions by mass-selective ion storage. (b) Timing sequence for the circuit.

however, Rettinghaus's system must be regarded as a success [see Section D.(1)]. From the foregoing, it can be seen that with mass-selective detection the ion trap is functioning as an ion source, mass filter and detector combined. In the alternative mode, *mass-selective ion storage* [see Chapter II D.(3)], the first two of these functions are retained and the presence of ions within the trap is demonstrated by pulsing them out through perforations in one of the end-caps on to a collector, usually the first dynode of an electron multiplier [19] or a "channeltron" [20]. The trap now operates in a manner entirely analogous to the mass filter, except that intermittent ion creation must be employed. A block diagram of a typical circuit is shown in Fig. 8.7 together with the timing sequence. In this system, ions are formed during the "ion storage period" by opening the "beam pulse" gate to admit electrons through a hole in the ring electrode. Mass selection then occurs during the "ion detection period" during the first part of which ions simply oscillate within the trap under the influence of the appropriate d.c./rf field. A short "cap pulse" is then applied to draw the ions from the trap into the multiplier and this is synchronized with a "gate pulse" applied to the signal processor so that the ion current recorded is that which only arises through deliberate ion ejection (in the absence of this precaution ions rejected by the trap during the mass-selection process would also be registered by the detection system). With a period extending to ca. 5 msec, the maximum repetition frequency is limited to ca. 200 Hz, and this means that unless some form of peak detection circuit is employed the long time constant associated with the recorder will produce a time-averaging effect on the signal as the d.c./rf potentials are scanned to produce a mass spectrum. The gate pulse and peak detecting function can be conveniently combined in a commercially available "boxcar detector" in which the gate pulse width and delay are variable with the leading edge of the cap pulse acting as a trigger. A further refinement which has been employed in a limited number of applications [20, 21] has been to relate the timing and duration of the ion creation and/or ejection pulses to the phase and periodicity of the rf drive potential. This requires some form of counting circuit, an example of which is shown as a block diagram in Fig. 8.8. A small

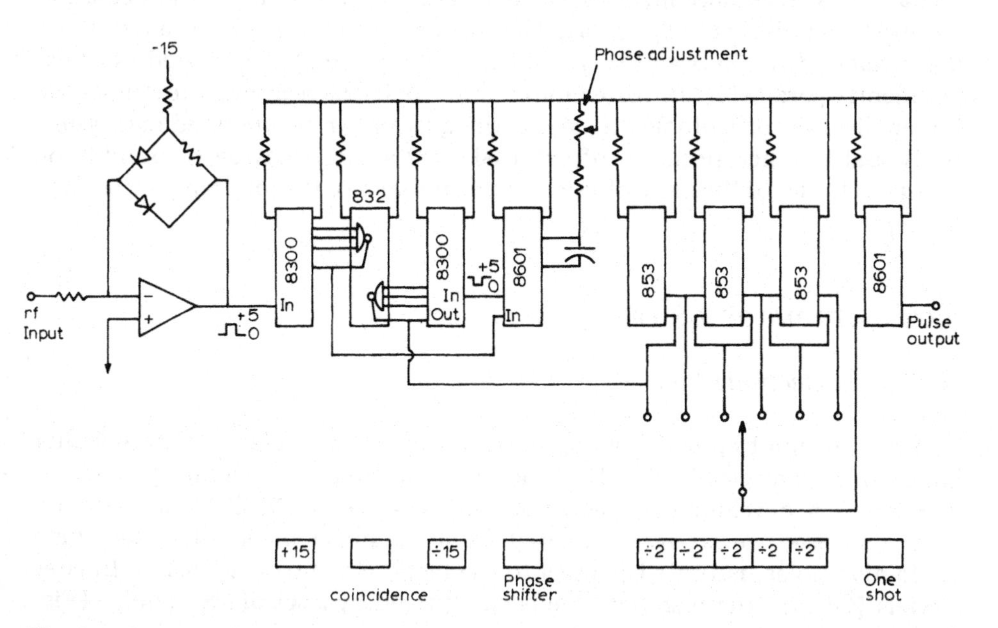

Fig. 8.8. Logic circuit employed by Dawson and Lambert [21] to synchronize the cap-pulse with the phase of the rf drive potential.

part (several volts) of the rf voltage applied to the trap is fed to an operational amplifier, the output of which is a square wave of 5 V amplitude. The frequency is reduced before the signal goes to the 8601 monostable multivibrator and the length of the inverted output pulse is controlled by a variable resistance. The 8300 shift register operates on the positive-going edge at the end of this pulse so that, in fact, the variable resistance controls the phase. The period between impulses is chosen by means of a multiposition switch connected to a series of flip-flops and in this particular system the storage time is adjustable between 0.22 and 7.75 msec for a drive frequency of 1 MHz. The final 8601 monostable produces a 10 μsec-long output pulse which initiates the beam, cap and gate pulse circuitry described above. An important result which comes from studies [21] on the phase synchronization of the ion ejection pulse is that mass discrimination effects can occur when ejection occurs at the wrong phase, particularly at a phase lag of π in the total pressure mode, and in certain situations ions can actually be ejected in the wrong direction.

The merits of the ion ejection technique have been discussed in Chapter II, but a particularly important feature is that during detection, ions of all masses are treated identically. A direct development from this approach has been the use of the ion trap as a "storage source" as described by Lawson et al. [22]. In this system, the conventional ion source of a mass filter is replaced by an ion trap (quistor) as shown in the block diagram in Fig. 8.9. The sequence of operations is essentially identical to that described above but the ion creation and ejection pulses are kept short. The former is used to trigger the latter and the variable delay between them (the ion storage time) can be used as an experimental parameter in, for example, ion–molecule reaction rate studies. By the application of suitable d.c. levels, the quistor can be operated mass selectively and the incorporation of external mass analysis has been found to be useful in studies of the basic physical characteristics of the ion trap [see Section D.(3)].

C. OTHER FORMS OF ION TRAP

(1) *The six-electrode trap*

A form of ion trap which is almost as old as the three-electrode trap is that based on a six-electrode structure and having a cubic-type geometry. This device was first developed by Langmuir and co-workers [3, 23] and their version comprised electrodes fabricated from six plane sheets of metal, some containing holes, mounted parallel to the faces of a cube (Fig. 8.10). In later models [24, 25] these sheets have been replaced by a set of six annulae (Fig. 8.11).

Ideally the electrode surfaces should be hyperbolic and in order to satisfy

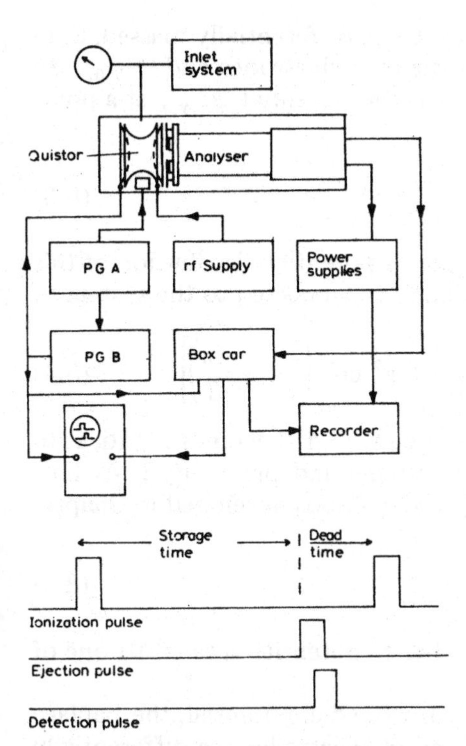

Fig. 8.9. Block diagram and timing sequence for the quistor/quadrupole combination employed by Todd and co-workers [40] for the study of ionic processes. Pulse generator PG A opens the electron "gate" and triggers PG B whose delay determines the storage time and thence the sampling time of the "boxcar detector".

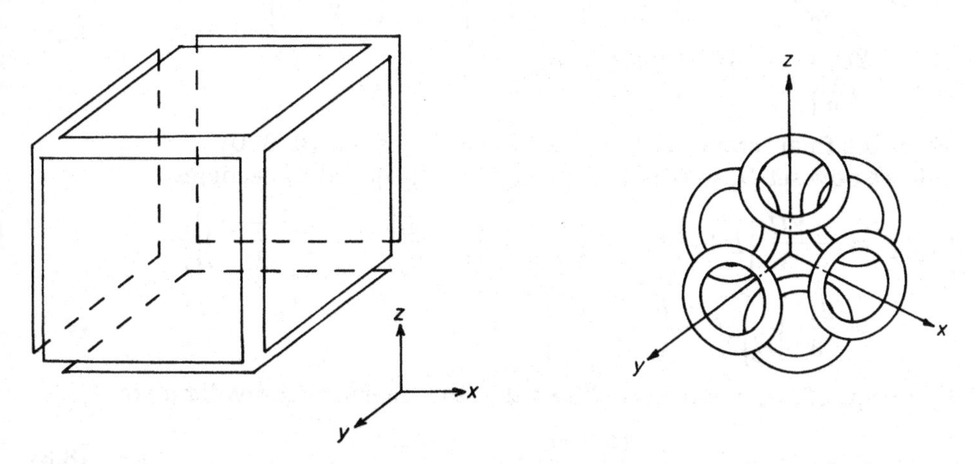

Fig. 8.10. Six-electrode trap of the type employed by Langmuir et al. [3, 23].
Fig. 8.11. Six-electrode trap of the type employed by Haught and Polk [24] and by Zaritskii et al. [25].

the condition $\nabla^2 \phi = 0$, the electrodes have to be differentially biassed. Suppose we apply a potential $+\psi_0$ to the pair of z electrodes and $-\psi_0/2$ to each of the pairs of x and y electrodes, then the potential $\psi_{x,y,z}$ at a point x, y, z is given by

$$\psi_{x,y,z} = \frac{\psi_0}{r_0^2}[z^2 - \tfrac{1}{2}(x^2 + y^2)] \tag{8.3}$$

where r_0 is the radius of the inscribed sphere. A potential distribution of this form can be achieved if a three-phase rf supply is connected to the system to give

$$V_{x,y,z} = \frac{V}{r_0^2}\left[z^2 \cos \omega t + x^2 \cos\left(\omega t + \frac{2\pi}{3}\right) + y^2 \cos\left(\omega t + \frac{4\pi}{3}\right)\right] \tag{8.4}$$

whence it can be seen that for, say, $t = 0$ eqn. (8.4) becomes identical to eqn. (8.3). Since the ion motion may be considered independently in each of the separate directions, it can be shown by the methods developed in Chapter II that

$$q_x = q_y = q_z = \frac{4eV}{mr_0^2\omega^2} \tag{8.5}$$

where V is the zero-to-peak amplitude of the rf potential applied to *one* of the pairs of electrodes.

We have already seen in eqn. (8.3) that for the d.c. potential, the Laplace condition can only be satisfied if the pairs of electrodes are differentially biassed. Suppose, in the general case, we applied the potentials U_x, U_y and U_z. Then at a point x, y, z we would have

$$U_{x,y,z} = \frac{2U_x}{3r_0^2}\left[x^2 - \left(\frac{y^2+z^2}{2}\right)\right] + \frac{2U_y}{3r_0^2}\left[y^2 - \left(\frac{x^2+z^2}{2}\right)\right] + \frac{2U_z}{3r_0^2}\left[z^2 - \left(\frac{x^2+y^2}{2}\right)\right] + \tfrac{1}{3}(U_x + U_y + U_z) \tag{8.6}$$

where the final term gives the potential at the centre (0, 0, 0).

If we now set $U_z = 0$ and $U_x = U_y = -U$, eqn. (8.6) becomes

$$\begin{aligned} U_{x,y,z} &= -\frac{2U}{3}\left\{1 + \frac{1}{r_0^2}\left[x^2 - \left(\frac{y^2+z^2}{2}\right)\right] + \frac{1}{r_0^2}\left[y^2 - \left(\frac{x^2+z^2}{2}\right)\right]\right\} \\ &= \frac{2U}{3}\left\{\frac{1}{r_0^2}\left[z^2 - \left(\frac{x^2+y^2}{2}\right)\right] - 1\right\} \end{aligned} \tag{8.7}$$

From eqn. (8.7), it is readily shown that the transformations for a are

$$a_z = -2a_x = -2a_y = \frac{16}{3}\frac{eU}{mr_0^2\omega^2} \tag{8.8}$$

In the devices which have been actually reported, the terms $1/r_0^2$ in eqns. (8.5)

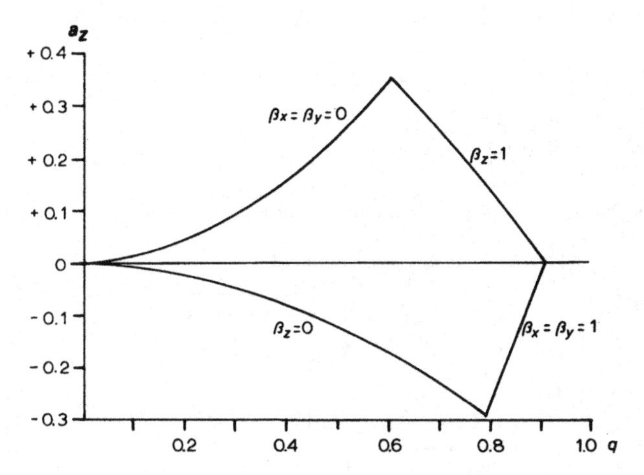

Fig. 8.12. Stability diagram for the six-electrode trap.

and (8.8) should be replaced by $5.15/d^2$ for the plane sheet electrodes and $8/d^2$ for the annular electrodes, where d is the distance between opposite faces of the cube.

The stability diagram which results from the q and a values given by eqns. (8.5) and (8.8) is shown in Fig. 8.12.

(2) *The storage-ring trap*

The storage-ring trap comprises a conventional quadrupole mass filter with the electrodes bent so as to form a closed loop. The two basic forms which have been reported are the circular configuration, by Drees and Paul [4] and by Church [5], and the "race-track" model, also by Church [5] (see Fig. 8.13). Because the rod electrodes are bent, this must necessarily introduce imperfections in the fields (cf. discussion in Chapter VI) and therefore the radius curvature of the ring, R, must be large compared with r_0. For the circular device, Church employed a ratio $R/r_0 = 45:1$ with $r_0 = 0.16$ cm. The field imperfections just mentioned can lead to non-linear resonances (see Chapter V) with a consequent possibility of losing otherwise stable ions. This effect can be reduced by making the rf drive frequency a large non-integral multiple of the secular oscillation frequency. Church chose a multiple in excess of 10 with a drive frequency of 52 MHz; ions were created by directing a gated beam of electrons between the quadrupole electrodes and detected by a resonance absorption technique. At a background pressure of 2×10^{-10} torr, $^3He^+$ ions were stored for periods in excess of 14 min.

(3) *The static ion trap*

The various types of trap so far described in this chapter have been "active"

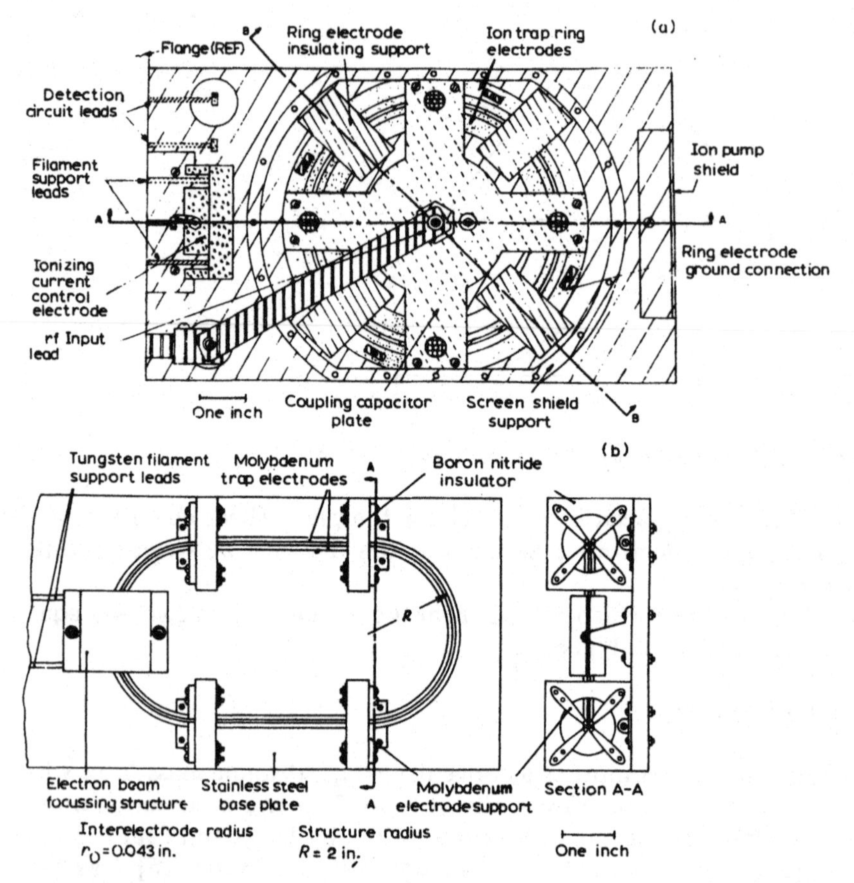

Fig. 8.13. Storage-ring traps developed by Church [5]. (a) Circular design; (b) race-track design.

devices in that they have relied upon the application of radio-frequency electric fields in order to confine the ions to bound trajectories, and we shall see in Section E.(3) that this can cause the ions to possess quite high kinetic energies. However, there are a number of instances (see Chapter X) where it is desirable to have the mean ion temperature as low as possible. This can be achieved in an instrument which is structurally identical to the three electrode trap but is "passive" in that a d.c. field only is applied between the end-caps and the ring and a homogeneous magnetic field, B, is applied along the z axis. A descriptive account of such a system has recently been given by Walls and Dunn [26]. If the end-caps are biassed positively with respect to the ring, the ions oscillate with a secular frequency, ω_z, given by

$$\omega_z = \left[\frac{eU}{\pi^2 m(r_0^2 + 2z_0^2)}\right]^{1/2} \tag{8.9}$$

Ions which attempt to escape along the r direction are turned back in a cyclotron orbit. As a result, the ions move in small circles at a frequency ν_c given by

$$\nu_c = \frac{eB}{2\pi m} \tag{8.10}$$

In addition, the centres of these circular orbits precess about the z axis with a frequency ν_m, and the frequencies of the three types of motion are related by

$$2\nu_m^2 - 2\nu_c\nu_m + \omega_z^2 = 0 \tag{8.12}$$

Typical quoted values for O_2^+ ions are $U = -1$ V (ring negative) and $B = 1.15$ Tesla ($\equiv 1.15 \times 10^4$ gauss) giving $\nu_c = 552$ kHz, $\omega_z = 67$ kHz and $\nu_m = 4.1$ kHz. At low pressures (10^{-10} torr), very long trapping times have been observed (3×10^6 sec for electrons, 10^5 sec for NH_4^+), the concentration of the ions being monitored by the noise power dissipated from the z motion of the ions into a $10^9\,\Omega$ resistor connected between the end-caps. This non-destructive detection has a high mass range, and by introducing power at the cyclotron frequency one can raise the noise power through an increase in the ion temperature. The method has been used to determine m/e values with a resolution of about two parts in 10^4, for example to distinguish between H_2O^+ ($m/e = 18.020$) and NH_4^+ ($m/e = 18.044$). A particular application of this system [27] has been in the determination of the total cross-sections for the recombination of NO^+ and O_2^+ ions with low energy electrons. This and other applications are discussed in Chapter X.

D. SURVEY OF APPLICATIONS OF ION TRAPS

In Section B, a survey of the construction and operation of ion traps has been made. Here we examine those applications where the results have cast light on ion trap behaviour; discussion of those applications in which the trap has been used merely as a tool for the study of a chemical or physical system is deferred to Chapter X.

(1) *Mass spectrometric applications*

The three-electrode ion trap was, of course, first developed for use as a mass spectrometer, and this is the chief application to which the device has been put. Despite this attention, however, no manufacturer has thought fit to develop the trap as a commercially available instrument.

The earliest published reports on the ion trap mass spectrometer are those by Berkling [16], Paul et al. [17] and Fischer [9]. Some aspects of the general technique, particularly detection, employed by this group have already been discussed (see Section B) and their work on the influence of space charge on ion trap behaviour is examined in Section E. (3). Here we consider some of the data relating to the performance of their instrument. As might be expected, the first mass spectra [9] obtained showed relatively poor resolution by modern standards but the ions H_2O^+, N_2^+ and CO_2^+ were observed, as were the ions m/e 38–43 from propane and the krypton isotopes at m/e 82, 83, 84 and 86. The best resolution obtained, $m/\Delta m = 85$, was with krypton at 3×10^{-6} torr, the resolving power being limited by the period between ion–molecule collisions relative to the period of the resonant field. Under the continuous ionization conditions employed in these experiments, the signal from the trapped ions was found initially to increase linearly with time and reach a saturation value after 5–20 μsec depending upon the electron beam current. From a signal of height h_∞ at saturation, Fischer [9] defined a mean ion lifetime $\bar{\tau}_i$ as

$$\bar{\tau}_i = \frac{h_\infty}{\dot{h}_0} \tag{8.13}$$

where $\dot{h}_0$ is the rate of increase of the signal with time over the initial linear portion. For N_2^+ at a measured nitrogen pressure of 6×10^{-6} torr, a value of $\bar{\tau}_i = 15$ msec was obtained which is approximately 40 times longer than the estimated period between ion–molecule collisions at the same pressure. Obviously, eqn. (8.13) is an approximation in that the ion–ion scattering effects, discussed later in Section E, are ignored.

The maximum ion densities calculated from the space charge-induced resonance shift [see Section E.(3)] were 2×10^6 and 4×10^6 ions cm^{-3} for nitrogen and krypton, respectively; the lowest detectable partial pressure was estimated as 1.8×10^{-8} torr.

The detection of low partial pressures using an ion trap was the objective of work reported by Rettinghaus [18] (see Section B). At a pressure of about 2×10^{-9} torr and over the mass range m/e 6–30, he achieved a maximum resolution of 300 with a half-life period of the ion concentration equal to 20 min. The estimated lowest partial pressure actually detectable was ca. 10^{-13} torr (corresponding to about four ions in the trap) provided that the total pressure did not exceed ca. 5×10^{-10} torr, the limit of detection being set by field distortions due to geometric defects and the space charge of the ions and the electron current. Field defects were also cited by Paul et al. [17] as being possibly responsible for "ghost" resonances. An interesting effect first noted by Rettinghaus was the increase in intensity of the peak at m/e 29 (assigned as COH^+) relative to the intensity of m/e 28 (CO^+), the assumption being that the latter species was undergoing an ion–molecule reaction with a hydrogen-containing molecule. This type of investigation is considered in greater detail in Section D.(3) and Chapter X.

The advantages to be gained by employing the method of mass selective storage coupled with ion ejection, developed by Dawson and Whetten, have

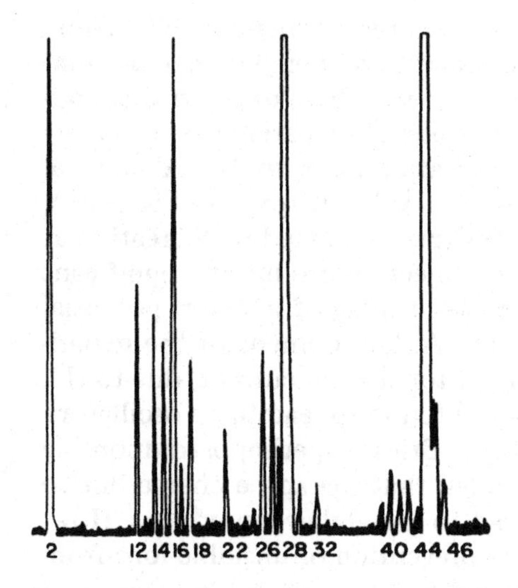

Fig. 8.14. Mass spectrum obtained with an ion trap made of stainless steel mesh [10] operating with a storage time of 1 msec in a pressure 10^{-6} torr.

already been discussed. This advance in technique opened up the way to improved performance and provided a greater understanding of the characteristics of the ion trap. In particular, the analogy with the mode of operation of the mass filter could be exploited to calculate theoretical ion trajectories [28] and examine effects such as the phase and position at ion creation on the expected peak shapes. A feature which emerged from the experimental work [13] was the phenomenon of peak splitting. This effect was later shown [29] to arise from non-linear resonances due to imperfect fields. Examples and a more complete account of this behaviour are given in Chapter V. Experimentally, it was found that the extent of peak splitting could be reduced by applying a d.c. bias between the end-caps such that the draw-out cap was the more negative. However, application of this bias did appear to limit the attainable resolution.

A typical mass spectral scan over the range m/e 2–50 taken with a mesh trap [10] is shown in Fig. 8.14; the pressure for this run was about 10^{-6} torr with a storage time of 1 msec and a resolution (FWHM) of 100. A large device with solid electrodes ($z_0 = 2.0$ cm) was operated at a maximum resolution of 1000 [19] over a very limited mass range. Other aspects of the work of Dawson and Whetten, including the study of ion loss kinetics, plasma oscillations and the phase synchronization of the ion ejection pulse are discussed elsewhere in this chapter.

The mass-selective detection technique was also employed in the mass spectrometer developed by Harden and Wagner [7, 20]. Their system, which was in many ways analogous to that of Dawson and Whetten, made provision for both axial and equatorial injection of the electron beam and also embodied an additional grid to gate the unwanted ions and prevent them from reaching the channeltron multiplier. Added to this was the refinement that the ionization and ion ejection pulses could be locked to the phase of the rf drive potential. This arrangement allowed a number of operating parameters to be examined. For example, using a very short ion creation pulse width and an axial electron beam, it was verified that, as expected from theory (see Fig. 2.34), the maximum number of ions retained in the trap coincided with creation at a phase lag of zero or π. Equatorial electron injection produced a significant level of ionization only at zero phase lag since at a lag of π, the rf potential applied to the ring repels the electron beam. A development of the experiment with axial electron injection was to monitor the ion current, due to N^+, as the ion creation pulse was widened. This was found to lead to an oscillatory fine structure, of the same frequency as the rf drive, superimposed upon the ion build-up curve, and the authors concluded that the space charge due to ions formed at an unfavourable phase acted to destabilize stored ions. However, such a curve might be expected if the prevention of unstable ions from reaching the detector was not 100% effective.

The effects of synchronization of the ion withdrawal pulse with the phase of the rf potential were less easy to rationalize: for $\omega = 220$ kHz, withdrawal was most effective when the amplitude of the rf was increasing, at a point corresponding to approximately half of its maximum value and at $\omega = 315$ kHz this position had advanced almost to the tip of the positive peak. Ion withdrawal appeared to be complete within ca. 5 μsec. From studies on the rate of ion loss, Harden and Wagner concluded that ion–molecule scattering was the dominant process at a pressure of ca. 10^{-6} torr. Some aspects of the work described above were repeated in a study by Mastoris [30] which included a verification of the delayed "warm-up" of the system before ions could be stored, the phase dependence of the efficiency of ion creation and the fact that scanning from low to high mass gave more reproducible but less intense peaks than scanning in the reverse direction.

A system similar to that of Harden and Wagner, having axial electron injection and a gated multiplier, was also employed by Sheretov and co-workers [31–34] in a series of studies which included an examination of the effects of the radius of the ionization zone and the phase of ion creation on the boundaries of the stability diagram and hence the resolution [33], the influence of unstable ions on the performance of the trap [34], and the effects of space charge [31] [see Section E.(3)]. The effect of unstable ions is an aspect of the system which does not appear to have been investigated explicitly by other workers but Sheretov et al. [34] argue that since the trap can be operated in a number of modes, these ions can be important. For example, in the "continuous" or "mixed" modes, ions are simultaneously created and selected according to their stability for the whole or a major fraction of the storage time, respectively. On the other hand, in the "pulsed" mode, ions are created during a short period and then selected during the storage period in a manner more analogous to the operation of an ion source and mass filter. From their analysis, the authors concluded that under ultrahigh vacuum conditions, the mixed mode provides a greater sensitivity than either of the other two methods, but that as the pressure is increased, the pulsed mode becomes optimal. In particular, because of the increased ion–molecule scattering which must occur at higher pressures, the storage time, and hence resolution, must be reduced if a linearity of signal with pressure is to be preserved.

(2) *The storage of microparticles*

One of the earliest papers on ion traps is that by Wuerker et al. [35] in which the authors described the trapping of ca. 20 μm-diameter aluminium dust particles. Their trap, which incorporated equatorial electron injection, was essentially the same as described previously except that provision was made for blowing in the dust and also for the illumination and observation of the particles. Typical operating conditions for $z_0 = 0.25$ in. and $e/m = 5.3 \times 10^{-3}$ coulombs kg^{-1} were $V = 500$ V (rms) and $\omega/2\pi = 200$ Hz. With this system, it was possible actually to observe the trajectories of single particles (see Fig. 8.15) or of arrays of particles (Fig. 8.16) and note how they were influenced by altering the parameters a and q. In particular, the effect of the constant force of gravity acting along the z axis could be examined. For such a force, F, we can write in a normalized form

$$A = 4F/m\omega^2 \tag{8.14}$$

for motion in the z direction, so that eqn. (2.30) becomes

$$\frac{d^2z}{d\xi^2} + (a_z - 2q_z \cos 2\xi)z = A \tag{8.15}$$

which, using the approximations developed in Section E.(2) [eqns. (8.32)–(8.38)] may be expressed as

$$\frac{d^2 Z}{d\xi^2} + \left(a_z + \frac{q_z^2}{2}\right)Z = A \tag{8.16}$$

for which the normalized solution is

$$Z = \left[B' \sin \left(a_z + \frac{q_z^2}{2}\right)^{1/2}\right] + \frac{A}{[a_z + q_z^2/2]} \tag{8.17}$$

Thus eqn. (8.17) shows that the uniform force displaces the centre of motion by an amount Δ proportional to its magnitude, and, since $\beta_z = [a_z + (q_z^2/2)]^{1/2}$ with $\omega_{0z} = \beta_z \omega/2$ [eqns. (8.41) and (2.33)], inversely proportional to the resultant frequency of motion. Using the above substitutions we find

$$\Delta = \frac{F}{m\omega_{0z}^2} \tag{8.18}$$

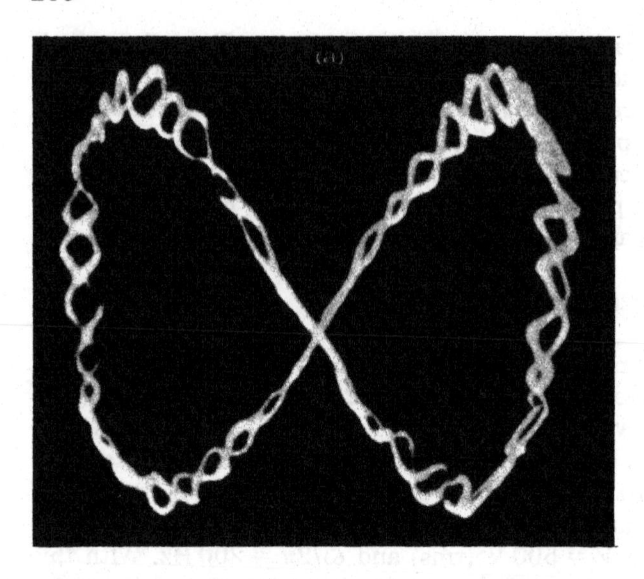
(a)

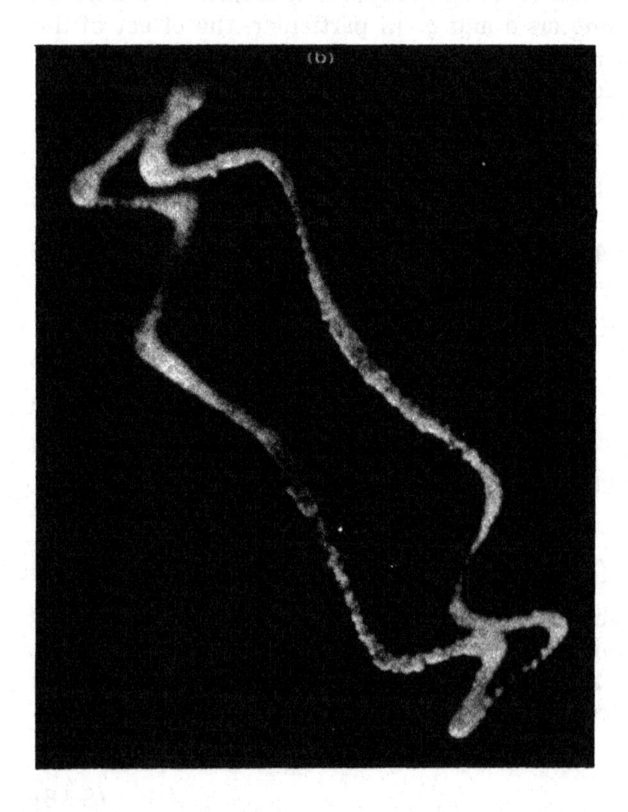
(b)

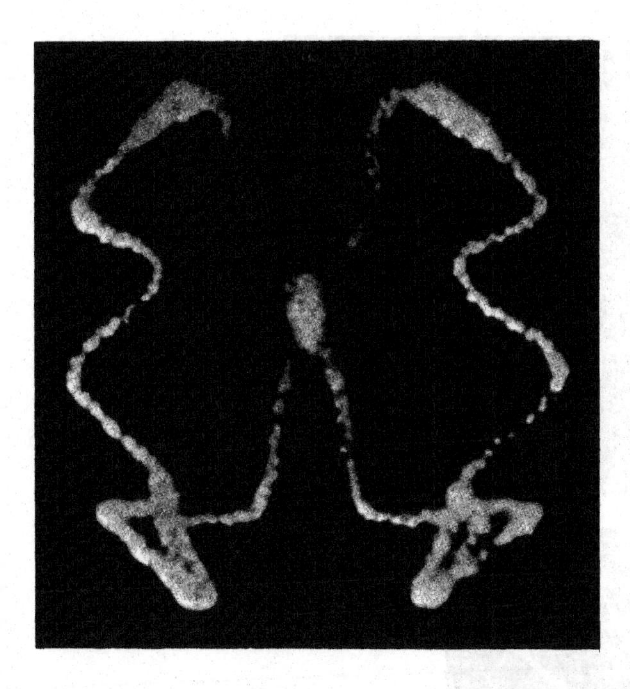

Fig. 8.15. Lissajous figures obtained [35] by viewing the trajectory of an aluminium microparticle trapped in the r–z plane. (a) $\beta_z/\beta_r = 2$, $a_z = 0$, $q_z = 0.232$; (b) $\beta_z/\beta_r = 1$, $a_z = -0.0643$, $q_z = 0.502$; (c) $\beta_z/\beta_r = 0.5$, $a_z = 0.102$, $q_z = 0.502$.

and, as ω_{0z} decreases, Δ approaches z_0 so that the particle "falls out" of the trap. Similarly, it may be shown by substitution for q^2 with $a = 0$ in eqn. (8.17) that the displacement due to gravity will increase with the square of the frequency of rf drive potential if V is held constant. This effect can be overcome by placing a d.c. bias between the end-caps with a polarity compatible with the sign of the charge on the stored particles.

One interesting feature which this work demonstrated was the extent of the validity of the approximate relationship

$$\beta_z = [a_z + (q_z^2/2)]^{1/2} \tag{8.41}$$

along the line $a_z = 0$. Thus direct observation of the value of β_z as a function of q_z indicated that the approximation holds to within 1% for $q_z < 0.4$.

The use of a.c. trapping fields has been extended to studies of charged species contained at ambient pressures. Thus Berg and Gaubler [36] have employed approximate quadrupole fields to investigate the evaporation and condensation of charged water droplets (see also Chapter X) and Whetten [37] has observed the trajectories of a variety of metallic and non-metallic materials in a quadrupole mass filter and in a three-electrode trap both operating in air at atmospheric pressure. As noted in Chapter III (p. 76), the particle motion is still governed by a Mathieu-type equation which incorporates an additional

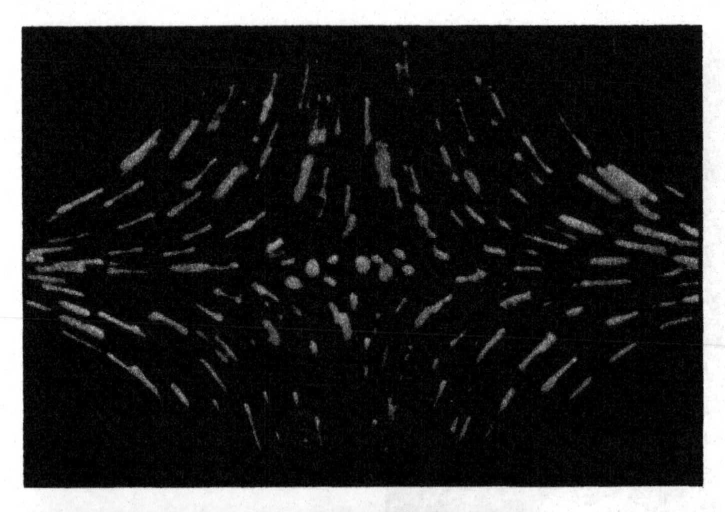

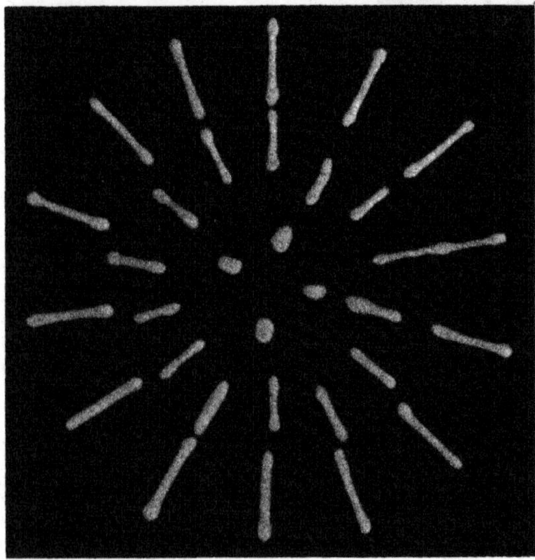

Fig. 8.16. (a) View of the r–z plane for a many-body suspension of charged aluminium microparticles [35]. The array has been "crystallized" by damping collisions with background gas at 10^{-3} torr. (b) View of r–θ plane for a "crystallized" suspension of 32 charged microparticles.

term arising from the velocity-dependent viscous drag (Stokes' law). With this system, it was possible to observe the effect of rf phase synchronization of the drawout pulse. In addition, a novel alternative technique of particle ejection was discovered. Thus suddenly reducing the rf voltage whilst maintaining the d.c. level applied to the end-caps resulted in the motion in the z direction becoming unstable so that the trapped species moved out through the end-caps.

Other workers have also experimented with stored microparticles, notably Wuerker et al. [3] and Zaritskii et al. [25], both groups apparently using this merely to characterize the six-electrode trap. In contrast, Waniek and Jarmuz [38] and Haught and Polk [24] have used the three-electrode and six-electrode traps, respectively, to investigate laser-induced evaporation or dissociation of stored particles. This work is discussed in Chapter X.

(3) *The quadrupole ion storage source (quistor)*

We have already seen from Section B.(4) that the ion trap can be used in place of the conventional ion source of a quadrupole mass filter to provide a means of delaying the period between ion creation and detection. Such a system can be employed for the study of unimolecular and bimolecular ionic processes such as metastable ion decay, charge-transfer, ion–molecule, and chemical ionization processes. A brief survey of this work has appeared [39] and a general description is given in Chapter X. In this section we concentrate on a set of results for the ammonia system [40] which provide a way of estimating the mean ion kinetic energies. We shall see later [Section E.(3)] how these values compare with those derived from theoretical models for the motion of the trapped ions.

The principal ion–molecule reactions occuring in ammonia are

$$NH_3^+ + NH_3 \xrightarrow{k_1} NH_4^+ + NH_2 \tag{8.19}$$

$$NH_2^+ + NH_3 \xrightarrow{k_2} NH_4^+ + NH \tag{8.20}$$

$$NH_2^+ + NH_3 \xrightarrow{k_3} NH_3^+ + NH_2 \tag{8.21}$$

and the data presented in Fig. 8.17 show the variation of yields for the primary and secondary ions as a function of storage time when the quistor ($r_0 = 1.0$ cm) was operating with $V = 400$ V and $\omega/2\pi = 2.75$ MHz. Because there was no means available for selecting a particular primary ion, the direct evaluation of the three rate constants was impossible. For example, the abundance of NH_4^+ must reflect contributions from both reactions (8.19) and (8.20). However, at short storage times, the level of NH_3^+ appears to remain constant so that we can equate $k_1[NH_3^+]$ to $k_3[NH_2^+]$, whereas for storage times in excess of 1 msec when the NH_2^+ has reacted completely, the value for k_1 can be found directly from the rate of decay of m/e 17. From this, a value for k_3 can be found using the equality assumed above. The combined effects of reactions (8.20) and (8.21) in removing NH_2^+ are represented by the decay in m/e 16 in Fig. 8.17 from which the sum of the combined rate constants ($k_2 + k_3$) may be found. From this, a value may be assigned to k_2.

A comparison between the data obtained in this way and that reported in the literature is presented in Table 8.1. Despite the variation between the

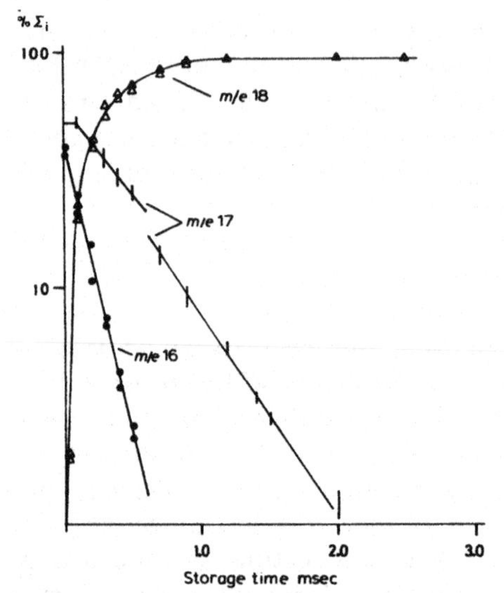

Fig. 8.17. Ion abundance plots showing primary and secondary ions formed from ammonia [40].

results of different groups of workers, there is clearly a strong indication that the primary ion kinetic energies are of the order 1–3 eV, a range which is compatible with that calculated from a model for the motion of the trapped ions (see p. 222).

E. EXPERIMENTAL AND THEORETICAL ASPECTS OF ION CONTAINMENT

(1) *Ion loss processes*

The minimization of unintentional ion loss is clearly of great importance for the successful functioning of the ion trap in any of the modes of operation considered so far. In this section, we consider how the problem may be quantified.

(a) *Mechanisms of ion loss*

Ion loss may occur through a number of different processes:

(i) intrinsically unstable trajectories;

(ii) quasi-unstable trajectories, where the limit of excursion of the ion exceeds the internal dimensions of the device;

(iii) ion–neutral molecule interactions;

(iv) ion–ion interactions.

(i) *Intrinsically unstable trajectories.* This category of ion loss simply arises from the basic principle of operation of the ion trap: if the a, q values for

TABLE 8.1

Values of rate constants[a] for reactions in ammonia

	This work	Ryan [41]			Huntress et al. [42, 43]		Marx et al. [44]		Derwish et al. [45]
		Thermal energies	1 eV	2 eV	Thermal energies	3 eV	Thermal energies	3 eV	
k_1	23.7 ± 2.6[b]					13.5	14.0 ± 1.5	9.0	13.0[e]
	10.7 ± 1.8[c]	18.1 ± 1.0[b]	8.1 ± 0.3[c]	7.5 ± 0.3[c]	15.5 ± 0.8[c]				
k_2	7.2 ± 2.6[d]	8.0 ± 2.0	< 0.5	7.2 ± 1.9	9.3 ± 0.5	8.8	9.0 ± 1.5	7.0	
k_3	27.3 ± 2.6[d]		10.9 ± 1.0	10.9 ± 1.5	13.8 ± 0.5	15.4	21.0 ± 4.0	30.0	
$k_2 + k_3$	35.0 ± 2.0[c]	15.0 ± 2.0[c]		18.1 ± 3.4[c]	24.2 ± 2.0[c]	24.2	30.0 ± 5.5[c]	37.0	18.0[e]

[a] Values expressed as $10^{10}k$ cm^3 molecule^{-1}sec^{-1}.
[b] Total appearance rate constant.
[c] Total disappearance rate constant.
[d] Values calculated assuming $k_1[NH_3^+] = k_3[NH_2^+]$.
[e] Values calculated from cross-section measurements.

a given ion place it outside the stability diagram, then the trajectory will be unbounded and the species will be rapidly removed from the trap, e.g. by charge neutralization at an electrode.

(ii) *Quasi-unstable trajectories.* The second type of ion removal is rather more critical to the operation of the ion trap and arises when ions are created outside the "initial ionization volume", a factor which has already been considered in detail in Chapter II (p. 54). Essentially, if an ion is formed at some distance from the centre and/or with a significant initial velocity, the maximum excursion of the ion, which is governed by the values of a and q and the phase of the rf field at creation, will result in the collision of the ion with an electrode. Thus the initial ionization volume may be a very small fraction of the inscribed volume, a situation which is more critical when the trap is employed in the mass spectrometric mode. In the mass filter, difficulties which result from the analogous effect are overcome by limiting the diameter of the entrance aperture; in the ion trap, suitable collimation of the electron beam coupled with phase-synchronized ion creation may be employed (see p. 198).

A practical consequence for the operation of the ion trap in either the mass spectrometric or total pressure modes is that sufficient time must be allowed for the unstable and quasi-unstable ions to be rejected before the detection or analysis process is applied. This time may be quite long (ca. 100 μsec) [46].

(iii) *Ion–neutral molecule interactions.* The study of ion–neutral interactions is, of course, an extensive subject in itself [47]. The processes which are relevant here include elastic scattering, inelastic scattering, charge-transfer and ion–molecule reactions. In the first two of these, one can visualize an ion with a stable trajectory being suddenly deflected such that its new situation is one of an ion having unfavourable initial starting position and/or velocity (cf. quasi-unstable ions above). In the case of charge-transfer, an ion in a stable trajectory is removed and a new ion, probably with effectively zero initial velocity, is created, possibly outside the initial ionization volume. If this happens, the total number of trapped ions will be reduced. On the other hand, it has been found [19] that when the pressure of a light gas, e.g. neon, is increased during the storage of heavy ions with which charge-transfer is unlikely, e.g. Hg^+, a stabilization effect appears to occur and the rate of Hg^+ ion loss at long storage times, ca. 10^3 sec, is appreciably reduced (see also p. 220).

When a mixture of sample gases in which charge-transfer can occur is present, entirely new ions may be formed, and the rate of this process may be a function of the kinetic energy of the original (primary) ion. Under "total pressure mode" operation, these new ions may be retained within the trap, and kinetic studies on argon/methane mixtures have shown that within experimental error the rate of loss of Ar^+ is exactly balanced by the rate of formation of CH_3^+ and CH_2^+ and further ion–molecule reaction products from these species [39]. A more detailed discussion on trapped-ion chemistry is given in Section D and Chapter X.

(iv) *Ion–ion scattering*. The problems associated with ion–ion interactions are rather less easy to discuss. Basically, two situations arise. One is the discrete event when a pair of like-charged ions collide leading to the disturbance, and possible instability, of one or both of the trajectories. This phenomenon will clearly become more significant as the number of ions contained within the trap increases and is thought to be responsible for establishing the kinetic energy distribution of the ion cloud (see p. 219).

The second difficulty is concerned with the modifications to the trapping field, and hence the ion trajectories, which occur through the space charge potential which develops as the number of stored ions increases. Under these conditions, the Laplace equation $\nabla^2\phi = 0$ no longer holds and one therefore expects the basic operational characteristics of the trap to change as the number of ions increases. In particular, one can envisage a "saturation" effect in which the linearity between the number of ions stored and the length of the ion-creation period disappears [10]. Indeed, under certain conditions, periodic instabilities, possibly due to plasma oscillations, have been noted [49], and these are considered in greater detail later (see p. 218).

We now examine the means of experimentally quantifying the rates associated with the ion loss processes which have just been discussed.

(b) *The kinetics of ion loss*

The earliest experiments in this area are those by Fischer [9] in which a "mean ion lifetime" was determined (see p. 196). Since then several authors [12, 13, 20, 22] have examined the problem in terms of a rate law analogous to those employed in chemical kinetics. Consider the situation where the trap contains N ions cm^{-3}, then during an ion creation period the rate of change of N with time must represent the difference between the rate of creation of ions and the rate of loss of ions

$$\frac{\mathrm{d}N}{\mathrm{d}t} = k_1 p - (k_2 N^2 + k_3 N p) \tag{8.22}$$

where p is the pressure of neutral molecules, k_1 is a "rate constant" for ion creation, k_2 the rate constant for loss by ion–ion scattering and k_3 the rate constant for loss by ion–neutral scattering. In practice, a plot of N versus t has the form shown in Fig. 8.18. Such a graph for a given pressure may be obtained by progressively increasing the length of the ionization period, allowing a short period for the quasi-unstable ions to be rejected and then ejecting the ions on to the detector by applying the cap pulse. These curves are similar to those obtained by Fischer [9] using the resonance absorption technique. At "saturation" $\mathrm{d}N/\mathrm{d}t = 0$ for $N = N_\infty$ so that

$$k_1 p = k_2 N_\infty^2 + k_3 N_\infty p \tag{8.23}$$

We can now proceed to evaluate k_1, k_2 and k_3. For example, at short times

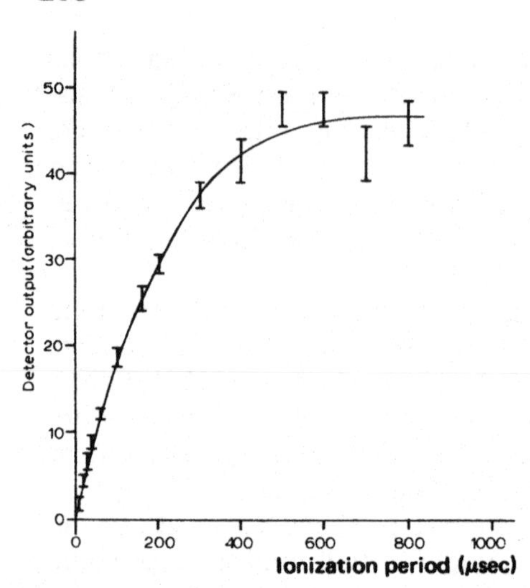

Fig. 8.18. Build-up curve for Ar^+ ions in a three-electrode ion trap.

over the linear portion of Fig. 8.18 we can make the approximation $N \simeq 0$ so that eqn. (8.22) becomes

$$\frac{dN}{dt} = k_1 p \tag{8.24}$$

from which k_1 is readily found for a given value of p.

If we now choose conditions so that $k_3 p \gg k_2 N$, then from eqn. (8.23) we have

$$k_3 = \frac{k_1}{N_\infty} \tag{8.25}$$

With a value of k_3 determined in this way one can then use eqn. (8.23) in a curve-fitting procedure on the first derivative of Fig. 8.18 to find a value for k_2.

The foregoing is clearly an approximate method and a more precise means of evaluating k_2 and k_3 is to simplify eqn. (8.22) by considering the situation where the electron beam has been suppressed so that no more ions are formed. We now have

$$\frac{dN}{dt} = -(k_2 N^2 + k_3 N p) \tag{8.26}$$

On rearrangement this becomes

$$\frac{dN}{N(k_3 p + k_2 N)} = -dt$$

which integrates to

$$\frac{1}{k_3p} \ln \left(\frac{k_2N + k_3p}{N}\right) = t + \text{constant} \tag{8.27}$$

The constant may be evaluated assuming an initial condition $N = N_0$ at $t = 0$ so that eqn. (8.27) becomes

$$\frac{1}{k_3p} \ln \left(\frac{N_0(k_2N + k_3p)}{N(k_2N_0 + k_3p)}\right) = t \tag{8.28}$$

which rearranges to

$$\frac{1}{N} = \left(\frac{k_3p + k_2N_0}{k_3pN_0}\right) \exp(k_3pt) - \frac{k_2}{k_3p} \tag{8.29}$$

Since eqn. (8.29) is of the general form

$$y = a \exp(bx) + c$$

with $a = (k_3p + k_2N_0)/k_3pN_0$; $b = k_3p$ and $c = -k_2/k_3p$, a standard exponential curve-fitting procedure can be employed on the decay data when plotted as $1/N$ versus t. Care should be taken, however, to ensure that the ion decay being monitored arises from genuine ion–ion and ion–molecule losses and not from the rejection of quasi-unstable ions.

Experimentally, the values for N, N_∞, N_0, etc. will be obtained as a signal in, say, millivolts. To convert this to absolute units and to determine the rate constants in equivalent units, requires that the transfer efficiency of ions from the trap to the collector and the gain of the multiplier, if used, be known. When the ion trap is employed as an ion source, the transfer efficiency of ions into and through the mass analyser must also be included. Inevitably, these corrections must introduce uncertainties. A further difficulty exists when it is desired to compare the results from ion traps of different dimensions, since the ion density is the only meaningful quantity. This may be estimated by assuming that the ions are contained within the inscribed oblate spheroid having a volume $4\pi r_0^2 z_0/3$ and then dividing the total number of ions by this quantity.

Typical values for the ion loss parameters for Ar^+ from a trap operating along the $a = 0$ line and acting as the source for a conventional mass filter are listed in Table 8.2. In order to estimate the relative importance of the two mechanisms of ion loss when the trap is completely filled, we can multiply the value of k_2 by N_∞^2 and k_3 by $N_\infty p$. Thus for the accompanying data we have

ion–ion scattering: $k_2N_\infty^2 = 2.9 \times 10^{10}\,\text{cm}^{-3}\,\text{sec}^{-1}$

ion–molecule scattering: $k_3N_\infty p = 4.2 \times 10^9\,\text{cm}^{-3}\,\text{sec}^{-1}$

This would suggest that under the conditions cited, ion–ion scattering is slightly the more dominant effect, although clearly as the number of trapped

TABLE 8.2

Typical ion-loss parameters for Ar^+ from an ion trap

r_0	1.0 cm
Pressure	2.4×10^{-5} torr
Drive frequency	0.762 MHz
V	140 V
q_z	0.59
N_∞	1.04×10^7 cm^{-3}
k_1	1.4×10^{15} cm^{-3} torr^{-1} sec^{-1}
k_2	2.7×10^{-4} cm^3 sec^{-1}
k_3	1.7×10^7 torr^{-1} sec^{-1}

ions falls, the second-order dependence of this factor on N would result in ion–molecule scattering becoming relatively more important giving an exponential decrease of trapped ion density with time. Dawson et al. [10] have attempted to assess the influence of these scattering processes by comparing their data with "theoretical" build-up and decay curves calculated assuming the dominance of one or other effect. They concluded that ion–molecule scattering was the major loss mechanism above 10^{-8} torr in the mass selective mode. As noted in Section D.(1) (see p. 196) the treatment of ion lifetimes developed by Fischer was based upon the assumption that only ion–molecule scattering is important. Comparing the two approaches we can write

$$\tau_i \equiv \frac{1}{k_3 p} \tag{8.30}$$

from which Fischer's data gives a value for N_2^+ of $k_3 = 1.11 \times 10^7 \text{ torr}^{-1} \text{ sec}^{-1}$, indicating close agreement with the result listed for Ar^+ in Table 8.2.

(2) *A theoretical model of ion trapping*

We have already seen in Chapter VII that the y motion of ions in a monopole may be represented as a ripple superimposed upon a "large" secular oscillation (see p. 154). In this section, we develop a slightly more rigorous treatment of this approach based upon work by Wuerker et al. [35] and Dehmelt and Major [48, 49] which provides the basis for a model with which one can estimate the space charge limited number of ions that can be contained within the trap as well as the mean kinetic energy of the secular motion of the ions.

For motion along the z axis of the trap, eqn. (2.30) may be re-written as

$$\frac{d^2z}{d\xi^2} = -(a_z - 2q_z \cos 2\xi)z \tag{8.31}$$

with $a_z = 4eU/mz_0^2\omega^2$, $q_z = 2eV/mz_0^2\omega^2$ and $\xi = \omega t/2$.

We can consider that the z motion is, in fact, made up of two components, a displacement, δ, due to the micromotion resulting from the high-frequency field and a larger displacement, Z, which describes the extent of the motion averaged over a period of the rf drive potential.

Then

$$z = Z + \delta \tag{8.32}$$

If we assume that the driving force, which is related to the value of q, is small, then we can write $\delta \ll Z$ but $d\delta/dt \gg dZ/dt$. With these approximations, we can substitute into eqn. (8.31) to get

$$\frac{d^2\delta}{d\xi^2} = -(a_z - 2q_z \cos 2\xi)Z \tag{8.33}$$

which, assuming $a \ll q$ and Z to be constant, integrates to

$$\delta = -\frac{q_z Z}{2} \cos 2\xi \tag{8.34}$$

Equation (8.34) indicates that the displacement due to the micromotion is out of phase with the rf potential by π and also increases linearly with the secular displacement Z [see Fig. 2.11 (b)].

The approximate value for δ from eqn. (8.34) can now be substituted into eqn. (8.32) to give

$$z = Z - \frac{q_z Z}{2} \cos 2\xi \tag{8.35}$$

whence the original Mathieu equation, eqn. (8.31), becomes

$$\frac{d^2z}{d\xi^2} = -a_z Z + \frac{a_z q_z Z}{2} \cos 2\xi + 2q_z Z \cos 2\xi - q_z^2 Z \cos^2 2\xi \tag{8.36}$$

Since the acceleration due to the rf drive, $d^2\delta/d\xi^2$, averaged over a period of the rf drive is equal to zero, we have that the acceleration of the secular motion averaged over the same period is given by

$$\left\langle\frac{d^2Z}{d\xi^2}\right\rangle_{av} = \frac{1}{\pi}\int_0^\pi \frac{d^2z}{d\xi^2}\,d\xi \tag{8.37}$$

so that the integral of eqn. (8.36) taken between these limits is

$$\frac{d^2Z}{d\xi^2} = -\left(a_z + \frac{q_z^2}{2}\right)Z \tag{8.38}$$

which, written in terms of time, becomes

$$\frac{d^2Z}{dt^2} = -\left(a_z + \frac{q_z^2}{2}\right)\frac{\omega^2}{4} Z \tag{8.39}$$

Equation (8.39) corresponds to simple harmonic motion of the secular motion of the secular component Z and is equivalent to

$$\frac{d^2Z}{dt^2} = -\omega_{0z}^2 Z \tag{8.40}$$

in which ω_{0z} is the secular oscillation frequency and is equivalent to that given by eqn. (2.33) if we write

$$\beta_z = \left(a_z + \frac{q_z^2}{2}\right)^{1/2} \tag{8.41}$$

[See also eqn. (3.15).]

If we now consider the case of $a_z = 0$ and substitute for q_z in eqn. (8.39), we get

$$\frac{d^2Z}{dt^2} = -\left(\frac{e^2V^2}{2m^2z_0^4\omega^2}\right) Z \tag{8.42}$$

The force on an ion of mass m and charge e is therefore given by

$$m\frac{d^2Z}{dt^2} = -e\frac{d\bar{D}_z}{dZ} \tag{8.43}$$

where

$$\frac{d\bar{D}_z}{dZ} = \frac{eV^2}{2mz_0^4\omega^2} Z \tag{8.44}$$

Integrating eqn. (8.44) between the limits $Z = 0$ and $Z = z_0$ gives

$$\bar{D}_z = \frac{eV^2}{4mz_0^2\omega^2} \tag{8.45}$$

The model we therefore arrive at is that proposed by Dehmelt [48], namely that when q_z is small, an ion can be regarded as oscillating in a parabolic potential well in the z direction with a frequency equal to that of the fundamental secular frequency ω_{0z}.

For motion in the r direction, eqn. (8.45) becomes

$$\bar{D}_r = \frac{eV^2}{4mr_0^2\omega^2} \tag{8.46}$$

and since $r_0^2 = 2z_0^2$, we have

$$\bar{D}_z = 2\bar{D}_r \tag{8.47}$$

This model is shown schematically in Fig. 8.19.

If a d.c. component is also applied and the condition $a \neq 0$ obtains, this will have the effect of increasing or decreasing the depth of the pseudo-potential well, depending upon the sign of the d.c. bias: for a negative bias, a point

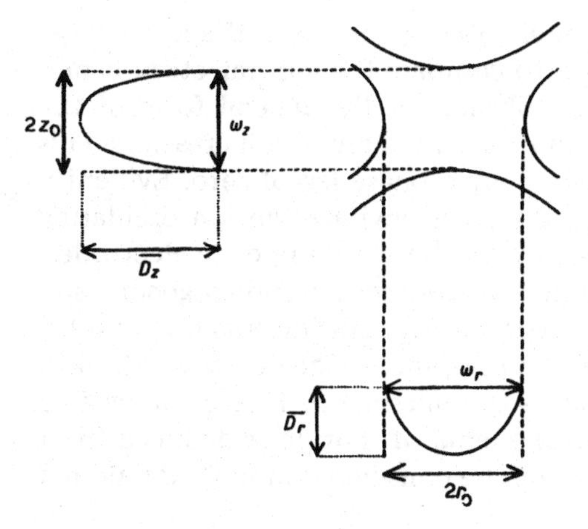

Fig. 8.19. Representation of the parabolic pseudo-potential wells, depths $\bar{D}_z$ and $\bar{D}_r$, retaining ions within a three-electrode trap. ω_{0z} and ω_{0r} are the respective secular frequencies of the ion motion within the wells.

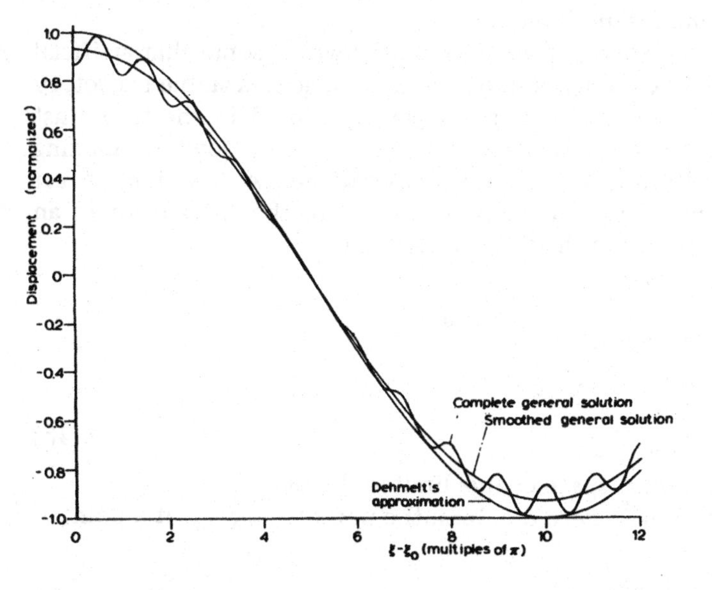

Fig. 8.20. Computer-plotted trajectories corresponding to ion motion for $\beta = 0.1$ showing the complete general solution, smoothed general solution and Dehmelt's approximation.

will be reached when the depth of the pseudo-potential well becomes zero and ions will no longer be retained. It should, however, be remembered that the central assumption behind this derivation is that $\delta \ll Z$, and from eqn. (8.34) we can see that as q_z increases for constant Z and ξ the fraction δ/Z becomes progressively larger so that the assumption ultimately breaks down, typically at a value of q_z equal to ca. 0.4.

Before we attempt to apply this model to considerations of space charge and ion kinetic energies, it is instructive to compare the ion trajectory it predicts with that obtained by analytical solution of the general form of the Mathieu equation for a given value of β and relative phase of ion creation. This is shown in Fig. 8.20 for $\beta = 0.1$ and a relative phase lag of zero. Evidently Dehmelt's pseudo-potential well approximation envisages an ion oscillating throughout the whole of the available distance between opposite electrodes. A possibly more realistic approximation is to use the "smoothed general solution" obtained from eqn. (2.32) by setting all the coefficients $C_{2n} = 0$ for $n \neq 0$: note how the micro-motion of the rf ripple is directed towards zero displacement as the secular displacement is at a maximum. This is compatible with the relative phases of the two components of motion as deduced from eqn. (8.34). These two approximations will be considered in more detail in a later section (see p. 222).

(3) *Space charge and ion kinetic energies*

(a) *Space charge and the Dehmelt model*

The model of ion trapping in a pseudo-potential well is somewhat artificial in that the well does not exist until an ion is actually held in a stable trajectory within the trap! Every ion which is subsequently placed in the trap must modify the fields experienced by those ions already present. However, making the assumption that with the r and z directions combined we have ions *at rest* (except for the micro-motion) held in a potential having the distribution of an oblate spheroid, we can invoke the Poisson relation

$$\begin{bmatrix} \text{electrostatic} \\ \text{potential, } \phi_i \end{bmatrix} + \begin{bmatrix} \text{pseudo-} \\ \text{potential, } \psi \end{bmatrix} = \text{constant}$$

such that

$$-\nabla^2 \phi_i = \nabla^2 \psi = 4\pi \rho_{max} \qquad (8.48)$$

where ρ_{max} is the theoretical space charge-limited density.

Now for a trap operating with zero d.c. bias, the potential $\psi_{x,y,z}$ at a point x, y, z is

$$\psi_{x,y,z} = \frac{\bar{D}_r}{r_0^2}(x^2 + y^2) + \frac{\bar{D}_z}{z_0^2} z^2 \qquad (8.49)$$

which on making the substitutions $\bar{D}_r = \bar{D}_z/2$ and $r_0^2 = 2z_0^2$ becomes

$$\psi_{x,y,z} = \frac{\bar{D}_z}{4z_0^2}(x^2 + y^2 + 4z^2) \qquad (8.50)$$

whence we have

$$\nabla^2 \psi = 3\bar{D}_z/z_0^2 \qquad (8.51)$$

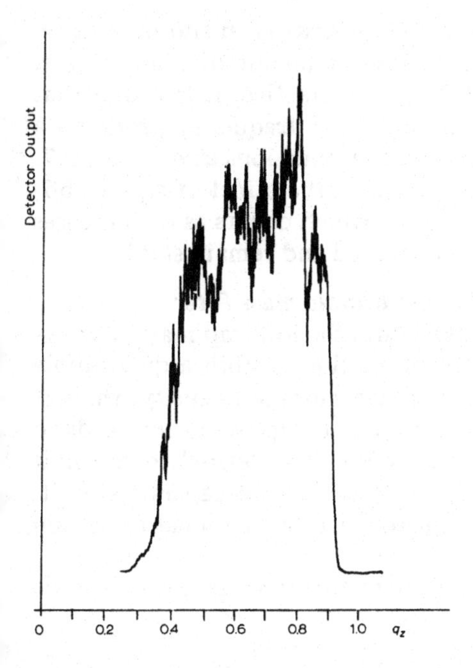

Fig. 8.21. Trace showing the number of Ar^+ ions stored in a three-electrode trap as a function of q_z. A storage time of 100 μsec was employed.

Equating this with eqn. (8.48) gives

$$\rho_{max} = \frac{3\bar{D}_z}{4\pi z_0^2} \tag{8.52}$$

or

$$N_{max} = \frac{3\bar{D}_z}{4\pi e z_0^2} \quad \text{ions per unit volume} \tag{8.53}$$

Substituting for $\bar{D}_z$ from eqn. (8.45) gives

$$N_{max} = \frac{3}{16\pi} \frac{V^2}{m z_0^4 \omega^2} \tag{8.54}$$

which, using the substitution $V^2 = (m^2 z_0^4 \omega^4 / 4e^2) q_z^2$ becomes

$$N_{max} = \frac{3}{64\pi} \frac{m\omega^2}{e^2} q_z^2 \tag{8.55}$$

Taking the data from Table 8.2 as an example, eqn. (8.55) gives a value $N_{max} = 3.5 \times 10^7$ ions cm^{-3}, a result which agrees well with observed value of N_∞.

Equation (8.55) indicates that N_{max} should increase with increasing m, ω and q_z, and decrease if multiply charged ions are stored. The increase with mass suggests that if the trap were operated at saturation in the mass spectrometer mode, the device should discriminate in favour of high masses. Furthermore, as q_z increases, so should N_{max}. In practice, it is found that scanning the rf amplitude for a fixed mass and fixed frequency produces a curve (Fig. 8.21) showing a maximum density of trapped ions at q_z = ca. 0.7. This would appear to be caused by a balance between the effect of eqn. (8.55) and the decrease in the initial ionization volume which occurs as q_z increases and so limits the region in which ions can be created and remain stable.

(b) *Space charge effects along the axis of a quadrupole mass filter*

The model of the pseudo-potential wells within the ion trap may be transferred to the motion of ions in the x and y directions within a quadrupole mass filter. In particular, one can estimate the maximum ion density which it is possible to contain within the filter before space charge repulsion leads to ion loss. This is an important consideration when it is desired to transmit large continuous beam currents, for example, in an isotope separation unit, or when a high density packet of ions is injected, e.g. from a quadrupole ion storage source (quistor).

Analogous equations to (8.49)–(8.52) apply so that with zero d.c. bias we now have

$$\rho_{max} = \frac{\bar{D}_y}{\pi r_0^2} \tag{8.56}$$

where $\bar{D}_y(=\bar{D}_z)$ is the depth of the pseudo-potential well in the y direction. Equation (8.56) gives a value for the maximum number of ions which can be contained within a unit length of the mass filter, and knowing the velocity with which ions are injected along the z axis, one can calculate the maximum rate of ion injection, and hence beam current, which is compatible with the value of ρ_{max}.

(c) *Fischer's treatment of space charge*

An alternative treatment of the problem of space charge within the ion trap pre-dates the approach described above and was first proposed by Fischer [9]. The model assumes that as the ion density builds up the resultant fields are equivalent to applying additional d.c. potentials $\Delta U/2$ to the end-cap and ring electrodes. Thus from eqns. (2.26) and (2.27) we have

$$-2\Delta U_z \cdot \frac{z}{r_0^2} = \frac{\rho z}{2\epsilon_0} \tag{8.57}$$

and

$$\Delta U_r \cdot \frac{r}{r_0^2} = \frac{\rho r}{4\epsilon_0}$$

giving

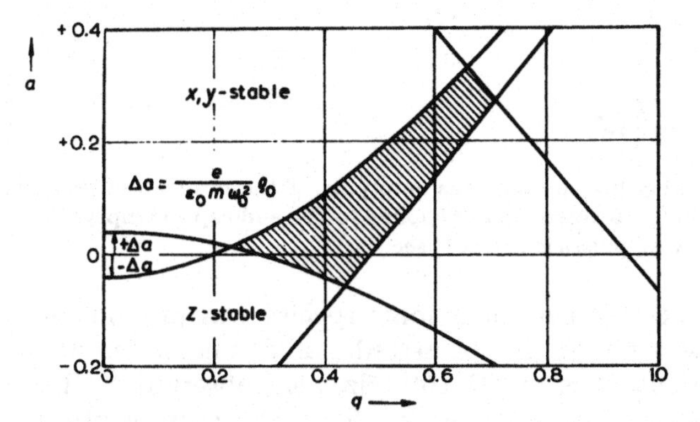

Fig. 8.22. The effect of space charge on the stability diagram (after Fischer [9]).

$$\Delta U_z = -\Delta U_r = -\frac{\rho r_0^2}{4\epsilon_0} \qquad (8.58)$$

where ϵ_0 is the permittivity.

The net effect is that the space charge acts as a de-focusing force in both the r and z directions: the d.c. potentials experienced by the ions are greater than they should be so that unstable trajectories will result. This may be viewed as an apparent shift in the boundaries of the stability diagram, as shown in Fig. 8.22 in which we combine the transformation for a, eqn. (2.28), with ΔU, eqn. (8.58), to get

$$\Delta a = \pm \frac{e}{\epsilon_0 m \omega^2} \rho \qquad (8.59)$$

From this equation, it appears that for a given ion density more favourable operation of the trap (i.e. lower values of Δa) should be obtained for high values of m and ω.

The space charge-induced shift of the stability diagram has also been examined experimentally and theoretically by Sheretov et al. [31] who found that, for a trap of normal geometry, as the ion density increased so did the resolution. For the shift observed, the authors were able to estimate the total number of charged particles trapped as 1.6×10^7 which accords well with the ion densities of 2×10^6 cm^{-3} and 4×10^6 cm^{-3} for nitrogen and krypton, respectively, reported by Fischer [9] who employed the same method. These appear to be the only examples of the use of the space charge-induced d.c. shift to evaluate ion densities; clearly, the method relies on being able to find experimentally a sharp boundary between having stable and unstable ions.

(d) *Oscillatory phenomena*

Two groups of workers have described conditions under which oscillatory behaviour of the ion concentration within the trap appears to be established.

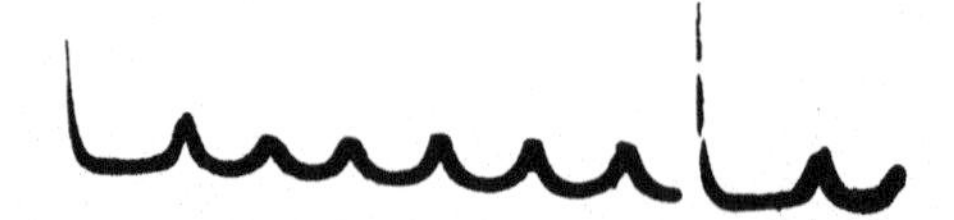

Fig. 8.23. Oscilloscope trace showing the spontaneous emptying of the ion trap in the z direction. The large pulses are due to a voltage pulse applied to the end-cap to empty the trap. The time between self-emptying pulses is 6×10^{-4} sec.

Fischer [9] noted that if, while a particular ionic species was just out of resonance, a foreign gas whose ions also possessed stable trajectories was added, then strong Kipp oscillations of 50–1000 Hz were observed at the output of the detection receiver, even though no form of sweep or other modulation was applied. The explanation offered for this was that as the space charge arising from the ions built up, it caused a shift in the resonance lines within the stability region (see also p. 217) so that ions were removed causing a reduction in the space charge and relaxation of the resonance line back to its former position. Because of the continuous ion creation, this cycle was then repeated.

The observation of the "self-ejection" of ions during continuous ionization at particular working points on the stability diagram was noted by Whetten and Dawson [51, 52] and was explained in a rather different way. Thus when the a, q values were close to the $\beta_z = 1/2$ and $2/3$ lines, the output from the detector had the form shown on the oscilloscope trace in Fig. 8.23. The low amplitude ripple had a period of ca. 6×10^{-4} sec which was inversely proportional to the ionization rate (which was changed by varying the gas pressure or electron current) and was proportional to the square of the frequency of the applied rf power. When a cap pulse was applied in order to eject the ions forcibly (see the larger amplitude pulses in Fig. 8.23), the period before the next self-ejection pulse was significantly longer.

In explaining these phenomena, the authors argued that when the secular oscillation frequency in the z direction, related to the value of β_z, was a subharmonic of that of the applied rf drive, the ions could easily gain extra energy and be ejected from the trap. Such energy could be gained from plasma oscillations, particularly when the plasma resonance frequency, ω_p, given [53] by

$$\omega_p = \left(\frac{4\pi Ne^2}{m}\right)^{\frac{1}{2}} \tag{8.60}$$

was equal to the applied drive frequency. Some sort of direct relationship between ω_p and ω appears to be reasonable in terms of the expression for N_{max} which we have already derived from the pseudo-potential well model. Thus rearranging eqn. (8.55) and replacing N by N_{max} in eqn. (8.60), we have

$$\frac{4\pi N_{\max} e^2}{m} = \frac{3}{16}\omega^2 q_z^2 = \omega_p^2 \qquad (8.61)$$

This expression accords with the observed relationship between the self-ejection period and ω^2 since an increase in ω^2 would demand a corresponding linear increase in $N_{\max}$ so that a longer ionization period would be required. On the basis of their original assertion, Whetten and Dawson replaced ω_p in eqn. (8.60) by ω ($2\pi \times 500$ kHz) to calculate an ion density of 1.6×10^8 ions cm^{-3}. Since the $\beta_z = 1/2$ and $2/3$ lines do not intersect the working apex of the stability diagram for operation in the mass spectrometric mode, plasma resonances should not be significant in this application, although they could affect operation in the total pressure mode, e.g. as an ion source.

(e) *Ion kinetic energies*

(i) *The energy distribution: ion–ion interactions.* When the ion trap is working as a mass spectrometer, there are no direct reasons for requiring a knowledge of the kinetic energy distribution of the ions. However, in applications where the trap is being used as a means of storing ions for the study of physical and chemical processes, e.g. ion–molecule and charge-transfer reactions, radiofrequency spectroscopy, ion–electron interactions (see Chapter X), a knowledge of ion energies is of paramount importance. We have already seen in Section D.(3) that the rate constants for ion–molecule reactions occurring within a quistor indicate mean primary ion energies in the range 1–3 eV. This problem has been discussed by Dehmelt [48] in terms of the pseudo-potential well model [see Section E.(2)] and the assumption that at low pressures the ion temperature, T_i, is established by ion–ion interactions according to the formulation originally developed by Spitzer [54] for a uniform plasma. Thus one can derive a "self-collision time", t_c, which provides a measure of the time required for the distribution of ion kinetic energies to become Maxwellian as a result of a sequence of two-particle interactions governed by the inverse square law. For an ion m/e 16 in a trap $z_0 = 0.707$ cm powered with a potential $V_0 = 400$ V oscillating at 2.75 MHz, we have $\bar{D}_z = 16.05$ eV. Assuming that for $N = N_{\max}$ we can write $3kT_i = e\bar{D}_z$, it has been shown [55] that $t_c = 0.424$ sec.

This means that in the above example, storage times in excess of 0.424 sec are required for the kinetic energy distribution to become Maxwellian; in fact, McDonald et al. [56] have suggested that an interval of some ten times this figure elapses before the high-energy tail of the distribution is fully established. Since most mass spectroscopic uses of ion traps involve storage periods of an order 10^{-3} to 10^{-2} times smaller than the above value of t_c, it may be assumed that ion–ion interactions can be ignored in calculating a value for the kinetic energy; we return to this problem below.

(ii) *Ion–neutral interactions.* The derivation which has just been outlined was based on the additional tacit assumption that the ensemble of ions is thermally isolated and that energy is neither lost to neutral molecules during collisions nor gained from the applied power by rf heating. Major and Dehmelt [49] have analysed this aspect of the problem in terms of elastic collisions of ions of mass m with particles of mass m_r at rest and have derived eqn. (8.62) for the average change in the energy of an ion taken over the phase of the rf field, $\langle \Delta\overline{w} \rangle_{av}$, during a collision.

$$\langle \Delta\overline{w} \rangle_{av} = m(1-\cos\theta)[m_r(m+m_r)^{-1}\langle v^2 \rangle_{av} - mm_r(m+m_r)^{-2}\langle u^2+v^2 \rangle_{av}] \qquad (8.62)$$

where θ is the scattering angle and u, v are the velocities of the secular motion and micro-motion, respectively.

Three particular cases have been examined in detail. When $m/m_r \ll 1$, the elastic scattering of the ions by essentially fixed centres gives rise to rf heating, leading to an exponential energy increase with time. For $m/m_r \gg 1$, the light neutral particles cause a viscous drag on the ions and so reduce the average energy with time. The experimental observation of the stabilization of heavy ions by the introduction of a light gas into the trap has already been noted (see p. 206) and Lambert [57] has reported an exponential decay in the intensity of N_2^+ ions trapped in the presence of heavier gases (argon and carbon dioxide), possibly reflecting an increase in ion energy as suggested above. In the special case of $m = m_r$, Dehmelt argues that when averaged over a secular oscillation period, there is zero change in the energy of the colliding ion since in such a collision the ion will be brought to rest and subsequently take up its original energy again from the rf field. The same argument applies to resonant charge exchange between an ion and its parent neutral species. However, at ion velocities of ca. 10^6 cm sec^{-1} the cross-section for resonant charge exchange increases with decreasing velocity [58] so that there is now a greater probability of new ions being formed at large values of z, i.e. at particular phases of the secular motion. Such an effect would lead to a net energy absorption.

(iii) *The estimation of ion kinetic energies.* Returning to the earlier discussion on the establishment of thermal equilibrium in the charge cloud at long storage times, one can imagine several different situations. For example, when the ion density is less than N_{max} and the ions are cold, they may be envisaged as partially filling up the potential well, like a fluid with a density N_{max}, up to a depth $\overline{D}_N$ given by

$$\left(\frac{\overline{D}_N}{\overline{D}_z}\right)^3 = \left(\frac{N}{N_{max}}\right)^2 \qquad (8.63)$$

As N approaches N_{max}, the ion temperature increases and ion–ion collisions may be expected to lead to loss at the electrodes: this is equivalent to heat loss from the ion cloud.

For trapping times which are short compared with t_c, one can make the assumption that ions are oscillating freely within the pseudo-potential well (see above) and find a "mean" kinetic energy by integrating over one secular period. For low pressure conditions, the total mean energy comprises components of motion in the r and z directions, however, when ion scattering becomes more important as the pressure is increased, Major and Dehmelt [49] have argued that the maximum permissible energy is determined by the lowest potential well, that given by $\bar{D}_r (= \bar{D}_z/2$ for $a = 0$ and $r_0^2 = 2z_0^2)$.

For the case of secular oscillation along the z axis only Todd et al. [55] have deduced a mean kinetic energy as follows. Consider simple harmonic motion with an amplitude z_0 and frequency ω_z, then we may write

$$z = z_0 \sin \omega_z t \tag{8.64}$$

and

$$\dot{z} = z_0 \omega_z \cos \omega_z t \tag{8.65}$$

Now by integrating over a time interval π/ω_z, the average velocity $\langle \dot{z} \rangle$ is given by

$$\langle \dot{z} \rangle = \frac{2z_0\omega_z}{\pi} \tag{8.66}$$

The kinetic energy corresponding to this mean velocity is therefore

$$\tfrac{1}{2}m\langle \dot{z} \rangle^2 = \frac{2mz_0^2\omega_z^2}{\pi^2} \tag{8.67}$$

which on substituting for ω_z^2 in terms of the depth $\bar{D}_z$ of the parabolic potential well

$$\omega_z^2 = \frac{2e\bar{D}_z}{mz_0^2} \tag{8.68}$$

gives

$$\begin{aligned} \tfrac{1}{2}m\langle \dot{z} \rangle^2 &= \frac{4}{\pi^2} e\bar{D}_z \\ &= 0.404e\bar{D}_z \end{aligned} \tag{8.69}$$

Thus for the well of depth $\bar{D}_z = 16.05$ V considered in the previous example, we have a mean kinetic energy along the z axis equal to 6.48 eV. Adopting the above argument about the maximum permissible energy, this reduces to 3.24 eV.

It should be remembered that, as indicated previously [Section E.(2)] the model used for this calculation assumes that an ion can oscillate throughout the whole distance $2z_0$ without striking the electrodes. In practice, the rf ripple superimposed upon such a trajectory would result in the ion being lost from the trap, so it is evident from eqn. (8.66) that using this value for the amplitude of the secular operation must necessarily give a high estimate of the mean ion kinetic energy. A more realistic approach may be to use the trajectory represented by the "smoothed general solution" of the Mathieu equation illustrated in Fig. 8.20.

This "smoothed" trajectory is obtained by equating the C_{2n} coefficients in eqn. (2.32) to zero for $n \neq 0$, giving a fundamental mode of the form

$$u(\xi) = A \cos \beta\xi$$

or

$$u(t) = A \cos (\beta\omega t/2) \tag{8.70}$$

Averaging the expression for $\dot{u}(t)$ over an interval $2\pi/\beta\omega$ gives a mean velocity

$$\langle \dot{u}(t) \rangle = \frac{A\beta\omega}{\pi} \tag{8.71}$$

in which A is a function of the initial conditions (see p. 16). To apply eqn. (8.71) to the calculation of a mean ion energy is a simple matter of evaluating $\langle \dot{u}(t) \rangle$ at the appropriate value of β and determining $\frac{1}{2}m\langle \dot{u}(t) \rangle^2$. For example, the conditions pertaining to the ion m/e 16 employed previously lead to "smoothed" mean energy along the z axis of 5.29 eV and along the r plane of 2.65 eV. These mean energies are clearly consistent with those suggested by the experimental data on ion–molecule reactions reported in Section D.(3) (see p. 203), and are in broad agreement with the results of recent calculations by Dawson [59] based upon the method of phase space dynamics.

REFERENCES

1 W. Paul and H. Steinwedel, Z. Naturforsch., 8a (1953) 448.
2 W. Paul and H. Steinwedel, U.S. Patent, 2, 939, 952 (1960).
3 R.F. Wuerker, H.M. Goldenberg and R.V. Langmuir, J. Appl. Phys., 30 (1959) 441.
4 J. Drees and W. Paul, Z. Phys., 180 (1964) 340.
5 D.A. Church, J. Appl. Phys., 40 (1969) 3127.
6 P.H. Dawson and N.R. Whetten, Advan. Electron. Electron Phys., 27 (1969) 59.
7 C.S. Harden and P.E. Wagner, EASP 100-93, Edgewood Arsenal, Maryland, 1971.
8 G. Lawson and J.F.J. Todd, Chem. Brit., 8 (1972) 373.
9 E. Fischer, Z. Phys., 156 (1959) 26.
10 P.H. Dawson, J.W. Hedman and N.R. Whetten, Rev. Sci. Instrum., 40 (1969) 1444.
11 R.F. Bonner, G. Lawson and J.F.J. Todd, Int. J. Mass Spectrom. Ion Phys., 10 (1972/73) 197.
12 J-P. Schermann and F.G. Major, NASA Rep. X-524-71-343, 1971.
13 P.H. Dawson and N.R. Whetten, J. Vac. Sci. Technol., 5 (1968) 11.
14 P.H. Dawson and N.R. Whetten, U.S. Patent, 3, 527, 939, (1970).

15 P.H. Dawson and N.R. Whetten, Brit. Patent, 1, 225, 272, (1971).
16 K. Berkling, Diplomarbeit, Bonn, 1956.
17 W. Paul, O. Osberghaus and E. Fischer, Forschungsberichte des Wirtschaft und Verkehrministeriums Nordrhein-Westfalen 415, Westdeutscher Verlag, Köln and Opladen, 1958.
18 G. Rettinghaus, Z. Angew. Phys., 22 (1967) 321.
19 P.H. Dawson and N.R. Whetten, Naturwissenschaften, 56 (1969) 109.
20 C.S. Harden and P.E. Wagner, EATR 4545, Edgewood Arsenal, Maryland, 1971.
21 P.H. Dawson and C. Lambert, Int. J. Mass Spectrom. Ion Phys., 14 (1974) 339.
22 G. Lawson, R.F. Bonner and J.F.J. Todd, J. Phys. E., 6 (1973) 357.
23 D.B. Langmuir, R.V. Langmuir, H. Shelton and R.F. Wuerker, U.S. Patent 3,065,640, (1962).
24 A.F. Haught and D.H. Polk, Phys. Fluids, 9 (1966) 2047.
25 A.A. Zaritskii, S.D. Zakharov and P.G. Kryukov, Sov. Phys. Tech. Phys., 16 (1971) 174.
26 F.L. Walls and G.H. Dunn, Phys. Today, 27(8) (1974) 30.
27 F.L. Walls and G.H. Dunn, J. Geophys. Res., 79 (1974) 1911.
28 P.H. Dawson and N.R. Whetten, J. Vac. Sci. Technol., 5 (1968) 1.
29 P.H. Dawson and N.R. Whetten, Int. J. Mass Spectrom. Ion Phys., 2 (1969) 45.
30 S. Mastoris, UTIAS Tech. Note 172, Institute for Aerospace Studies, University of Toronto, 1971.
31 É.P. Sheretov, V.A. Zenkin and V.F. Samodurov, Sov. Phys. Tech. Phys., 18 (1973) 282.
32 É.P. Sheretov, V.A. Zenkin and O.I. Boligatov, Prib. Tekh. Eksp., (1971) 166.
33 É.P. Sheretov and V.A. Zenkin, Sov. Phys. Tech. Phys., 17 (1972) 160.
34 É.P. Sheretov. V.A. Zenkin and V.F. Samodurov, Sov. Phys. Tech. Phys., 18 (1973) 262.
35 R.F. Wuerker, H. Shelton and R.V. Langmuir, J. Appl. Phys., 30 (1959) 342.
36 T.G.O. Berg and T.A. Gaubler, Amer, J. Phys., 37 (1969) 1013.
37 N.R. Whetten, J. Vac. Sci. Technol., 11 (1974) 515.
38 R.W. Waniek and P.J. Jarmuz, Appl. Phys. Lett., 12 (1968) 52.
39 R.F. Bonner, G. Lawson, J.F.J. Todd and R.E. March, Advan. Mass Spectrom., 6 (1974) 377.
40 G. Lawson, R.F. Bonner, R.E. Mather, J.F.J. Todd and R.E. March, J. Chem. Soc. Faraday I, (1976) in the press.
41 K.R. Ryan, J. Chem. Phys., 53 (1970) 3844.
42 W.T. Huntress Jr., M.M. Moseman and D.D. Elleman, J. Chem. Phys., 54 (1970) 843.
43 W.T. Huntress Jr. and R.F. Pinizzotto Jr., J. Chem. Phys., 59 (1973) 4712.
44 R. Marx and G. Mauclaire, Int. J. Mass Spectrom. Ion Phys., 10 (1972/73) 213.
45 G.A.W. Derwish, A. Galli, A. Giardini-Guidoni and G.G. Volpi, J. Chem. Phys., 39 (1963) 1599.
46 G. Lawson, R.E. Mather and J.F.J. Todd, to be published.
47 See, for example, P. Ausloos and S.C. Lias (Eds.), NATO Advanced Study Institute on Ion–Molecule Interactions, Plenum Press, London, in the press.
48 H.G. Dehmelt, Advan. At. Mol. Phys., 3 (1967) 53.
49 F.G. Major and H.G. Dehmelt, Phys. Rev., 170 (1968) 91.
50 G. Lawson, J.F.J. Todd and R.F. Bonner, in D. Price and J.F.J. Todd (Eds.), Dynamic Mass Spectrometry, Vol. 4, Heyden, London, in the press.
51 N.R. Whetten and P.H. Dawson, J. Vac. Sci. Technol., 6 (1969) 100.
52 P.H. Dawson and N.R. Whetten, AFCRL-69-0185, General Electric Company, Schenectady, 1969.
53 D. Pines, Solid State Phys., 1 (1955) 368.
54 L. Spitzer, Physics of Fully Ionized Gases, Wiley-Interscience, New York, 1956.

55 J.F.J. Todd, R.F. Bonner and G. Lawson, to be published.
56 W.M. McDonald, M.N. Rosenbluth and W. Chuck, Phys. Rev., 107 (1957) 350.
57 C. Lambert, M.Sc. Thesis, Laval University, Quebec, Canada, 1974; P.H. Dawson and C. Lambert, J. Vac. Sci. Technol., 12 (1975) 941.
58 D. Rapp and W.E. Francis, J. Chem. Phys., 37 (1962) 2631.
59 P.H. Dawson, Int. J. Mass Spectrom. Ion Phys., 17 (1976) to be published.

CHAPTER IX

TIME-OF-FLIGHT SPECTROMETERS

J.P. Carrico

A. INTRODUCTION

It is well known that the performance capabilities of mass spectrometers are limited by the initial conditions of the ions. Multiple field regions are used in deflection instruments to deal with energy and divergence aberrations. Time-lag focusing [1] and focusing fields [2, 3] are used in time-of-flight (TOF) devices for the same purpose. A different approach to this problem is to base the measurement of mass on a property of the ion motion in the analyzer field which is independent of the trajectory and, therefore, independent of the initial conditions. One approach is the mass filter where operation is based on the stability of an ion in the inhomogenous, oscillatory electric field (IOEF). As discussed elsewhere in this book, stability in the IOEF depends on the field parameters and the mass-to-charge ratio (m/e) of the ion but not on its initial conditions.

The flight time of an ion in the IOEF is another m/e-dependent property of this field which is independent of the initial position and velocity. This independence can be understood by realizing that, since the IOEF is a linear function of the coordinates, the system is similar to a harmonic oscillator. In fact, the d.c. component of the equation of motion for an ion in the IOEF describes a harmonic oscillation of the type experienced by a pendulum. As is well known, the period of the pendulum is a fundamental constant which is independent of the initial conditions for small oscillation amplitudes. Efforts have been made to develop a mass spectrometer based on the measurement of this period [4]. Unfortunately, several problems limit the usefulness of this approach, namely, the saddle-point nature of the potential distribution which causes ions to disperse in the transverse direction, difficulty in introducing and extracting ions from the field, and difficulty with interpretation of the spectra. The latter difficulty arises when ions oscillate for many periods with one species overtaking another.

The fundamental character of the motion in the static field is preserved with the addition of an oscillatory component although the motion may no longer be truly periodic. The resulting period or "quasi-period", whichever the case may be, remains independent of the initial velocity and position; a

References pp. 239—240

physical analogy is a pendulum with an arm which periodically varies in length. This period or "quasi-period" is used in the TOF mass spectrometer [5–7] described in this chapter. The principle of operation is described and analytical as well as experimental results are reported. Also, several new applications of this novel TOF principle are described.

B. PRINCIPLE OF OPERATION

We consider the ion dynamics for the IOEF produced by the potential distribution expressed by eqn. (2.6). The equation of motion for an ion in the field produced by this distribution can be expressed in the canonical form of the Mathieu equation [see eqn. (2.21)].

For simplicity, we consider a single component of the motion, say in the z direction. The solution of the component of the equation of motion for this direction is written as

$$z = z_0 G_1(a_z, q_z, \xi_0, \xi) + \dot{z}_0 G_2(a_z, q_z, \xi_0, \xi) \qquad (9.1)$$

where z_0 and $\dot{z}_0$ are z components of the initial position and velocity, respectively (the dot denotes differentiation with respect to ξ) and the functions G_1 and G_2 are two independent solutions of the equation of motion. The functions G_1 and G_2 obey

$$\begin{aligned} G_1(a_z, q_z, \xi_0, \xi_0) &= 1 \quad G_2(a_z, q_z, \xi_0, \xi_0) = 0 \\ \dot{G}_1(a_z, q_z, \xi_0, \xi_0) &= 0 \quad \dot{G}_2(a_z, q_z, \xi_0, \xi_0) = 1 \end{aligned} \qquad (9.2)$$

If G_2 has other zeros ξ_j than ξ_0, an ion with initial position (in the z direction) appears at $z = z_0 G_1(\xi_j)$ at the time $t_j = 2\xi_j/\omega$. The arrival time, t_j, obviously does not depend on the initial velocity of the ion but only on a, q, and ξ_0. For fixed field parameters, the arrival time depends only on m/e. This dependence is the basis for determining the mass-to-charge ratio of an ion by measuring its flight time in the IOEF. Since there is no dependence on the initial velocity, the arrival time will not be sensitive to variations in the kinetic (thermal and recoil) and potential (acceleration) energies of the ions at time t_0. Of course, the arrangement of the configuration is important in order to measure the correct arrival time. If the detector is not located at $z = z_0 G_1(\xi_i)$ then the ion is not measured at time t_i and the velocity independence of the measurement is lost.

In addition to the requirement that the function G_2 possess zeros other than ξ_0, feasibility of the IOEF TOF principle depends on the requirement that ion displacement in the field be less than the dimensions of the field region during the interval $\xi_i - \xi_0$. It is important to note that stability (see Chapter II) is not a necessary condition for meeting this requirement. The major portion of our analytical work, however, was performed for values of a and q belonging to the stability region shown in Fig. 2.30(b). This region was obtained by overlapping the stability regions for all three coordinate directions where $a_z = -2a_y = -2a_x$ and $q_z = -2q_x = -2q_y$. Rearranging the solution given in eqn. (3.19) in the form of eqn. (9.1), one obtains

$$G_1(a, q, \xi, \xi_0) = \frac{1}{W_0}[\dot{\mathrm{F}}_2(\xi_0)\,\mathrm{F}_1(\xi) - \dot{\mathrm{F}}_1(\xi_0)\,\mathrm{F}_2(\xi)]$$

and (9.3)

$$G_2(a, q, \xi, \xi_0) = \frac{1}{W_0}[\mathrm{F}_1(\xi_0)\,\mathrm{F}_2(\xi) - \mathrm{F}_2(\xi_0)\,\mathrm{F}_1(\xi)]$$

where F_1 and F_2 are defined by eqns. (3.5) and (3.19) and W_0 is given in eqn. (3.21). The functions $\mathrm{F}_1(\xi_0)$ and $\mathrm{F}_2(\xi_0)$ are the same as $\mathrm{F}_1(\xi)$ and $\mathrm{F}_2(\xi)$, respectively, except ξ_0 is substituted for ξ. The dots denote differentiation with respect to the indicated functional argument. The parameter β is a function of a and q (see Chapter II). The analytical properties of the solutions of the Mathieu equation are discussed in the literature [7–9].

From the definitions of a and q, the line $a = 2(U/V)q$ in the space of a and q is the family of points corresponding to all values of U and V for which their ratio U/V is a constant and to all values of m/e, ω, and weighting factors [see eqn. (2.1)]. This line is commonly referred to as an operating line and a, q points on it which correspond to given m/e ratios are called operating points. Since, for given operating conditions, operating points for all m/e ratios fall on the line having slope $\epsilon_1 = 2U/V$, it is convenient to present analytical data for points on a given operating line. Also there is a close connection between the parameter β and the flight time. Thus, it is convenient to use β to indicate the position of a stable operating point on the operating line. This is possible because a given iso-β curve intersects a given operating line at a single point. In most of our work we used ϵ_1-β space.

C. ION DISPLACEMENT

The determination of the conditions for which the ion displacement remains less than the dimensions of the field region is a formidable problem because the displacement depends on the initial position and velocity, the injection phase, the operating point, and the zero of G_2 which is utilized. The calculations are simplified considerably if the coordinate of the initial position is zero for the nominal direction in which the time-of-flight is measured, i.e. $z_0 = 0$. In this case, the measured arrival time corresponds to one of the times t_i at which the ions reappear at $z = 0$. From eqn. (9.1), the z component of the displacement at time $t = 2\xi/\omega$ is then

$$z(\xi) = \dot{z}_0 G_2(a_z, q_z, \xi, \xi_0) \tag{9.4}$$

Since the initial velocity $\dot{z}_0$ can be viewed as a scaling factor, we need consider only the behavior of G_2 in determining the displacement.

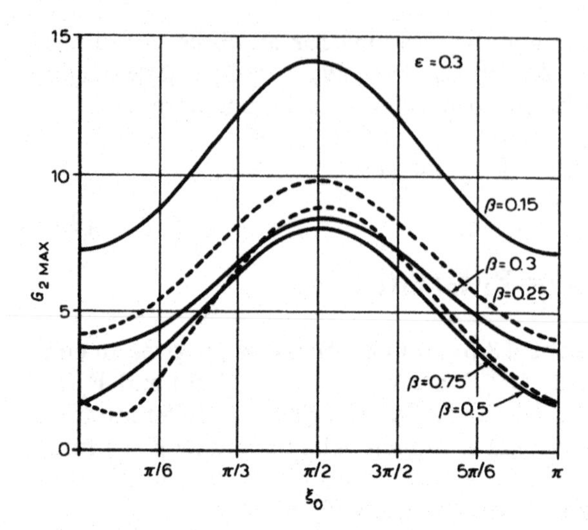

Fig. 9.1. Dependence of displacement on injection phase ξ_0.

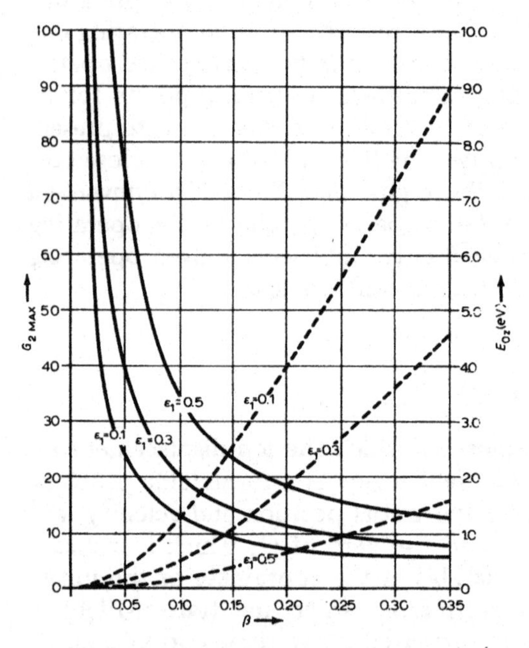

Fig. 9.2. Solid curves, $G_{2\max}$ vs. β for $\epsilon_1 = 0.1$, 0.3, and 0.5. Broken curves, E_{0z} vs. β for $\epsilon_1 = 0.1$, 0.3, and 0.5.

Using a computer, we determined the maximum value of the function G_2 in the interval $\xi_1 - \xi_0$ for operating points belonging to the stability region of Fig. 2.32(b) where ξ_1 is the first zero of G_2. Figure 9.1 shows the maximum value of G_2 as a function of ξ_0 for several values of β on the operating line $\epsilon_1 = 0.3$. The function $G_{2\max}$ generally increases as β decreases and the maximum displacement increases with ϵ for a given β. Also, the minimum and maximum values of $G_{2\max}$ occur in the neighborhood of $\xi_0 = 0$ and $\pi/2$, respectively. Some results [10] on the determination of the maximum allowed energy of injection using the energy constraint

$$E_{0z} < \frac{5ML^2f^2}{G_{2\max}^2} \tag{9.5}$$

are shown in Fig. 9.2 where the mass M is in amu, f is the frequency (in MHz) of the applied a.c. voltage, L is the maximum allowed penetration (in cm) of the ion in the z direction which is assumed to be the electrode spacing for this direction, and $G_{2\max}$ is the maximum value of the function G_2 in the interval $\xi_1 - \xi_0$. The solid curves show the behavior of G_2 vs. β for several values of ϵ_1. The broken curves illustrate the allowed energy values E_{0z} vs. β when $f = 1$ MHz, $M = 50$ amu, and $r_0 = 1$ cm.

It is important also to consider the ion displacement in the transverse directions x and y. The difficulty is that it is no longer possible to assume that the initial position is zero. This prevents us from studying the behavior of the x and y displacements as a function of a, q, and ξ_0 without first assigning values for the initial position and velocity. However, as discussed later, electrode configurations are possible where the predominant displacement effect is associated with the z direction. The only concern then is that the size in the transverse direction of the ion pulse exiting from the field be less than the exit aperture otherwise a loss of signal is incurred. Work is under way to identify operating conditions, if any, for solutions which exhibit transverse focussing and suitable TOF properties. Of particular importance are the periodic solutions discussed by Lever [11] (see Chapter II, p. 57). It is expected that conditions will be found for higher transmission than for the mass filter because of different restrictions on the values of a and q used in the two types of spectrometers. In the mass filter, it is essential that the operating point be close to a boundary of the stability region associated with the given direction; resolution depends on the proximity of the operating point to the boundary. Since the functions G_1 and G_2 in eqn. (9.1) increase rapidly as an edge of the stability region is approached, ion transmission in the mass filter decreases with the resolution. However, in an oscillatory field time-of-flight spectrometer, either values of a and q remote from the stability boundaries or dwell times t_j and t_0 in the oscillatory field can be used which ensure that the displacement is less than the field dimensions. This suggests that transmission of the ions in the time-of-flight spectrometer could be less sensitive to the initial conditions.

D. ARRIVAL TIME

The arrival time, t_i, is calculated from

$$G_2(a_z, q_z, \xi_i, \xi_0) = 0 \qquad (9.6)$$

i.e. t_i is the ith zero of the function G_2. In terms of the trigonometric expansion [eqn. (9.3)], eqn. (9.6) becomes

$$\mathrm{F}_1(\xi_0)\,\mathrm{F}_2(\xi_i) - \mathrm{F}_2(\xi_0)\,\mathrm{F}_1(\xi_i) = 0 \qquad (9.7)$$

Formally, eqn. (9.7) is solved for ξ_i in terms of a_z, q_z, and ξ_0, i.e.

$$\xi_i = \xi_i(a_z, q_z, \xi_0) \qquad (9.8)$$

Equation (9.8) expresses the fact that for fixed values of the field parameters and ξ_0, the arrival time depends only on m/e.

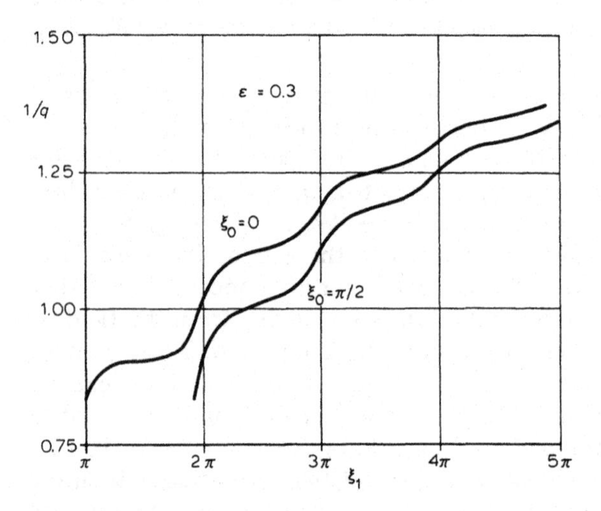

Fig. 9.3. $1/q$ vs. ξ_1 for $\xi_0 = 0$ and $\pi/2$ and $a/q = 0.3$. $1/q$ is proportional to m/e and ξ_1 is the normalized flight time.

The particular arrival time t_i used in an IOEF TOF device depends on the configuration of the apparatus. We have used a computer to determine the dependence of the first zero ξ_1 of G_2 on the position of the operating point for a fixed injection phase ξ_0. The equation of motion [eqn. (2.21)] was integrated using four-point Runge–Kutta methods and converging to $z = 0$. This method was cross checked by a trial-and-error method in which values of ξ were substituted into the right-hand side of eqn. (9.7) until the value producing zero was identified.

Figure 9.3 shows the behavior of $1/q$ as a function of ξ_1 for given values of ξ_0. Values of q belonging to the stability region of Fig. 2.32(b) were used.

Since $1/q$ and ξ_1 are proportional to m/e and the arrival time t_1, respectively, the curves in Fig. 9.3 depict the behavior of m/e as a function of the arrival time; for fixed operating conditions, the correspondence is unique. We note that the arrival time increases with increasing m/e and ϵ_1. Also, the difference in arrival times, i.e. the dispersion, increases with m/e and ϵ_1 (see also Fig. 2.40). The stepwise behavior is of particular interest since the dispersion is larger where the curves are flat. Since the operating point for a given m/e depends on the field parameters and m/e, the flight time and dispersion can be controlled electronically. The IOEF TOF spectrometer appears to be one of the few instances of a device in which the mass dispersion can be electronically controlled.

The largest value of ξ which can be used in a practical apparatus is limited by the condition that the size of the ion trajectory must be less than the dimensions of the analyzer region. The maximum value of the function G_2 increases with m/e along a given operating line. The rate of this increase is greater than the rate of decrease of velocity with mass for a fixed injection energy. Thus, for fixed field parameters and injection energy, the penetration in the z direction for ions of some mass-to-charge ratio will be greater than the length of the field region in that direction. Ions having this or greater m/e ratios will be lost to the electrodes. Thus, stability is not a sufficient condition for ion displacement to be less than the dimensions of the field region. Therefore, the largest value of ξ_1 as well as the largest dispersion which can be used will depend on the field parameters, mass-to-charge ratio, injection energy, insertion phase, and the length of the field region in the z direction.

E. RESOLUTION

An expression for the mass resolution is obtained by the usual method of equating the mass dispersion and peak width. The time domain is used in describing both of these parameters since the m/e of an ion is determined by measuring the ion's flight time in the IOEF. The variation δt_a in the arrival time due to a small change δm in the mass is written as

$$\delta t_a = \frac{\partial t_a}{\partial m} \delta m \tag{9.9}$$

Assuming that the only contribution to the peak width Δt_a is the variation Δt_0 in the injection time, we have

$$\Delta t_a = \frac{\partial t_a}{\partial t_0} \Delta t_0 \tag{9.10}$$

Equating δt_a and Δt_a, solving for δm, and defining the resolution as $R = m/\delta m$, the expression for the resolution associated with measurement of the

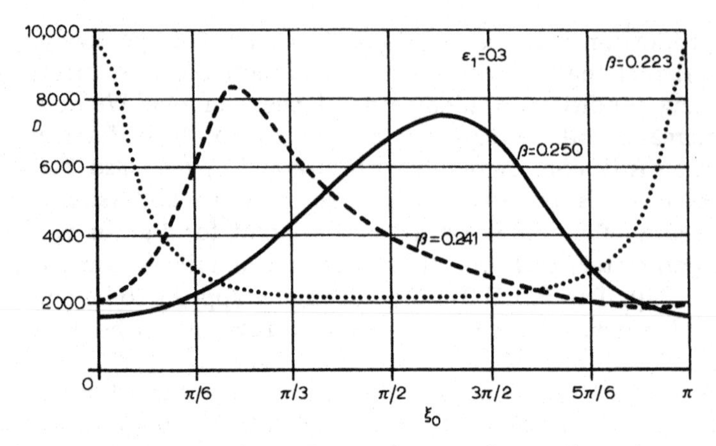

Fig. 9.4. Dispersion factor D vs. ξ_0 for several operating points.

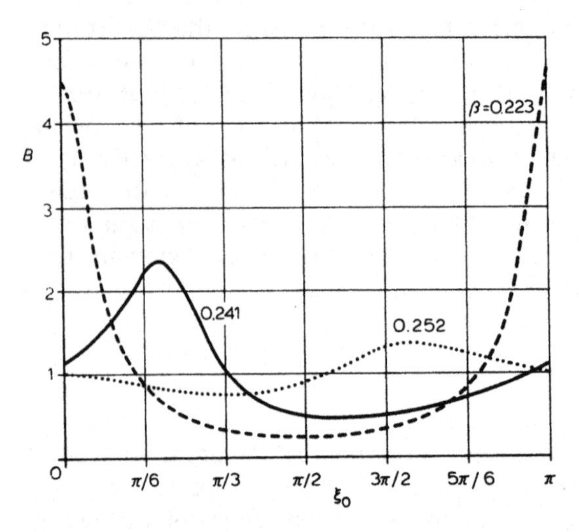

Fig. 9.5. Bunching factor B vs. ξ_0 for operating points of Fig. 9.4.

first zero of G_2 is, in terms of the canonical parameters

$$R = \frac{q(\partial \xi_1/\partial q)}{(\partial \xi_1/\partial \xi_0)\,\Delta\xi_0}. \tag{9.11}$$

The ξ_0 dependence of the dispersion factor $D \equiv \partial\xi_1/\partial q$ is shown in Fig. 9.4 for several operating points. The dependence of the bunching factor $B = \partial\xi_1/\partial\xi_0$ on ξ_0 is shown in Fig. 9.5. The spread in arrival times for ions of a given m/e is less than the spread in injection times when $B < 1$. The variation in resolution as a function of ξ_0 is shown in Fig. 9.6 where $R/\Delta\xi_0$ vs. ξ_0 is plotted for the same operating points used in Figs. 9.4 and 9.5. All three

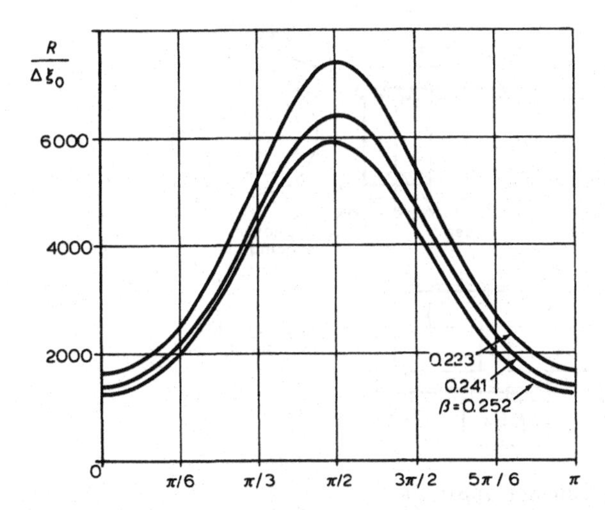

Fig. 9.6. Resolution factor $R/\Delta\xi_0$ vs. ξ_0 for operating points of Fig. 9.4.

factors, dispersion, bunching, and resolution depend on the injection phase ξ_0 and the operating point.

F. EXPERIMENTAL RESULTS

A diagram of the experimental apparatus [5] used to test the IOEF TOF principle is shown in Fig. 9.7. An approximation to the potential distribution

$$V = -\frac{1}{L^2 - (R_1^2/2)}(V_{\mathrm{d.c.}} + V_{\mathrm{a.c.}} \cos \omega t)\left[z^2 - \frac{(x^2+y^2)}{2}\right] \tag{9.12}$$

was obtained using a series of rings connected to a potential divider. The length, L, of the electrode structure was ~ 36 cm and the inside radius, R_1, was ~ 2 cm. The voltage ranges were $0 \leqslant V_{\mathrm{a.c.}} \leqslant 1500$ V and $0 \leqslant V_{\mathrm{d.c.}} \leqslant 300$ V. The frequency range of the a.c. component was 20–50 kHz.

The electron beam was pulsed on for ~ 100 nsec at a preselected phase of the a.c. voltage in order to ionize neutral molecules in the source region. Then, by pulse-biasing grid G_2, positive ions were accelerated in the z direction through a 1.3 cm diameter hole in the grounded, cone-shaped electrode. In this way, the ions were injected into the analyzer region at zero field, i.e. $z_0 = 0$. Upon proper adjustment of the a.c. and d.c. voltages, injected ions returned to $z = 0$ where they were detected by a channel electron multiplier positioned on the z axis and shielded from the source region by grid G_2. With air as a source gas, N_2^+ and O_2^+ were easily resolved as shown in Fig. 9.8. Also, the isotopes of xenon were resolved when this gas was introduced into the source region [5]. Ion residence times on the order of 400 μsec were used in these experiments.

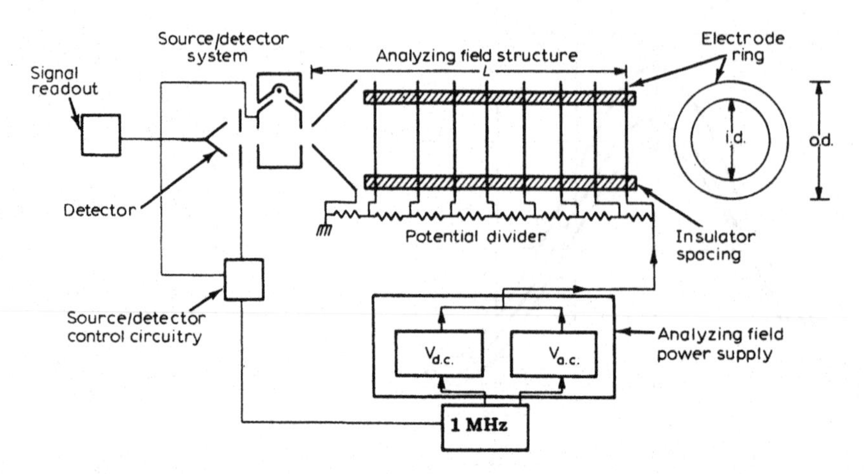

Fig. 9.7. Schematic drawing of experimental apparatus.

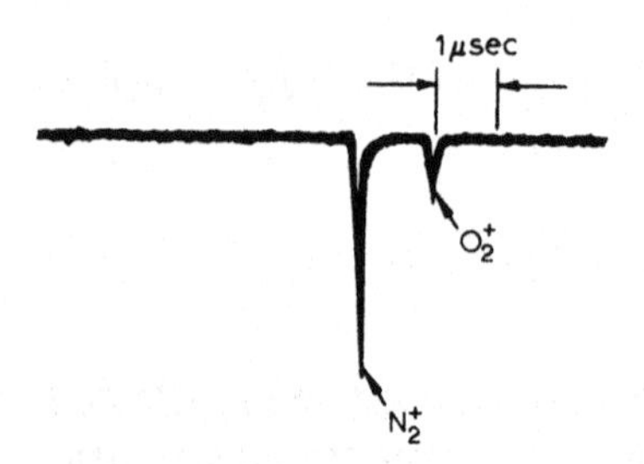

Fig. 9.8. Resolved air peaks.

G. PROPOSED NEW DESIGN

The resolution achieved with the apparatus shown schematically in Fig. 9.7 was limited by the relatively poor approximation to the desired potential distribution [eqn. (9.12)] provided by the ring-electrode structure. The resulting field probably contained non-linear terms which would introduce an energy-dependent arrival time as well as a dependence on $x(\xi_0)$ and $y(\xi_0)$. The electrode configuration [7] shown schematically in Fig. 9.9 is expected to overcome this problem. The cone and hyperboloidal-shaped field electrodes are designed to produce the potential distribution.

$$V(x, y, z, t) = \frac{1}{4L^2}(x^2 + y^2 - 2z^2)(V_{\text{d.c.}} + V_{\text{a.c.}} \cos \omega t) \tag{9.13}$$

where $L^2 = r_0^2/2$. Using the parameters of Fig. 9.2 for this device, namely $f =$ 1 MHz, $M = 50$ amu, $L = 1$ cm, and the operating point $a = -0.228$ and $q =$ 0.759 (i.e. $\epsilon_1 = 0.3$ and $\beta = 0.241$), the injection energy for the z direction

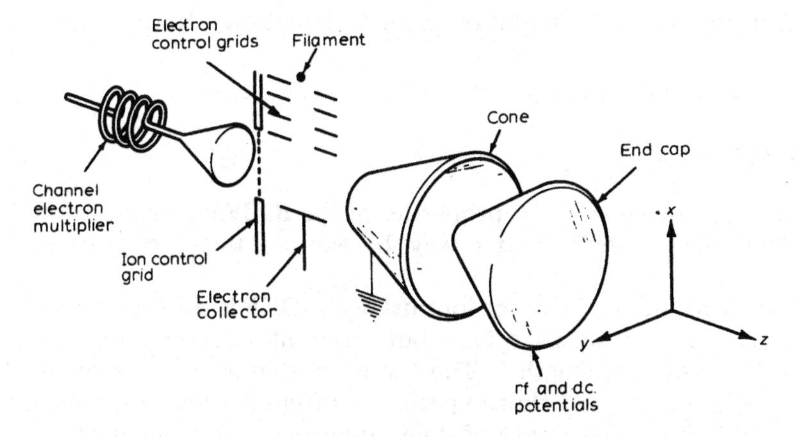

Fig. 9.9. IOEF TOF spectrometer with hyperboloidal electrodes.

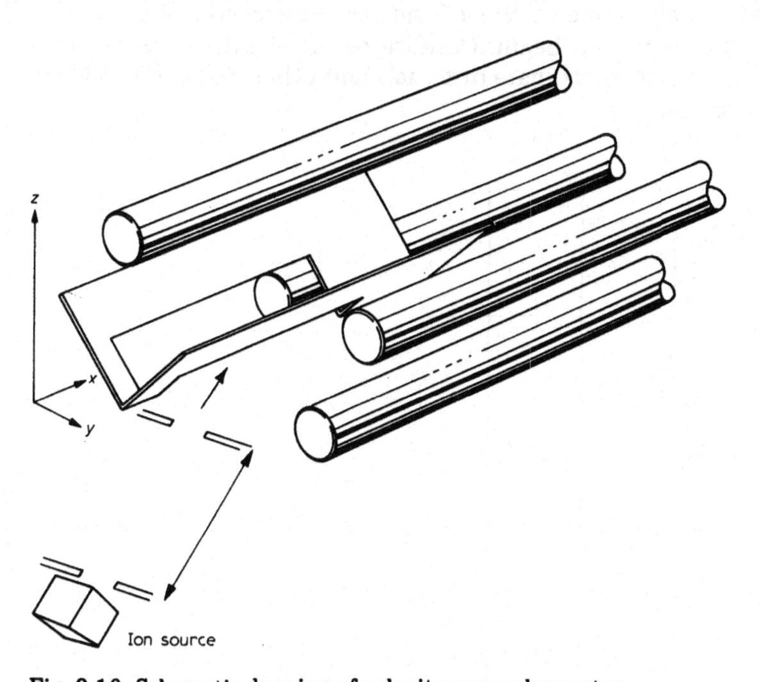

Fig. 9.10. Schematic drawing of velocity monochromator.

must be less than 3 eV for $\xi_0 = \pi/2$ (see Fig. 9.2). From Fig. 9.7, the resolution factor for these operating conditions is 6400. If the resolution is to be 1000, the injection interval must be $\sim 2\,\mu$sec. Assuming a source acceleration energy of ~ 3 eV, this injection interval requirement restricts the thickness of the usable ionization region to 1 mm. Of course, a larger ionization region is possible if we use a higher acceleration voltage for removing the ions from the source region followed by retardation prior to injection into the oscillatory field.

H. NEW MODES OF OPERATION

Several other applications of TOF properties of the IOEF are proposed in this section. Analytical work is under way to explore their potential as feasible devices.

A velocity monochromator [12] for ions using an IOEF is shown schematically in Fig. 9.10. Ions of the same mass but different velocities are pulse injected into a drift region of length L. The fastest ions enter the subsequent oscillatory field region first and thus experience different forces than slow ions entering later. The ξ_0 dependence of the normalized z component of the velocity $\dot{G}_2 = \dot{z}/\dot{z}_0$ at the time corresponding to the first zero ξ_1 of G_2 is shown in Fig. 9.11. By operating on the left side of these curves, it is possible to obtain a monochromatic ion beam. Conditions for simultaneous velocity focusing in more than one coordinate direction and other electrode configurations are being considered.

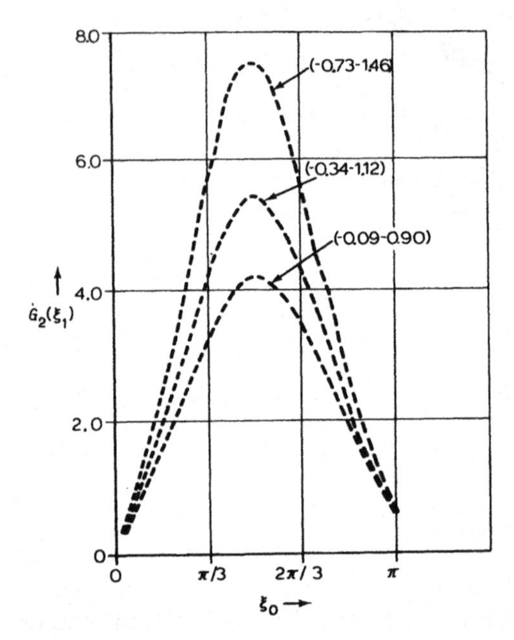

Fig. 9.11. Dependence of normalized exit velocity $\dot{G}_2(\xi_1)$ on ξ_0.

The second proposed application of the IOEF is an ion emission microscope [13]. At first sight it does not appear possible to use the mass filter in

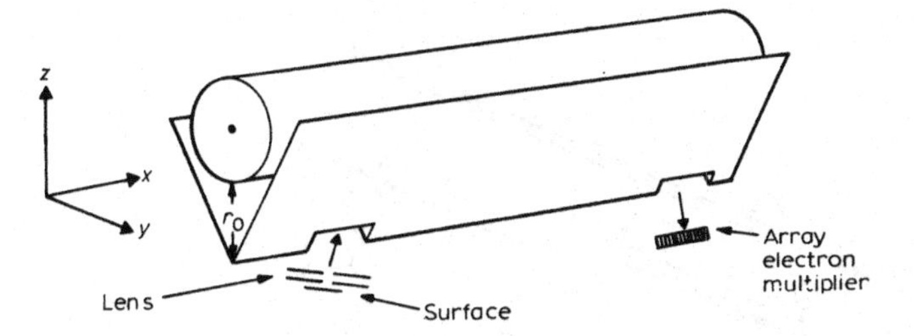

Fig. 9.12. Schematic drawing of proposed ion microscope.

this application because of the "scrambling" of the ion trajectories by the fringing fields. However, the focusing configuration first proposed by Lever (see Fig. 9.12) would alleviate this problem. Several operational aspects being analyzed are given below.

(1) A priori, the requirement for focusing of rational values of the β parameter does not seem justified. Rather, equality of the flight times for the involved directions appears sufficient. On the other hand, rational values for β appear necessary in order that the flight time is independent of the injection phase for proper imaging. This independence requires that $\partial \xi_1/\partial \xi_0 = 1$. Calculations for the IOEF TOF spectrometer show that this condition is met for certain rational values of β. For operating points other than these selected rational values, pulse injection and TOF operation are required.

(2) The requirement $z > 0$ for the monopole electrode configuration restricts the choice of β_z values for operation. However, the combination electrode structure shown in Fig. 9.13 removes this restriction. Ions, injected into the oscillatory field through the apex of a short tripole or monopole structure, enter the adjoining quadrupole structure before returning to $z = 0$. The ions finally leave the oscillatory field region through the apex of a second tripole or monopole structure.

(3a) The energy of injection into the oscillatory field affects performance in two ways. First, this energy is limited by the constraint that the size of an ion's trajectory be less than the finite dimensions of the oscillatory field region. This constraint depends on the operating point and, hence, the mass resolution. It appears possible to relax the energy constraint by electrically isolating the entrance and quadrupole sections and operating the former at low resolution and the latter at the desired resolution. However, this mode of operation is likely to impair image quality unless the ions are pulse-injected into the oscillatory field at a particular phase of the rf, because the necessary condition $\partial \xi_1/\partial \xi_0 = 1$ is not expected to hold for the transition region between the two sections.

(3b) The other aspect of the injection energy problem is the deleterious effect on image quality of the spread δv_x in the x component of the velocity. A simple calculation shows that the velocity spread must be less than 0.05% for a 5 cm long device and 25 μm spatial resolution prior to magnification.

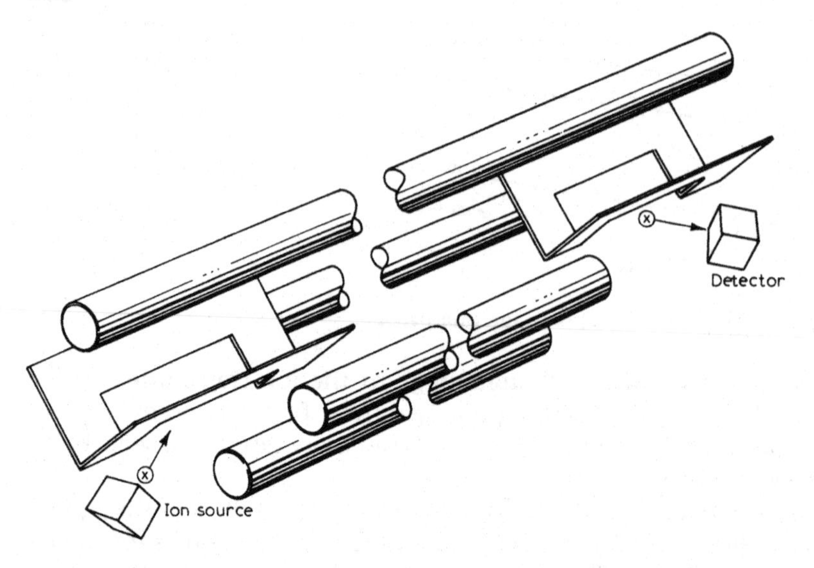

Fig. 9.13. Schematic drawing of combination-structure mass filter.

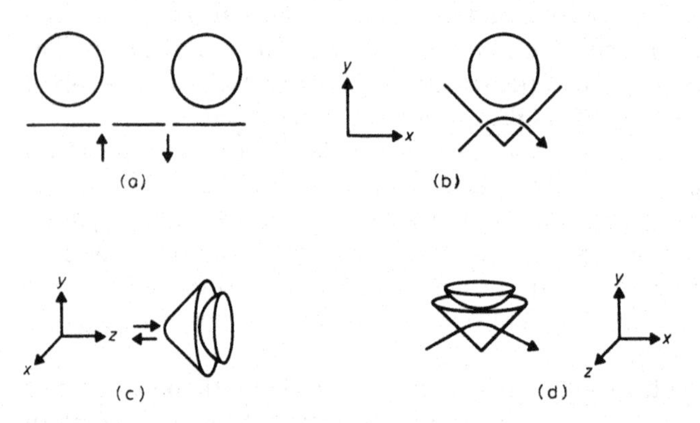

Fig. 9.14. Some proposed electrode configurations for an ion microscope.

This requirement will be very difficult to achieve even with a reduction in the spread through use of a collimator or velocity selector.

As noted in our discussions, it may be necessary to use pulse injection and a concomitant TOF measurement scheme to alleviate potential problems. Also to this end, other configurations such as those shown in Fig. 9.14 are being considered [13, 14].

The third area of new applications of the IOEF pertains to some recently discovered TOF properties of this field [15]. Returning to the complete solution [eqn. (9.1)] for the z component of the equation of motion, we note that for times corresponding to zeros (ξ^*) of the functions G_2, the solution reduces to

$$z(\xi^*) = z_0 G_1(\xi^*) \qquad (9.14)$$

In this case, the position of an ion at time $t^* = 2\xi^*/\omega$ is proportional to its initial position and is independent of its initial velocity. We are investigating the possibility of using this independence in an imaging device. On the other hand, for zeros of the function G_1, the position depends on the initial velocity. This property might be used in an energy analyzer.

The z component of velocity is obtained by differentiation of eqn. (9.1) with respect to ξ. We have

$$\dot{z} = z_0 \dot{G}_1 + \dot{z}_0 \dot{G}_2 \qquad (9.15)$$

For zeros of $\dot{G}_1$, the velocity is independent of the source geometry and for zeros of $\dot{G}_2$, it is independent of the initial velocity. These properties may have application in charged-particle scattering experiments requiring monochromatic particles. Sources having either a large energy or spatial variation might be accommodated.

Realization of the aforementioned properties requires TOF measurements. Also, it is necessary to identify suitable electrode configurations and convenient methods for introducing and removing the charged particles from the IOEF.

I. CONCLUSIONS

Discussions in the previous sections underscore the potential applications of the TOF properties of the IOEF. Although work to date has been mainly analytical, limited experimental success with the IOEF TOF spectrometer has demonstrated the feasibility of this concept. The promise in future work is the realization of relatively high mass spectrometer performance in a small device pending the identification of optimum operating points and electrode tolerances. A great deal of analytical work remains to be performed in this area as well as for the other proposed TOF concepts. Also, there is obviously a need for experimental work to demonstrate feasibility and determine performance limitations.

REFERENCES

1 W.C. Wiley and I.H. McClaren, Rev. Sci. Instrum., 26 (1955) 1150.
2 J.M.B. Bakker, Int. J. Mass Spectrom. Ion Phys., 6 (1971) 291.
3 W.P. Poschenrieder, Int. J. Mass Spectrom. Ion Phys., 9 (1972) 357.
4 For example, see E.W. Blauth, Dynamic Mass Spectrometry, Elsevier, Amsterdam, 1966.
5 J.P. Carrico, L.D. Ferguson and R.K. Mueller, Appl. Phys. Lett., 17 (1970) 146.
6 J.P. Carrico in D. Price (Ed.), Dynamic Mass Spectrometry, Vol. 2, Heyden, London, 1971.

7 J.P. Carrico in D. Price (Ed.), Dynamic Mass Spectrometry, Vol. 3, Heyden, London, 1972.
8 N.W. McLachlan, Theory and Applications of Mathieu Functions, Oxford University Press, New York, 1947.
9 F.M. Arscott, Periodic Differential Equations, MacMillan, New York, 1964.
10 J.P. Carrico, Int. J. Mass Spectrom. Ion Phys., 12 (1973) 309.
11 R.F. Lever, IBM J. Res. Develop., 10 (1966) 26.
12 J.P. Carrico, J. Phys. E., 8 (1975) 717.
13 J.P. Carrico, J. Phys. E., 8 (1975) 18.
14 P.H. Dawson, Int. J. Mass Spectrom. Ion Phys., 12 (1973) 53.
15 J.P. Carrico and R.K. Mueller, Phys. Lett. A, 47 (1974) 427.

CHAPTER X

APPLICATIONS IN ATOMIC AND MOLECULAR PHYSICS

J.F.J. Todd

In this and the three succeeding chapters, an attempt is made to review the principal fields in which quadrupole mass spectrometry has been applied. Probably the most extensive area of application has been that of atomic and molecular physics: it was for this market that the commercial instruments were first introduced and the variety of investigations which have consequently been made possible provides an obvious basis for illustrating the unique features possessed by the mass filter.

The account which follows is divided into two main sections. The first deals with general applications of the quadrupole, in which the instrument is used essentially as an analyser for neutral or ionic species, e.g. the monitoring of residual gases, reaction products, etc.; the fields of vacuum technology, surface studies and gas phase studies are considered in turn. Section B is devoted to an account of the "special applications" of quadrupole fields in which use is made of properties such as ion containment, thereby providing access to the study of ionic processes which might otherwise be impossible.

A. GENERAL APPLICATIONS

(1) *Vacuum technology*

The current utilization of high vacuum techniques ranges from the simulation of space conditions for astronaut training to the freeze-drying of foods. Frequently it is very important to know, at least qualitatively, not only the ultimate pressure but also the composition of the residual gases in the vacuum system both in the laboratory and on a production line basis, particularly in the manufacture of semiconductors and high power valves (e.g. klystrons). The selection of proper pumping equipment for a particular application requires information on the partial pressures of expected gases since the efficiency of most pumps depends on the composition of the gas to be removed. At this point it should be noted that most pressure measuring devices exhibit different sensitivities for various gases and, as a consequence, the total pressure is often expressed as the "nitrogen equivalent pressure" which, in extreme cases, may lead to an order of magnitude error. Another very important area of investigation is the study of adsorption, desorption and permeability for constructional materials, particularly stainless steel and glass.

In principle, any mass spectrometer can be used as a residual gas analyser, whereas, in fact, the quadrupole because of its size and performance parameters, has now virtually supplanted all other instruments in this field. A revealing discussion of partial pressure analysis has been given by Huber [1, 2] in which it is implied, as the most important conclusion, that the analyser must be regarded as an integral part of the vacuum chamber. This consideration includes full bake-out capability with the analyser mounted in the "nude configuration" inside the apparatus. Failing this, as in the case of some magnetic instruments, the interconnection must be by means of a high conductance bakeable tube to reduce pressure differentials.

Comparisons of the performance of the omegatron, farvitron, radio frequency, time-of-flight and quadrupole analysers have been published [3–5], all suggesting that, in the main, the quadrupole is the best general purpose residual gas analyser, although for some more specialized applications they are unsuitable. This degree of utility can be readily understood in terms of the nude insertion availability, relatively fast scanning and the ease with which an electron multiplier detection can be incorporated, leading to a reduction of the minimum detectable partial pressure (MDPP) (see Chap. VI, p. 138).

The quadrupole [6] and to a lesser extent the monopole [7, 8] and ion storage devices (see Chapter VIII) are now frequently used to monitor vacuum chamber atmospheres. The instruments currently employed in RGA applications may be divided into two groups: those used for extremely low level detection incorporating a multiplier, and much smaller instruments with a Faraday cup detector designed to replace the Bayard–Alpert gauge. Both types of quadrupole are now commercially available with the performance characteristics of the larger instrument well documented in the literature [9–13]. The analytical ion gauge being a more recent development has not yet received the same attention [14–16] although the number of applications is potentially quite large [17–19]. The construction and general performance of these two types of instrument are compared in Table 10.1; however, some variation between this data and individual instruments is inevitable.

Attempts to improve the performance of these RGA instruments have progressed along two separate paths: the use of discharge sources to eliminate spurious peaks caused by reaction on hot filament wires and, to a lesser extent, the application of ion counting techniques. Böhm and Günther [20, 21] have reported the use of Penning type discharge sources for a quadrupole, a combination which is useful above 10^{-6} torr but is inoperable in the UHV region. The limit for partial pressure measurements is usually set by the background current in the electron multiplier induced by fast ions, excited neutral particles and short wavelength photons, all of which may be reduced to virtually zero by off-axis positioning of the detector (see Chapter VI).

TABLE 10.1

Comparison of quadrupoles for residual gas analysis

	Mass analyser	Analytical ion gauge
Construction		
Ionizer	Electron impact Penning discharge	Electron impact
Analyser	12 cm stainless steel rods	5 cm rods or tubes
Detector	Electron multiplier	Faraday cup
Performance		
Mass range	0–200	2–70
Resolution*	200	40
Sensitivity	10–100 A $torr^{-1}$	10^{-4} A $torr^{-1}$
Scan rate	1 msec–10 min/amu continuously variable	10 msec–3 sec/amu 3 preselected ranges
MDPP (Scan rate dependent)	3 sec/amu 5×10^{-13} torr 10 msec/amu 5×10^{-11} torr 1 msec/amu 5×10^{-7} torr	3 sec/amu 1×10^{-9} torr 10 msec/amu 1×10^{-7} torr

*Defined as $M/\Delta M$ where ΔM is the peak width in amu measured at half-peak amplitude.

In a counting system with a properly adjusted discriminator the noise pulses can be eliminated, whereas the number of ion pulses measured is to a large extent independent of the multiplier gain and consequently its ageing. Such a method obviously removes the mass discrimination effects occuring when multiplier output currents are measured directly. Very similar work has been reported by Torney and Roehrig [22, 23] and the general conclusion is the extension of the MDPP to the 10^{-14} to 10^{-16} torr range. The latest advances in the use of quadrupoles for residual gas analysis have been reported by Fite and Irving [24, 25] in which an instrument consisting of 8 in. × 3/4 in. rods achieved a resolution of ca. 2500 (10% valley).

An ionization gauge proposed by Schwarz [26] represents a departure from conventional ideas since the operating conditions are such that electrons rather than ions are confined between the rods. Electrodes at either end of the rod structure repel the electrons back and forth, thus increasing the path length. The ions produced are collected by a cylindrical screen surrounding the rod assembly.

(2) *Surface studies*

Quadrupole mass filters have found wide application in the study of interactions between gases and solid surfaces, particularly metals. The apparatus usually involves a bakeable ultrahigh vacuum chamber with the ion source of the analyser mounted in close proximity to the surface under investigation. An experimental technique (Fig. 10.1) in which both the kinetics of

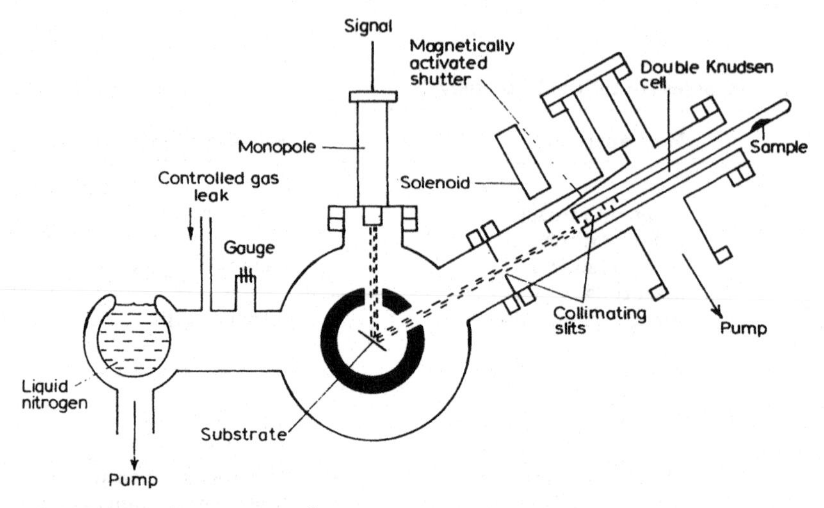

Fig. 10.1. Representation of the apparatus reported by Hudson and Sandejas [27] for the observation of adsorption and crystal nucleation.

nucleation of vapour deposits and certain of the related physical processes, such as thermal accommodation of vapour at the substrate surface, has been described by Hudson and Sandejas [27]. The measurements made included the interactions between cadmium vapour and atomically clean tungsten foil. Flash desorption techniques have been employed in the study of the interaction of ammonia with tungsten surfaces [28], hydrogen with 304 stainless steel [29] and oxygen and hydrogen atoms with silver and gold [30].

The interactions between modulated molecular beams and solid surfaces have been fairly extensively investigated [31–33]. Colthorp et al. [34] have published an account in which a chopped molecular beam of nitrous oxide molecules was fired at a tungsten ribbon filament heated to 1800–2500 K. The analyser, incorporating phase sensitive detection capabilities, could be positioned to receive scattered molecules at different angles with respect to the line of incidence of the nitrous oxide beam. West and Somorjai [35] have described similar experiments between nitrous oxide molecules and clean and carbon-covered platinum (100) single crystal surfaces.

Further stimulation of interest in gas/solid surface interactions has been generated by the advent of such new techniques as electron induced desorption (EID) [36], field ion mass spectrometry (FIMS) [37] and secondary ion mass spectrometry (SIMS). These techniques all fall into the excitation–emission category and are summarized in Table 10.2. The use of a charged species as the stimulant virtually precludes any instrument likely to introduce stray magnetic fields, thus making the quadrupole the ideal analyser. Sandstrom et al [38], in monitoring both ionic and neutral species formed in the electron induced desorption of carbon monoxide and hydrogen from

TABLE 10.2

Excitation/emission techniques for surface studies

Emission	Stimulation				
	Photons	Electrons	Ions	Heat	Electric field
Photons					
Electrons	ESCA	AES			
Ions		EID	SIMS		FIMS
Neutrals				Thermal desorption	

Abbreviations: EID, electron induced desorption; ESCA, electron spectroscopy for chemical analysis; AES, Auger electron spectroscopy; SIMS, secondary ion mass spectrometry; FIMS, field ionization mass spectrometry.

tungsten, noted that the wide angle of acceptance for ions formed outside the source is a major attribute of the quadrupole in this application. Huber and Rettinghaus [39] have employed a quadrupole to investigate the ions formed in the electron induced desorption of water, carbon monoxide and oxygen from platinum, molybdenum and platinum/iridium and have discussed the possible effects such desorption would have on the operation of an ion gauge. The adsorption and desorption of gases on tungsten in particular has received a great deal of attention [40–44] and other metals studied include nickel [45], silver [46] and platinum [47].

The SIMS technique involves the analysis of secondary ions, characteristic of the surface species, which are ejected as a result of bombardment with a beam of fairly high kinetic energy primary ions [48]. The use of quadrupoles [49] is particularly advantageous for this analysis since the performance is largely unaffected by the wide range of energies of the secondary ions, thus replacing the large and expensive double focussing magnetic instruments employed previously.

Experiments employing SIMS can be conveniently divided into two classes [50]. In "static" SIMS which requires a vacuum sufficient to maintain a "clean" surface even under stand-by conditions, high detector sensitivity is needed in order to minimize the primary ion current to ensure as little surface disruption as possible. Static conditions are employed in the study of the SIMS process itself and particularly in investigating oxidation effects [50–52]. "Dynamic" SIMS employs a much higher primary ion current since the sputtering effect is relied upon to maintain a clean surface. This process is usually used to study the composition of surfaces and thin films. The etching away of the surface leads to a depth profile, provided that ions only from the centre of the crater reach the detector [53].

In the static analyses reported by Benninghoven [50, 54, 55] the quadrupole, with off-set multiplier, was placed close to the target to maintain high sensitivity. Crude ion energy analysis was accomplished by means of a pole bias facility enabling the rod assembly to be "floated" between 0 and 30 V. Under negative ion monitoring conditions, a weak magnetic field is employed [51] to prevent electrons reaching the electron multiplier (see p. 140).

Trace analysis capability, desirable in dynamic SIMS applications, can only be achieved by preventing the higher energy secondary ions entering the quadrupole thereby eliminating peak tailing and poor resolution. Huber et al. [56] have described a SIMS apparatus (Fig. 10.2) which utilizes a deflection plate between the quadrupole and multiplier to provide limited energy selection. More sophisticated ion energy analysis has been reported by Wittmaak et al. [57] using a parallel plate capacitor in front of the analyser whereas Schubert and Tracy [58] have employed cylindrical mirror energy analysers, again in front of the quadrupole. A very different approach has been reported by Thomas and de Kluizenaar [59] in which one of the electrodes in the quadrupole ion source was modulated at 770 Hz with a sine wave between 2 and 10 V peak-to-peak. The detector system incorporated a "lock-in" amplifier set to respond to 770 Hz modulation which ignored all the "fast" ions since the applied sine wave was insufficient to affect their motion.

The advantages of the SIMS technique have been enumerated by Benninghoven and Loebach [60] along with accounts of the surface analyses of such widely divergent systems as stainless steel [61], molybdenum foil and

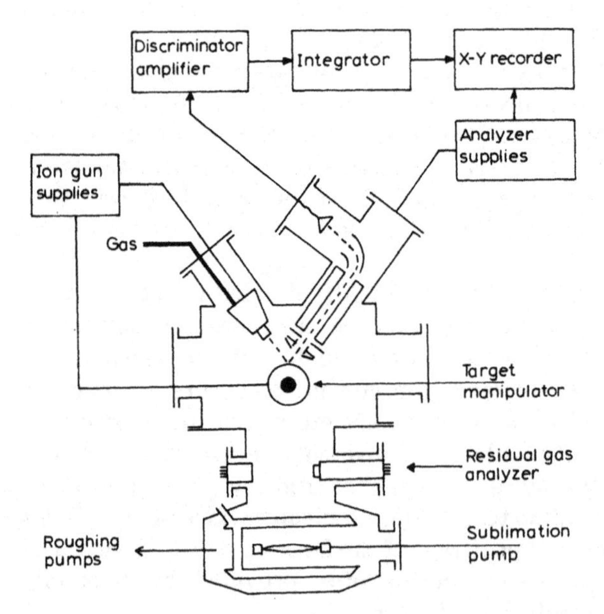

Fig. 10.2. Schematic representation of a typical SIMS apparatus.

crystals [54, 60], surface oxidation of chromium [50], other transition metals [62], and adsorbed species on the surface of crown glass [56]. A sophisticated system incorporating simultaneous SIMS, EID and flash heating techniques has been employed by Benninghoven [63] to investigate the interaction of gases with tungsten surfaces. Dawson [64] has recently provided a comprehensive analysis of the suitability of different kinds of energy analyser when coupled with a quadrupole mass filter in a typical SIMS application using the method of phase space dynamics (see also p. 99).

TABLE 10.3

Performance comparison for AES, ESCA, and SIMS techniques

	AES	ESCA	SIMS
Information depth	15 Å	50 Å	1 Monolayer
Detection of			
elements	Yes	Yes	Yes
isotopes	No	No	Yes
compounds	For special cases	Chemical shift	Yes
hydrogen	No	No	Yes
Sensitivity (typical values)			
monolayers	10^{-3}	10^{-2}	10^{-6}
$g\,cm^{-2}$	10^{-10}	10^{-9}	10^{-13}
variation for different elements	10	10	10^{3}

A comparison of the information and sensitivity available from SIMS, AES and ESCA techniques (Table 10.3) indicates that all are able to supply information on the composition of a solid surface but vary in their average information depth. Elements are traced by all three whereas only SIMS is capable of giving isotopic abundances. With ESCA, chemical compounds can be identified through the chemical shift, and with SIMS by means of either the molecular ions emitted or fragmentation products. Only under certain conditions can the chemical shift in the AES technique give compound analysis. Hydrogen and its compounds can only be detected by SIMS. All the techniques exhibit a high sensitivity, the disadvantage of the SIMS sensitivity being the relatively large variation (three orders of magnitude) for some elements.

(3) *Reaction studies involving solids or surfaces*

A number of investigations have been reported in which a quadrupole mass filter has been used to monitor the partial pressure of the species formed when involatile substances are heated, and to determine some of the properties of these species. Such experiments usually employ a quadrupole fitted with a Knudsen cell and have included continuous analysis of metallic vapours [65], the determination of some of the thermodynamic properties of the vapour species issuing from aluminium [66], and an effusion–mass spectrometric study of the vaporization properties of vanadium and vanadium nitride by Farber and Srivastava [67], who determined the dissociation energy of the latter compound. Other investigations include the determination of the latent heat of vaporization of lead dichloride and rubidium chloride [68], the formation of PbClBr in the vapour phase from a melt of lead dichloride and lead dibromide [69], the activity of lead dibromide in a melt of lead dibromide and potassium bromide [70] (indicating the formation of the $KPbBr_3$ complex) and a study of molten mixtures of lead dichloride with sodium, potassium, rubidium and caesium chlorides [71].

The interposition of an inhomogeneous magnetic field between the Knudsen cell and the ion source of the analyser allowed Meschi and Searcy [72] to determine the magnetic moments of S_2, Se_2, Te_2, Se_6 and Se_5 in the vapour phase by monitoring the variation of peak intensity with magnetic field.

Thermogravimetric–mass spectrometric techniques have been shown to be capable of providing accurate analytical data for the identification of components released from geochemical samples. Some of the systems studied include the analysis of Orgueil carbonaceous chondrite [73], the analysis of an oil shale containing Dawsonite [74] and the investigation of minerals and lunar samples by simultaneous thermal and X-ray analysis [75].

Lever [76] has described the use of a monopole in monitoring germanium oxide (m/e 90) formed in the reaction

$$Ge(solid) + O_2(gas) \rightarrow GeO(gas)$$

between oxygen (10^{-8} to 10^{-5} torr) and germanium heated in the range 500–900°C. The production of germanium oxide was observed to be proportional to the oxygen partial pressure and sticking coefficients were evaluated.

The changes in chemical composition of gases present at pressures of 10^{-9} to 10^{-4} torr above metal oxide catalysts have been measured immediately following exposure to 30 μsec pulses of UV radiation [77, 78]. Time profiles of flash-assisted desorption of oxygen, nitrogen and carbon dioxide have been investigated for zinc oxide and titanium dioxide surfaces. The introduction of methanol, ethanol or methyl iodide to these surfaces prior to flashing yields flash initiated time profiles for species which must have been produced by fast photo-assisted reaction of the components over the illuminated surface [79].

The use of quadrupoles to analyse the products from a low pressure catalytic reactor have been reported by Soloman et al. [80] and Bradley et al. [81, 82] have monitored the decomposition of ammonia flowing over a heated platinum spiral. The reactivity of low index (111 and 100) and stepped platinum single crystal surfaces for reactions between hydrogen and deuterium and the dehydro cyclization of n-heptane have been described by

Somorjai et al. [83]. The catalytic decomposition of hydrazine at low pressures, again on platinum surfaces, to give di-imide has been investigated by Willhoft and Robertson [84].

(4) *Gas phase studies*

We have already seen in the previous sections how quadrupole mass spectrometry has aided in the solution of a wide range of problems involving solids, surfaces and gas–surface interactions. This diversity of application is equalled, if not out-shone, by the enormous wealth of gas phase investigations which have been made possible because of those properties of the quadrupole which generally make the instrument more ideally suited than any other type of mass spectrometer for this kind of work. It is impossible and inappropriate in the present context to give a completely comprehensive review of all the applications in this area, and indeed most specialists will be aware of the numerous texts and articles pertaining to important instrumental techniques in their own fields. Here the aim has been to provide a brief general survey of the uses of mass filters for the analysis of ions and neutral species formed, for example, in flames, afterflows, drift tubes, and crossed beams, and to indicate the particular advantages which quadrupoles possess.

(a) *The monitoring of ions and neutral species formed in swarms and plasma*

One of the currently most fruitful areas of research is the investigation of the reactions of ions formed in gases at pressures where the motion of the species is diffusion-controlled: of particular interest is the determination of rate constants and cross-sections for ion–molecule reactions, cluster formation and ion–electron recombination, all of which are relevant to our understanding of the physics and chemistry of the atmosphere and upper atmosphere.

Early experiments [85] were limited to the measurement of ion mobilities and diffusion coefficients without the provision of means for identifying the species involved, and although a few experiments were reported [86, 87] in which mass spectrometric monitoring was employed, it was not until the mid-1960's that the use of the quadrupole-type mass spectrometer became widely established. The particular advantages of quadrupoles for this type of application are evident from two recent reviews, by Smith and Plumb [88] and by Hasted [89], which provide a critical appraisal of the problems associated with the sampling of ions from afterglows and plasma. Where the bulk of the gas is only weakly electronegative, the principal species in a plasma are positive ions and electrons. This means that at the walls of the containing vessel, the (generally) faster moving electrons will be preferentially discharged, giving rise to a positive space charge which acts to oppose further discharge of electrons. In this way, a "plasma sheath" is established across which is a potential gradient arising from the difference between the wall potential and the potential within the bulk plasma. For the extraction of ions, an orifice must be provided, but with a diameter sufficiently small (a few microns) compared with the thickness of the sheath that the latter does not "collapse" leading to non-uniformities in the plasma. Ideally, the thickness of the orifice should be less than the diameter and be of the order of the mean free path of the gas molecules.

Generally, the orifice plate is electrically isolated from the surrounding walls so that it may be biassed relative to the bulk plasma potential. When ions pass from the field in the sheath (E_1) into the external accelerating field (E_2) the orifice behaves as a single aperture lens of focal length f given by $1/f \simeq -(E_2 - E_1)4V$ where V is the orifice potential relative to that of the bulk plasma. For convenience, E_2 should be small, and this is readily achieved with a quadrupole which has an ion injection energy of a few volts (see Chapter VI), in contrast to a magnetic sector instrument requiring ion acceleration potentials of several kilovolts. Indeed, quadrupole instruments are commercially available in which "pole biassing" is possible, a feature which allows the whole analyser system to operate at a reference potential other than ground.

Following from this there are other features which make quadrupoles advantageous. For example, it is to be expected that without further refinement the simple extraction system described above would lead to an appreciable spread in the kinetic energy of the emerging ion beam. In magnetic sector or time-of-flight instruments, which rely on a velocity term in the equations for ion focussing, an inevitable discrimination and loss in sensitivity will occur. For the quadrupole, however, this consideration is far less critical since there is no exact focussing requirement and, for the resolution normally required, the energy spread is not likely to be sufficient to cause an undesirable variation in the times taken for the ions to traverse the rf field. Related to this is the less critical requirement of the injection geometry for the mass filter resulting in the necessity for only a crude form of ion collimation. Provided the energy and angular spread of the ion beam are contained within tolerable limits, the quadrupole can be operated with a high percentage ion transmission which is essential for sampling the low ion currents employed in drift tube work. The ease which the quadrupole can be switched from positive to negative ion detection is desirable in this type of work and operation in the "total pressure mode" is very useful in the initial setting up of the experimental system.

A further very distinct advantage of the quadrupole is the absence of a magnet, which might otherwise influence the behaviour of the ions under observation, and which results in light weight and compactness. This means that the analyser may be mounted nude within, say, a flow system. Furthermore, the short path length of the analyser means that even at low kinetic energies the flight time is such that ion–molecule scattering collisons are less likely so that a higher working pressure (ca. 10^{-4} torr) may be tolerated.

From the earlier discussion, it may be concluded that the actual sampling of ions from swarms and plasma is a somewhat complex process. For example, if the gas flow through the orifice is too high, the resulting expansion causes cooling which may lead to ion populations, including cluster formation characteristic of a lower temperature than that of the bulk plasma [89]. On the other hand, weakly bound clusters formed in the plasma may be collisionally dissociated in the region of comparatively high pressure near the orifice under the influence of the accelerating field of the lens system, the potential on which should be maintained at such a value that the energy gained by the ions between collisions is less than the binding energies involved [88]. Numerous general accounts of the theory and experimental techniques employed in the study of ion behaviour under swarm conditions have appeared, notably those by Ferguson and co-workers on flowing afterglow techniques [90–93], McDaniel et al. on drift tube experiments [93–95], Kebarle on stationary afterglows [96] and Studniarz [97] on electrical discharges. Parkes has provided a useful survey of the work relating to negative ion–molecule reactions occurring under swarm conditions [98] *.

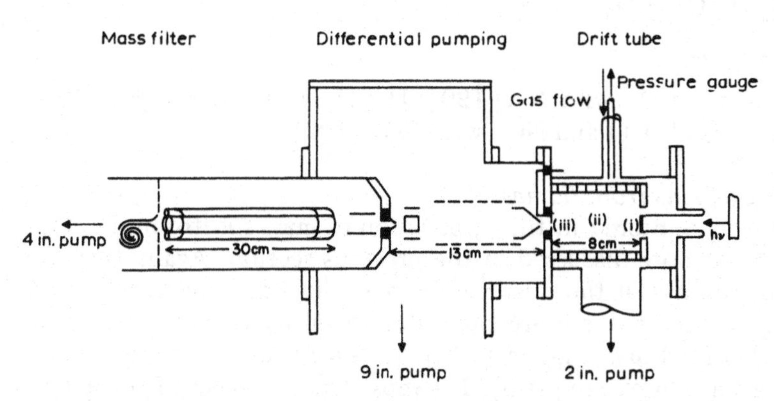

Fig. 10.3. Schematic diagram of the apparatus used by Parkes [99] for studying the behaviour of negative ions.

As an illustation of the techniques employed in ion sampling, we consider the apparatus (Fig. 10.3) described by Parkes [99] for the study of electron attachment and negative ion–molecule reactions in oxygen. Slow electrons formed by emission from a near-illuminated photocathode (i) undergo attachment in the drift tube (ii), operating at a pressure of ca. 1 torr, from which the ions are sampled through a 250 μm hole (iii) in a 12.5 μm thick platinum foil. Before entry to the mass filter, the ions traverse a differentially pumped section (ca. 10^{-4} torr) containing a conical skimmer biassed to + 300 V (this

* The author is indebted to Dr. D.A. Parkes for supplying a copy of his review prior to publication.

preferentially attracts the negative ions and sweeps neutrals aside), a cylindrical lens at + 100 V and a second skimmer at + 40 V. Correct alignment of the beam is achieved by appropriately biassing the deflection plates and further focussed by a cylindrical lens (+ 100 V) positioned immediately before the mass filter, the poles of which are biassed with + 60 V and provided with pre- and post-filter sections to which the rf potential only is applied (see Chapter VI).

The data are obtained as a variation of signal arising from particular ionic species with the reduced field ($V\,cm^{-1}\,torr^{-1}$) in the drift tube, but are subject to corrections arising from possible mass discrimination which can arise from preferential sampling through the hole from the drift tube, ion–molecule reactions occurring on the differentially pumped side of the sampling hole where the pressure is relatively high and mass-dependent transmission of the ion optics and the mass filter.

Using this technique, it has been possible to determine the rates of reactions involving O_3^- and O_4^-, viz.

$$O_2^- + 2\,O_2 \rightleftharpoons O_4^- + O_2$$

$$O_4^- + H_2O \rightleftharpoons O_2 \cdot H_2O^- + O_2$$

$$O^- + 2\,O_2 \rightarrow O_3^- + O_2$$

as well as reactions of O^- with N_2O [99], H_2, D_2, C_2H_2 and C_2H_4 [100], O^{2-} with CO and CO_2 [101] and NO^- with CO_2 [102].

(b) *The sampling of ions from flames*

The problems encountered in this application of mass spectrometry are in many ways akin to those discussed in the previous section. Again, this is an obvious area for the use of the quadrupole mass filter and the first systems to have been described in detail are those developed by Calcote and Reuter [103] and by Hayhurst and Sugden [104]. A general survey of the field has recently been given by Calcote [105]. The most critical aspect of the experimental system relates to possible processes occurring at the sampling orifice, which usually takes the form of a conical "skimming" nozzle. Because the orifice is in contact with the flame, it must necessarily be hot and this can lead to chemical reactions giving spurious products which were not directly formed in the combustion process or to the removal of active species which one might otherwise expect to detect. For example, Morley [106] has reported that in flames containing an excess of oxygen ("lean" flames) the negative ion current as sampled through a nickel cone decreased slowly with time, with the signals from lighter ions falling more rapidly, whereas the positive ion signals remained at a constant level. This was thought to be due to oxidation of the cone causing a negative charge to build up which then opposed the tendency for negative ions to be carried into the hole by the gas flow.

A more serious effect which must be taken into account is the adiabatic cooling which occurs as the sampled gases expand through the nozzle. The rapid decrease in density that occurs is desirable in that the chemical reactions are quenched so that the populations of the various species sampled accord with those formed in the flame, but because many of the intermediates are present in equilibrium proportions, the rate in the endothermic direction falls faster. For example in CO/O_2 flames containing traces of C_2H_2 the equilibrium

$$CHO^+ + H_2O \rightleftharpoons H_3O^+ + CO \qquad \Delta H = -25 \pm 12\,\text{kcal mole}^{-1}$$

is rapidly established [106] so that on sampling, the equilibrium shifts to the right-hand side making the observed ratio $[H_3O^+]/[CHO^+]$ greater than that in the flame. It has been shown that this effect [106] leads to an apparent variation in the equilibrium constant as the concentration of water in the flame is changed. The related problem of the hydration of ions during expansion through a hole [see also Section A.(4)(a)] has been considered in detail by Hayhurst and Telford [107].

(c) *Beam studies*

One of the most exacting applications of a mass spectrometer is as a detector for the products formed in the reactions of ions or neutral species collimated into well-defined beams. The experiments may take the form of a single beam of ionic or neutral species projected into a scattering chamber, or merged or crossed with a further beam of neutral molecules or even photons.

(i) *Ion–molecule reactions.* A comprehensive review of this rapidly expanding field has been presented by Herman and Wolfgang [108]. The particular advantages of quadrupoles in this application are the high percentage ion transmission and the non-critical dependence on energy spread in the ion beam. For scattering chamber experiments, the requirements for high efficiency differential pumping are also less stringent. An early single beam system incorporating a quadrupole mass filter but providing velocity analysis only was described by Ding et al. [109] and a sophisticated version of this general method, "the guided beam technique", employing rf octopole fields, is described in Section B.(1)(b) (see p. 258).

Further information on the dynamics of the reaction may be gained from an angular analysis of the ionic products and this was achieved by rotating the mass filter with respect to the scattering centre, in the systems described by Champion et al. [110] and by Mahan and co-workers [111].

Crossed-beam techniques, although requiring a greater detector sensitivity and involving other experimental difficulties, become essential for studies at low relative kinetic energies, and results using mass filters with such systems have been obtained by Turner et al. [112] and by Fink et al. [113, 114].

Two recent accounts, by Klein et al. [115] and by Vestal et al. [116] both describe crossed-beam apparatuses employing a magnetic spectrometer for primary ion selection and a quadrupole for charged product analysis. In each case, product energy analysis is provided, but the former employs a "monoplasmatron" ion source of the type described by Menzinger and Wåhlin [117] whilst the latter system incorporates a high pressure (ca. 1 torr) chemical ionization source such that the primary ions are collisionally de-excited before reaction.

(ii) *Neutral beam reactions.* The difficulties encountered in this type of study are, in general, even more severe than those experienced in ion beam research; a detailed account of the theory and techniques involved has been presented by Fluendy and Lawley [118]. A major problem which arises is that, even though a beam of neutral species may be detected by secondary electron emission from a suitable dynode, qualitative analysis is only possible if some form of ionization process is involved. Initially, products and reactants were distinguished by selective surface ionization, in which readily ionizable species, such as alkali metal atoms, lost electrons on collision with a pretreated heated wire and gave rise to positive ions which were collected on an adjacent electrode [119]. Subsequently, Herschbach and co-workers [120] replaced the simple ion collector by a quadrupole mass filter which not only provided mass analysis but, through the use of an electron multiplier, also improved the senstivity by a factor of ca. 10^3, the whole system having a detection efficiency of ca. 50%.

Conventional electron impact ionization has also been employed, but this brings with it a number of problems: the ionization efficiency is generally low so that some form of oscillating electron source [121] is desirable and the background gases may also be ionized so that modulation of one or both of the primary neutral beams is required [122]. McFadden et al. [123] have recently described experiments in which chopped molecular beams of CH_3 or C_2H_5 were crossed with Cl_2, Br_2, I_2 and ICl, the effect of background ionization being reduced by surrounding the Brink-type ionizer with a liquid nitrogen cooled liner, and a detailed description of a modulated beam system incorporating a quadrupole mass spectrometer has been given by Jenkins and Voisey [124]. A further difficulty is that the detection efficiency will be related to the rate at which the neutral species pass through the ionization region: very fast neutrals may escape ionization altogether and have to be removed by placing an electrostatic deflector between the source and the analyser [125].

The measurement of molecular flow rate (beam intensity), flow density and velocity distribution has been described by Heald and Powell [31, 126] and Siekhaus et al. [127] have noted that the extraction of ions from a cross beam ionizer into a quadrupole depends upon the velocity of the molecules before ionization. Fluendy and Lawley [118] have given an extensive treatment of the theory of detection systems under various conditons of operation.

(iii) *Photofragment spectroscopy.* Quadrupole mass spectrometers have also found use in the relatively new technique of photofragment spectroscopy [128–130]. Here a molecular beam is crossed with a (polarized) photon beam having sufficient energy to photodissociate the molecules. The fragments formed are then identified by analysis with a quadrupole mass filter placed orthogonal to the plane containing the two incident beams. A velocity analysis of the products is achieved through a time-of-flight technique which involves relating the arrival times of particular products to pulses of light in the photon beam, and from this the partitioning of the relative kinetic energy between the recoiling fragments is determined. A further refinement of the technique is to examine the product yields and energetics as a function of the rotation of the plane of polarization of the light in order to gain information on the orientation of the transition dipole in the molecular frame. To date, the method has been restricted to studies on relatively simple molecules such as I_2, Cl_2, NO_2, ICN, NOCl and alkyl iodides.

(d) *Photoionization and photodetachment studies*

The relatively non-critical ion entry conditions for the mass filter together with its high sensitivity have made the instrument a natural choice as the analyser for processes involving the photo-ejection of electrons from molecules and ions. For example, Jones and Bayes [131] have been able to study electronic energy transfer from $^{32}O_2(a\ '\Delta g)$ to $^{36}O_2(X\ ^3\Sigma \bar{g})$ by using the fact that argon resonance radiation will ionize only the former species. Selective photoionization of NO by krypton resonance radiation was used by the same workers to analyse the products of the photolysis of nitrogen dioxide [132], and also by Bone [133] who has examined cluster formation by NO^+ created in a mixture of less easily ionized molecules such as H_2O, CH_3OH and H_2S. The ionization limits, as indicated by the use of argon, krypton and xenon resonance radiation, were used in the detection of steady state free radical concentrations during the oxidation of hydrocarbons [134]. Truby [135–137] has employed pulsed photoionization in He/I_2 and He/Br_2 mixtures as a means of forming electrons and complex positive and negative ions whose time-dependent behaviour could then be monitored.

The determination of the electron photodetachment cross-sections of negative ions is an important adjunct to the study of the physics and chemistry of the upper atmosphere. One experimental system utilizing a quadrupole mass filter as the analysing element has been employed by Burt for the measurement of the photodetachment cross-sections of O_2^-, [138], CO_3^- and $CO_3^- \cdot H_2O$ [139] and O_4^- [140] at pressures of ca. 3 torr. The ionic species from a glow discharge are admitted to a differentially pumped chamber which contains a beam of xenon resonance radiation brought to a focus immediately in front of a skimmer having a 0.1 mm diameter sampling orifice for a separately pumped mass filter. With the spectrometer output phase-locked to the incident chopped photon beam, the reduction in ion intensity resulting from detachment can be monitored directly.

(e) *Electron impact induced ionization and excitation*

Although the sources of the earliest quadrupoles employed for residual gas analysis tended to have somewhat crude electron guns, mass filters generally have certain advantages over magnetic instruments for the determination of ionization efficiency data. For example, if automated scanning of the electron energy is required, the high potentials necessary for operation of sector mass spectrometers necessitates the use of complex interfacing circuits [141], whereas quadrupole sources operate at very low potentials and external control systems can be coupled directly. A further advantage is that the increased sensitivity of the mass filter permits the use of lower electron beam currents. Morrison and co-workers have described an elaborate system [142] in which a computer was used to generate the electron energy scan and provide direct time-averaging and deconvolution of the output data. The samples studied include the rare gases [143], H_2O and H_2S [144], NH_3 and PH_3 [145] and CH_4 and SiH_4 [146]. In addition to the above-mentioned attributes, the authors [142] cite as further advantages of the quadrupole for this application, the absence of discrimination of the ions caused by initial kinetic energies; the trapezoidal peak shape, enabling stable ion currents to be maintained; the low ion extraction energies required, which reduce perturbing affects on the electron beam; and the relative cheapness of the instrument. Marchand et al. [147] have described a crossed beam system comprising a cylindrical electron energy selector and a large quadrupole (length 1.1 m, $r_0 = 2.0$ cm, $\omega = 1$ MHz) capable of accepting ions with radial energies of up to 1 eV. Ion counting was used with a specially constructed large aperture multiplier.

There has been a limited number of other applications in which quadrupoles have been employed in the investigation of excitation and dissociation processes resulting from electron impact. Thus Fite and co-workers have described experiments on the polarization of Lyman alpha radiation emitted in electron collisions with hydrogen atoms and molecules and the excitation of atomic hydrogen to the $2(2)S^{1/2}$ state by electron impact [148]. Of more immediate interest to mass spectroscopists is the recent system described by Reeher and Svec [149] in which a pulsed dual ionization chamber source is used in conjunction with a quadrupole mass filter for the identification and characterization of neutral fragments produced during the electron bombardment of complex molecules.

B. SPECIAL APPLICATIONS

In this section, we examine those applications in which quadrupole-type devices have been employed for purposes other than the simple detection and identification of products. In the main, these have been experiments in which there has been a need to "guide" a beam of ions between a widely separated source and detector or to confine an ensemble of charged particles in order to permit the observation of the properties of such systems.

(1) *The use of rf fields for "beam guides"*

(a) *The study of metastable ion lifetimes*

Because the mass filtering action of the quadrupole is based on the path stability of ions with a given m/e value, unlike the momentum focussing which obtains in magnetic sector spectrometers, the decay of "metastable" ions between the source and the analyser is not manifest as broad peaks falling at non-integral mass numbers, as in spectra recorded on the latter type of instrument. Such ions would be transmitted with the normal "daughter" ions but possibly with a higher resolution resulting from their lower velocity. In the mass spectrometric mode, ions formed through dissociation during the flight through the mass filter would simply be lost; however, in the total pressure mode, provided that both the parent and daughter masses fall within the pass band of the system and that the kinetic energy released is small, then both types of ion are retained within the beam. For this reason, von Zahn and Tatarczyk [150, 151] used a 314 cm-long quadrupole ($r_0 = 1.5$ cm), powered with an rf potential only, oscillating at 2.6 MHz, as a beam guide positioned between a combined ion source and small Nier–Johnson double focussing sector mass spectrometer, and a second quadrupole acting as a mass analyser (Fig. 10.4). A mass-selected monoenergetic beam of ions from the source, such as m/e 72 from n-pentane, was passed through the beam

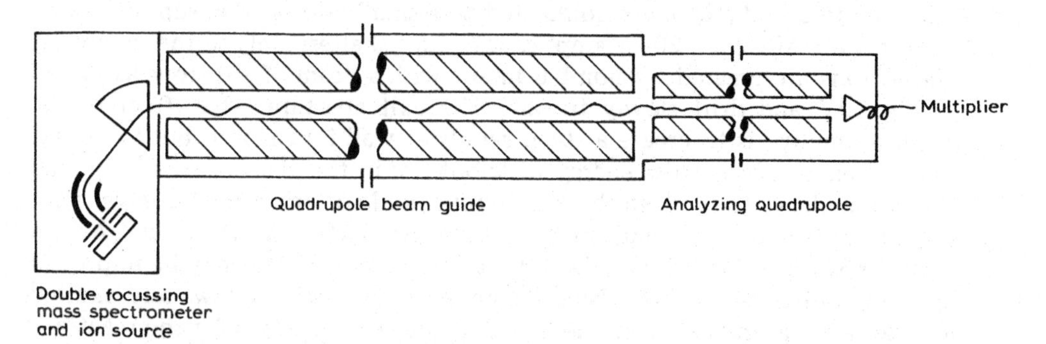

Fig. 10.4. Schematic diagram of the apparatus employed by von Zahn and Tatarczyk [150] for the study of long-lived metastable ions.

guide and the relative abundances of the parent and daughter ions, e.g. m/e 56 and 42, recorded on the analyser. Defining $t = 0$ as the time of ionization, with t_1 and t_2 as the times at which ions left the source and entered the analyser, respectively, it was possible to determine the percentage decay of the parent ion into the daughter species over time intervals $(t_2 - t_1)$ in the range 0.1–1 msec. A correction had to be made for further ion decay occurring during the flight time $(t_3 - t_2)$ through the analysing spectrometer, and a check was made to assess the importance of collision-induced metastable decay processes. Using this technique, the authors found that for selected primary species formed from *n*-butane, *n*-pentane and *n*-hexane, between 10 and 20% decayed during the interval 10–150 μsec after ionization.

(b) *Determination of the integral cross-sections for ion–molecule reactions*

The use of quadrupole mass filters as detectors in the study of ion–molecule reaction rates has already been discussed in Section A.(4)(c)(i). Here we consider an experimental system, developed by Teloy and Gerlich [152] in which the confining properties of quadrupole electric fields have been used to overcome some of the disadvantages associated with conventional systems employing tandem mass spectrometers. These problems include the fact that at low energies space charge,and spurious electric and magnetic fields inhibit the preparation of stable, well-defined intense ion beams; a fraction of the primary ions may be in excited electronic, vibrational or rotational states; and the collection and detection efficiencies of the secondary ions may depend on their mass, energy, scattering angle and position in the scattering region where they were formed. The modified system, shown schematically in Fig. 10.5, incorporates inhomogeneous rf oscillatory electric fields at three separate points: an ion storage source, a mass and velocity filter and an rf octopole to guide the ions through the collision chamber. The ion storage source, which is shown in greater detail in Fig. 10.6 consists of a stack of 10 molybdenum plates into which U-shaped holes have been cut, and two solid end plates. Three of the central plates are slotted so that the ions may drift out. The two end-plates are connected to a small positive d.c. supply and an rf supply (10 MHz, 30–200 V amplitude) coupled alternately to the remaining plates. Ions are created by bombardment with a beam of electrons as shown in Fig. 10.6 and drift slowly towards the exit aperture. The effect of the above geometry is to create a steep repulsive wall for the ions but to impart to the ions a very narrow energy distribution; in the absence of the applied rf, no ions leave the cell. In the continuous mode of operation, storage times range from 0.1 to 1 msec and for beam currents of 10^{-12} A for Ne^+ and Ar^+ an energy resolution of 0.1 eV (FWHM) can be obtained: this may be improved by using pulsed operation. Using a source of this design allows any excited ions to relax, as indicated by the fact that the cross-sections for the reactions $N^+ + CO \rightarrow C^+ + NO$ and $H_3^+ + He \rightarrow HeH^+ + H_2$ were found to depend on the source pressure and residence times of the primary species.

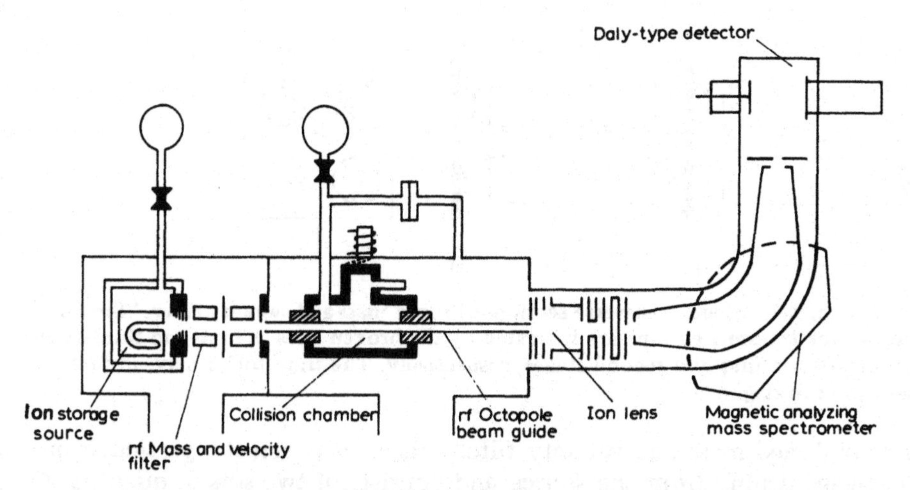

Fig. 10.5. Schematic diagram of the "guided-beam" apparatus employed by Teloy and Gerlich [152] for the measurement of integral cross-sections for ion–molecule reactions. Ions emanate from the storage source (see Fig. 10.6), are selected by the rf mass and velocity filter (see Fig. 10.7) and are guided through the collision chamber with an rf octopole before being analysed in the magnetic sector instrument.

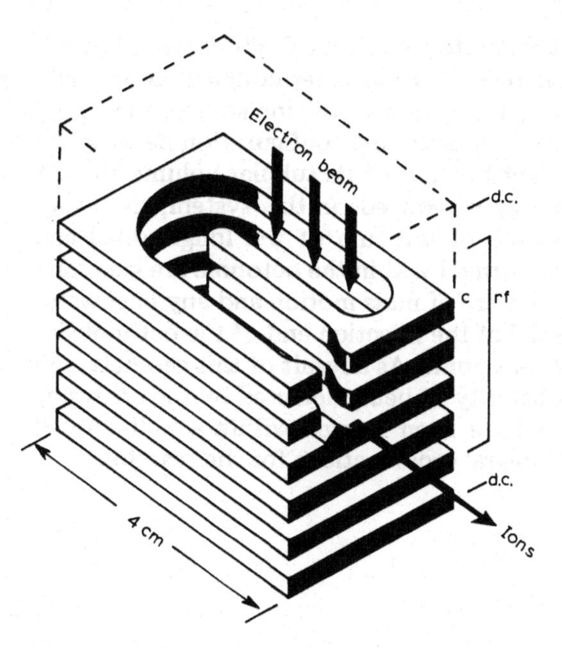

Fig. 10.6. Electrode arrangement of the storage ion source (simplified). The device comprises 5 plates positioned above and 6 plates mounted below the "central" plate c. The end plates (d.c. only) have slits for the electron beam instead of U-shaped holes; the two outputs of an rf supply are connected alternately to the remaining plates.

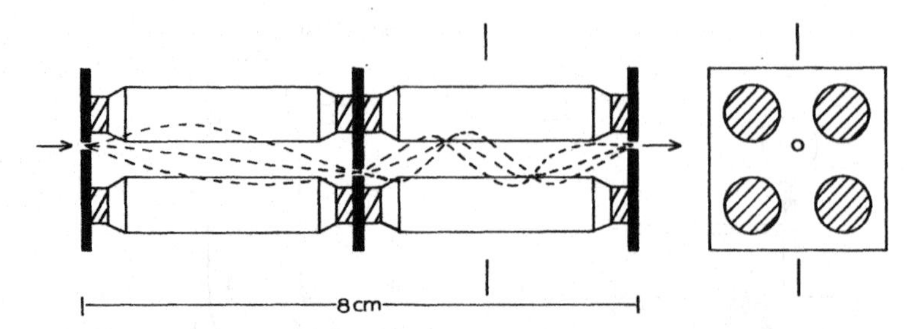

Fig. 10.7. Longitudinal and transverse sections of the rf mass and velocity filter. "Smooth" trajectories for transmitted ions are indicated by the broken lines as first- and third-order trajectories in the first and second stages, respectively. The trajectories may, in fact, be the same in both stages.

The combined mass and velocity filter acts so as to select a particular primary species issuing from the source and consists of two sets of quadrupole rods coupled in tandem but separated by an off-axis aperture, (Fig. 10.7). The device makes use of mass-dependent focussing (cf. Chapters II and III) rather than path stability, and ions enter and leave the filter at off-axis positions having described discrete numbers of secular oscillations (which may be different) in the two filters.

The rf octopole consists of eight accurately parallel cylindrical rods equally spaced in an octagonal array with neighbouring poles coupled to opposite phases of the oscillating supply potential. As with the ion source, the use of a multipole array was preferred over a quadrupole configuration because the oscillatory field is weak over a larger portion of its tubular volume so that the distribution of the kinetic energy is reduced: in this system, the transverse kinetic energy was less than 3% of the sum of the longitudinal and transverse energies. Secondary ions formed within the octopole are generally directed forward according to the centre of mass motion and any backward-moving ions are likely to be reflected at the injection end of the octopole by the electrode potentials applied at that point. As a result of this geometry, all the secondary ions have a high probability of being detected by the analysing mass spectrometer (magnetic) and have been formed within a well-defined region of the scattering chamber. Integral cross-sections for the reactions

$$Ar^+ + D_2 \rightarrow ArD^+ + D$$

$$Ne^+ + CO \rightarrow C^+ + Ne + O$$

$$\left.\begin{aligned} H^+ + D_2 &\rightarrow D^+ + HD \\ &\rightarrow D^+ + H + D \end{aligned}\right\}$$

$$\rightarrow HD^+ + D$$

$$\rightarrow D_2^+ + H$$

have been measured [152, 153] using this technique.

(2) *Applications of charged-particle traps*

(a) *Experiments with microparticles and droplets*

We have already seen in Chapter VIII that early experiments with illuminated stored microparticles provided confirmation of the theory of operation of three- and six-electrode ion traps. In general, however, little practical use appears to have been made of this technique. One application, using the six-electrode trap, has been described by Haught and Polk [154] and relates to the dissociation of individual 10–20 μm diameter particles of lithium hydride by pulsing them with a 20 MW Q-spoiled ruby laser. This gave rise to a high temperature, high density plasma, the rate of expansion of which could be measured by positioning charge collectors along different directions, relative to the axis of the laser beam. In this way, it was estimated that there was total ionization of all the atoms in the original particle, and that the mass and energy densities of the expanding plasma were isotropically distributed, with plasma energies in excess of 100 eV. The use of a three-electrode trap for the storage of single aluminium particles of 25 μm diameter has been reported by Waniek and Jarmuz [155]. In these experiments, the object was to explore the use of laser evaporation to cause acceleration to meteoritic velocities. To do this, the individual particles, charged negatively to $e/m =$ ca. 3×10^{-6} C g^{-1}, were held by the trap such that the amplitude of the motion was less than the diameter of the focal spot of the beam from a pulsed ruby laser operating at 6943 Å. By photographing the trajectories illuminated through the use of second, helium–neon CW, laser the terminal velocity after each ruby laser pulse could be calculated. Having allowed for the mass of the particle lost by evaporation the resulting vapour velocity was computed to be $(4.2 \pm 1) \times 10^4$ cm sec^{-1}. In order to achieve a comparable result by electrostatic means, one would need a 10 MeV accelerator.

A totally different area of application which has relied on methods of particle suspension similar to those described above, has been in cloud physics, where interest is centred on the stability of charged droplets [156], evaporation and condensation phenomena [156] and thermal effects [157]. The apparatus used in this study [158, 159] was extremely simple in design and consisted of two 0.75 in. diameter spheres, between which the oscillating potential (60 Hz) was applied, mounted vertically 10.1 mm apart within a 1.75 in. diameter cylinder. Glass windows were provided for illuminating and viewing the droplets which were charged by injection from a syringe needle maintained in a strong electric field.

The motion of the droplets over quite long time periods was recorded using high speed photography and the rate of evaporation monitored by noting the change in diameter of the droplet with time or the upwards drift of the particle resulting from the loss in mass. From a study of the temperature dependence, an activation energy for evaporation equal to 7.6 kcal mol^{-1} was deduced. When the radius, r, of the droplet carrying charge q reached a critical value such that a stability criterion of the form $\gamma > q^2/16\pi r^2$ (γ = surface tension) no longer held, the particle exhibited a sudden downwards motion corresponding to a loss of charge. Under normal conditions no new droplets were observed, but in the presence of saturated water vapour, condensation occurred suggesting that the charge lost by the unstable particle was some form of gaseous ion. Other relevant studies included the rate of heat exchange between strongly charged droplets and the surrounding air and interaction of a silver iodide particle and moist air.

(b) *Radiofrequency spectroscopy of trapped ions*

We have seen from Chapter VIII that the trap may be used for the study of bimolecular processes involving ions; with a suitable colliding partner, such interactions may be employed to study the spectroscopic properties of stored ions. The most comprehensive use of this technique has been by the group at the University of Washington and Dehmelt [160] has presented a survey of this work. Both radiofrequency and static (Penning-type) three-electrode traps have been employed.

One set of experiments using an apparatus described by Dehmelt and Major [161, 162] was based upon spin exchange with a polarized atomic beam. Thus He^+ ions were created in an rf trap at 3×10^{-8} torr by an 80 msec duration pulse at a repetition rate of 1 Hz which was fast compared to the mean lifetime (measured by a resonance technique) of the ions in the trap under those conditions ($\gtrsim$ 8 sec). However, when a beam of polarized caesium atoms, created by optical pumping with circularly polarized light from a caesium resonance lamp, was admitted, the resulting collisions quickly ($\sim$ 100 msec) caused the stored ions to assume a polarization through

$$Cs\uparrow + He^+\downarrow \rightarrow Cs\downarrow + He^+\uparrow$$

In addition, rapid He^+ ion-loss processes occurred, principally through the spin-dependent charge transfer process

$$Cs + He^+ \rightarrow Cs^+ + He^* + \Delta E$$

with He^* in the 2^1S_0 or 2^3S_1 states.

A charge-exchange collision of an He^+ ion with a Cs atom in the $m_s = +1/2$ state was considered as being equivalent to an observation of the He^+ ion spin state, the difference in He^+ ion loss rates depending upon whether the ions had spins parallel or antiparallel to that of the atoms. In practical terms, therefore, the measurement consisted in observing the difference (about 2%) in the He^+ ion loss rate when the Cs beam was polarized and when the polarization had been destroyed by magnetic resonance disorientation.

The hyperfine magnetic resonance spectrum of the hydrogen-like species $^3He^+$ has been observed by Schuessler et al. [163] using an improved version of this technique: a magnetic field of 7.13 G was maintained across the trap by Helmholtz coils and microwave power in the frequency range 8 GHz fed through a $\lambda/2$ slot antenna cut radially into one of the cap electrodes. The signal-to-noise ratio of the ion counting signal was improved by applying an auxiliary oscillating potential of frequency $(\omega_z + \omega)$ between the end-cap electrodes and monitoring the signal induced in the detection circuit tuned to the secular frequency ω_z. The measurements were then taken by comparing the rate of decay of $^3He^+$ with the microwave power on and off resonance.

A spin-dependent excitation process has also been studied in a spin-resonance experiment by Graeff et al. [164] using a static trap of the Penning type (see Chapter VIII). Electrons were held in the trap and allowed to interact with a beam of polarized sodium atoms (created by passage through a field generated by a 6-pole magnet). As with the previous studies on He^+, rapid spin-exchange occurred between the electrons and the alkali metal atoms, but this was followed by the spin-dependent excitation process

$$e + Na(3s) \rightarrow e + Na^*(3P) \qquad \Delta E = +2.1\,\text{eV}$$

As a result, electron cooling occurred, and this was monitored by momentarily reducing the applied d.c. voltage to a lower value so as to decrease the depth of the trapping well for the faster electrons and then, with the d.c. restored to its original value, determining the remaining number of trapped electrons using auxiliary excitation at the frequency ω_z. The cyclotron resonance and spin resonance lines were then observed by noting the effect of radiating power into the system at the appropriate frequencies. This technique has subsequently been extended to the measurement of the anomalous magnetic moment of free electrons utilizing an rf magnetic field gradient.

In a more recent report, Church and Mokri [166] have described a different approach to the study of free electron resonances which, whilst making use of the interaction with polarized atoms as above, employs a different "bolometric" technique. This provides a means of monitoring the temperature of stored electrons [167] or ions [168] by observing the noise induced in a suitably coupled resonance-tuned circuit by the ion motion along the z axis of the trap, a process which may also be employed to "cool" the ion motion by dissipating the energy as Joule heating of a resistor (see also Chapter VIII, Section C.(3)].

A variant on the method involving collisions with polarized atomic particles was the study of the rf spectrum of H_2^+ which was monitored by the selective dissociation with linearly polarized photons [169, 170] to form H^+, a process which depends upon the angle between the electric vector and the internuclear axis. Initially, the ions were detected by resonant absorption, but in a refined experiment to observe the rotation hyperfine spectrum of H_2^+ [171], the ionic species were monitored by selective ejection and counting by an electron multiplier.

(c) *Electron–ion recombination studies*

With the increasing interest in the rates of reactions of ions present in the upper atmosphere (cf. Section A.(4)(a)), measurement of the cross-sections for ion–electron recombination has become essential in order to quantify the charge-removal processes and provide realistic data for use in models of the overall system. A new approach to this has been developed by Walls and Dunn [172] using the static trap described in Chapter VIII, Section C.(3) [173] in a method similar to that reported by Byrne and Farago [174]. The ions under investigation, NO^+ or O_2^+, were created by low energy electron bombardment at pressures of ca. 10^{-10} torr. At an ion density of ca. 10^6 cm^{-3}, a Maxwellian distribution of energy was established within approximately 10 msec and the number of ions remaining in the trap was monitored by the noise power technique [167] over a period of 500 sec, sufficient for excited states to decay. This established the "natural" rate of loss of the ions. A beam of electrons, variable over the energy range 0.045–10 eV, was then admitted from one of two guns, the "high resolution" model giving an axial distribution 0.03 eV wide, and the increased rate of decay of the ions monitored. The cross-section, σ, for recombination at each energy was then evaluated from a knowledge of the electron current, its duration, the number of ions in the trap before and after introducing the electrons and the natural rate of ion decay. The data were then plotted in the form of σ vs. electron energy, or rate coefficient as a function of temperature; discontinuities in the former were taken as evidence that despite the lengthy initial decay period a number of final state channels appear to be involved in recombination.

(d) *Heavy ion plasma confinement*

In providing the atomic frequency standard the aim is to employ the hyperfine resonance exhibited by an atomic system isolated as far as possible from perturbations arising from the surroundings. The ion trap appears to have advantages in this respect, and one possible standard is the 40.7 GHz line emitted from $^{199}Hg^+$. It is clearly desirable to optimize the signal by maximizing the density of the trapped ion, and one possible way of achieving this seems to be through the reduction of space-charge repulsion [see Chapter VIII, Section E.(3)] by the simultaneous storage of positive and negative ions. In an exploratory study, Schermann and Major [175] investigated the simultaneous storage of Tl^+ and I^- ions produced by the photodissociation of thallium iodide. The appropriate stability diagram for this system is shown in Fig. 10.8. In this plot, the a, q values for I^- have been reduced by the ratio of the atomic masses m_I/m_{Tl}, and the a values reversed in sign, whence it can be seen that in order to confine the two species simultaneously the choice of operating points is restricted to the shaded area of overlap. Experimentally, the ions were detected by resonant excitation of the secular motion and the rate of ion production was evaluated from separate measurements involving a UV iodine vapour lamp and a specially designed ionization chamber. Ion

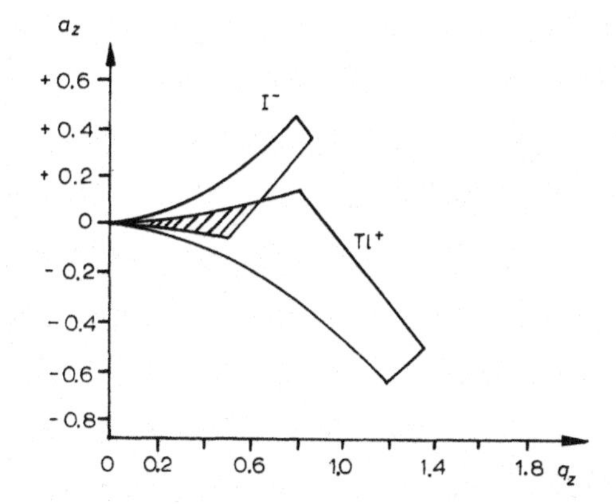

Fig. 10.8. Composite plot of the stability regions for Tl^+ and I^- ions in an rf quadrupole trap. The numerical values of a_z, q_z refer to the Tl^+ ions only, the co-ordinates for the I^- region being scaled by the factor (m_{I^-}/m_{Tl^+}). The shaded area then represents the conditions under which both species should be simultaneously stable.

decay experiments with the trap were then performed using intermittent radiation at a thallium iodide pressure of 4×10^{-7} torr in the presence of helium at 10^{-4} torr acting as a buffer. A comparison was then made with the rate of loss of Tl^+ ions when both Tl^+ and I^- ions were stored and when only Tl^+ ions were stable. The measurements indicated that for simultaneous storage both the Tl^+ and I^- ion densities decreased at the same rate, suggesting that recombination was the most important ion-loss mechanism, but nevertheless under these conditions the density of Tl^+ was higher than when it was the only species in the trap.

(e) *The quadrupole ion storage source (quistor)*

We have already seen in Chapter VIII [Sections B.(4) and E.(3)] that the three-electrode trap may be employed as a storage source in conjunction with a conventional mass filter. This system has been used by the University of Kent group in the study of unimolecular and bimolecular ionic processes [176]. For example, at low pressures the decay of metastable ions may be monitored by varying the delay time between ion creation and ejection for the reactions

$$CF_3^+ \rightarrow CF_2^+$$
$$\rightarrow CF^+$$

occurring in hexafluoroethane. In this experiment, the trap is behaving in a manner analogous to the quadrupole "beam guide" employed by Tatarczyk and von Zahn [150] which has already been described [see Section B.(1)(a)]. If a similar experiment is performed but at higher pressures, bimolecular ion–molecule interactions predominate. These reactions have already been discussed in Chapter VIII [Section D.(3)] in the context of providing

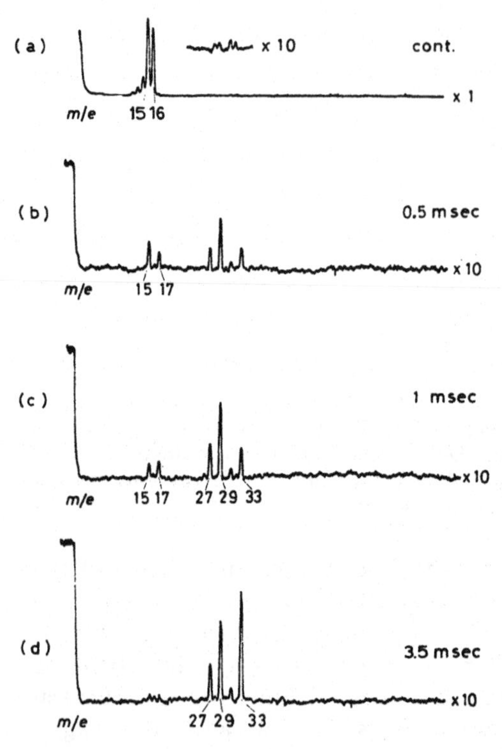

Fig. 10.9. A series of mass spectra of a methane–methanol (400 : 1) mixture at a total pressure of 10^{-4} torr obtained with the quistor/quadrupole combination. Trace (a) corresponds to continuous ionization with no storage whilst traces (b)–(d) demonstrate the effects of pulsed ionization and increased storage times as indicated.

experimental evidence for the mean primary ion kinetic energy. Other systems which have been investigated include methane and ethylene [177] and water and ammonia [178]. The data for methane show that the abundances of the secondary species CH_5^+ and $C_2H_5^+$ are broadly similar to those observed by Munson and Field [179] in the high pressure mass spectrum of this compound, from which it has been concluded that chemical ionization should be observable in the quistor, but at pressures some 10^4 times lower than in conventional CI sources. Such an effect is demonstrated [180] for a (400 : 1) methane–methanol mixture in Fig. 10.9. Other acid–base systems have been investigated, such as those involving H_3O^+ [178]: apparently cluster ions are not formed in the quistor, presumably because either the primary ions are too energetic for the weakly exoergic association products to form, or, if formed, the products rapidly undergo collisional dissociation. Other types of investigation with the quistor–quadrupole combination include charge-transfer [176], "spectrum simplification" (through the occurrence of consecutive ion–molecule reactions in a pure sample at ca. 10^{-4} torr and storage times of 3.5 msec) and the sequential ionization of trapped ions using a double electron pulse technique.

REFERENCES

1 W.K. Huber, Vacuum, 13 (1964) 399.
2 W.K. Huber, Vacuum, 13 (1964) 469.
3 A.P. Averina, L.N. Linnik and G.I. Nikitina, Instr. Exp. Tech. USSR, (1965) 373.
4 G.I. Slobodenyuk, Sov. Phys. Tech. Phys., 16 (1971) 798.
5 K.G. Günther, Vacuum, 10 (1960) 293.
6 G.A. Hofman, Vacuum, 24 (1974) 65.
7 M.A. Richardt, Vide, 22 (131) (1967) 272.
8 F. Lah and R. Zavasnik, Inst. Phys. Phys. Soc. London Conf. Ser. 6, Part 2, 1968, p. 703.
9 G.I. Slobodenyuk, A.I. Titov, V.S. Voronin and V.I. Ivashkin, Zavod. Lab., 36 (1970) 745.
10 G.I. Slobodenyuk, A.I. Titov, V.S. Voronin and V.I. Ivashkin, Inst. Exp. Res., (1968) 650.
11 H. Gentsch, Messtechnik, 79 (1971) 74.
12 H. Gentsch, J. Vac. Sci. Technol., 6 (1969) 93.
13 C. Von Brunnee, L. Delgmann and K. Kronenberger, Vak. Tech., 13 (1964) 35.
14 A.P. Averina, V.I. Vinogradov and T.G. Grinchenko, Inst. Exp. Technol., (1966) 907.
15 C.J. Bargery, Vacuum, 18 (1968) 160.
16 G. Rettinghaus and W.K. Huber, Vak. Tech., 20 (1971) 65.
17 D.L. Swingler, Vacuum, 18 (1968) 669.
18 G.W. Ball, Vacuum, 19 (1969) 331.
19 J.Yinon and F.S. Klein, Vacuum, 21 (1971) 379.
20 H. Böhm and K.G. Günther, Z. Angew. Phys., 17 (1964) 553.
21 H. Böhm and K.G. Günther, Vak. Tech., 14 (1965) 192.
22 F.L. Torney, Jr. and J.R. Roehrig, J. Vac. Sci. Technol., 6 (1969) 906.
23 F.L. Torney, Jr. and J.R. Roehrig, NASA Rep. CR-1747, Norton Research Corp., Cambridge, Mass., 1971.
24 W.L. Fite and P. Irving, Ind. Res., (Sept.) (1973) 60.
25 W.L. Fite and P. Irving, J. Vac. Sci. Technol., 11 (1974) 351.
26 H. Schwarz, Inst. Phys. Phys. Soc. London Conf. Ser. 6, Part 2, 1968, p. 685.
27 J.B. Hudson and J.S. Sandejas, J. Vac. Sci. Technol., 4 (1967) 230.
28 Y.K. Peng and P.T. Dawson, J. Chem. Phys., 54 (1971) 950.
29 L.C. Beavis, J. Vac. Sci. Technol., 10 (1973) 386.
30 B.J. Wood, J. Phys. Chem., 75 (1971) 2186.
31 J.H. Heald, Vacuum, 17 (1967) 511.
32 L.A. West, E.I. Kozak and G.A. Somorjai, J. Vac. Sci. Technol., 8 (1971) 430.
33 R.H. Jones, D.R. Olander, W.J. Siekhaus and J.A. Schwartz, J. Vac. Sci. Technol., 9 (1972) 1429.
34 R.N. Colthorp, J.T. Scott and E.E. Muschlitz, Jr., J. Chem. Phys., 51 (1969) 5180.
35 L.A. West and G.A. Somorjai, J. Vac. Sci. Technol., 9 (1972) 668.
36 K.W. Ashcroft, J.H. Leck, D.R. Sandstrom, B.P. Stimpson and E.M. Williams, J. Phys. E., 5 (1972) 1106.
37 T. Utsumi and O. Nishikawa, J. Vac. Sci. Technol., 9 (1972) 477.
38 D.R. Sandstrom, J.H. Leck and E.E. Donaldson, J. Appl. Phys., 38 (1967) 2851.
39 W.K. Huber and G. Rettinghaus, J. Vac. Sci. Technol., 6 (1969) 89.

40 D.R. Sandstrom and J.H. Leck, Vacuum, 18 (1968) 126.
41 T.E. Madeley, J.T. Yates, D.A. King and C.J. Uhlaner, J. Chem. Phys., 52 (1970) 5215.
42 D.A. King, T.E. Madeley and J.T. Yates, J.Chem. Soc. Faraday I, 68 (1972) 1347.
43 D.A. King, C.G. Goymour and J.T. Yates, Proc. Roy. Soc. (London) Ser. A, 331 (1972) 361.
44 G.C. Goymour and D.A. King, J. Chem. Soc. Faraday I, 69 (1973) 736.
45 M. Onchi and H.E. Farnsworth, Surface Sci., 11 (1968) 203.
46 C. Corotte, P. Ducros and D. Lafeuille, C.R. Acad. Sci. Ser. B, 265 (1967) 1040.
47 W.K. Huber and G. Rettinghaus, J. Vac. Sci. Technol., 7 (1970) 289.
48 M.W. Siegel, R.H. Krauss and J.W. Boring, J. Chem. Phys., 57 (1972) 3576.
49 A. Benninghoven, Appl. Phys., 1 (1973) 3.
50 A. Benninghoven, Surface Sci., 28 (1971) 541.
51 A. Benninghoven, C. Plog and N. Treitz, Int. J. Mass Spectrom. Ion Phys., 13 (1974) 415.
52 A. Benninghoven and A. Müller, Surface Sci., 39 (1973) 416.
53 J. Maul, F. Shulz and K. Wittmaak, Phys. Lett. A, 41 (1972) 177.
54 A. Benninghoven and E. Loebach, Rev. Sci. Instrum., 42 (1970) 49.
55 A. Benninghoven, Z. Phys., 230 (1970) 403.
56 W.K. Huber, H. Selhofer and A. Benninghoven, J. Vac. Sci. Technol., 9 (1972) 482.
57 K. Wittmaak, J. Maul and F. Schulz, Int. J. Mass Spectrom. Ion Phys., 11 (1973) 23.
58 R. Schubert and J.C. Tracy, Rev. Sci. Instrum., 44 (1973) 487.
59 G.E. Thomas and E.E. de Kluizenaar, Rev. Sci. Instrum., 45 (1974) 457.
60 A. Benninghoven and E. Loebach, J. Radioanal. Chem., 12 (1972) 95.
61 R. Schubert, J. Vac. Sci. Technol., 11 (1974) 903.
62 A. Benninghoven and A. Müller, Thin Solid Films, 12 (1972) 439.
63 A. Benninghoven, J. Vac. Sci. Technol., 9 (1972) 600.
64 P.H. Dawson, Int. J. Mass Spectrom. Ion Phys., 17 (1975) 423, 447.
65 P.H. Tyon, Advan. Mass Spectrom., 5 (1971) 551.
66 M. Farber, R.D. Srivastava and O.M. Uy, J. Chem. Soc. Faraday I, 68 (1972) 249.
67 M. Farber and R.D. Srivastava, J. Chem. Soc. Faraday I, 69 (1973) 390.
68 J.W. Hastie and D.L. Swingler, High Temp. Sci., 1 (1969) 46.
69 H. Bloom and J.W. Hastie, J. Phys. Chem., 71 (1967) 2360.
70 H. Bloom and J.W. Hastie, Aust. J. Chem., 21 (1968) 583.
71 H. Bloom and J.W. Hastie, J. Phys. Chem., 72 (1968) 2706.
72 D.J. Meschi and A.W. Searcy, J. Chem. Phys., 51 (1969) 5134.
73 E.K. Gibson and S.M. Johnson, Thermochim. Acta, 4 (1972) 49.
74 M. Muller-Vonmoos and F. Back, Thermal Analysis, Vol. 2, Academic Press, New York, 1969, p. 1229.
75 H.G. Wiedemann and G. Bayer, Z. Anal. Chem., 266 (1973) 97.
76 R.F. Lever, Surface Sci., 9 (1968) 370.
77 J. Cunningham, J.J. Kelley and A.L. Penny, J. Phys. Chem., 74 (1970) 1992.
78 J. Cunningham and A.L. Penny, J. Chem. Phys., 76 (1972) 2353.
79 J. Cunningham and N. Samman, in D. Price and J.F.J. Todd (Eds.), Dynamic Mass Spectrometry, Vol. 4, Heyden, London, 1976, p. 247.
80 A.J. Soloman, A.C. Zettlemoger and R.D. Iyengar, J. Catal., 10 (1968) 304.
81 J.N. Bradley, J.R. Gilbert and A.J. Park, Trans. Faraday Soc., 65 (1969) 2772.
82 J.N. Bradley, J.R. Gilbert and A.J. Park, Advan. Mass Spectrom., 4 (1968) 669.
83 G.A. Somorjai, R.W. Jayner and B. Lang, Proc. Roy. Soc. (London) Ser. A, 331 (1972) 335.
84 E.M.A. Willhoft and A.J.B. Robertson, Mass Spectrometry, Proc. Symp. Mass Spect., Enfield College of Technol., 1967, Butterworths, London, 1968, p. 161.
85 See, for example, H.S.W. Massey, Electronic and Ionic Impact Phenomena, Volume III, Oxford University Press, London, 2nd edn., 1971, Chap. 19.

86 R.L.F. Boyd, Nature (London), 165 (1950) 142.
87 P.F. Knewstubb and A.W. Tickner, J. Chem. Phys., 37 (1962) 2941.
88 D. Smith and I.C. Plumb, J. Phys. D., 6 (1973) 1431.
89 J.B. Hasted, Advan. Mass Spectrom., 6 (1974) 901.
90 E.E. Ferguson, Advan. Electron. Electron Phys., 24 (1968) 23.
91 E.E. Ferguson, F.C. Fehsenfeld and A.L. Schmeltkopf, Advan. At. Mol. Phys., 5 (1969) 1.
92 E.E. Ferguson, in J.L. Franklin (Ed.), Ion–Molecule Reactions, Vol. 2, Butterworths, London. 1972. Chap. 8.
93 E.W. McDaniel, V. Čermák, A. Dalgarno, E.E. Ferguson and L. Friedman, Ion–Molecule Reactions, Wiley-Interscience, New York, 1970.
94 E.W. McDaniel, Collision Phenomena in Ionized Gases, Wiley, New York, 1964.
95 E.W. McDaniel and E.A. Mason, The Mobility and Diffusion of Ions in Gases, Wiley, New York, 1973.
96 P. Kebarle, in J.L. Franklin (Ed.), Ion–Molecule Reactions, Vol. 1, Butterworths, London,1972, Chap. 7.
97 S.A. Studniarz, in J.L. Franklin (Ed.), Ion–Molecule Reactions, Vol. 2, Butterworths, London, 1972, Chap. 14.
98 D.A. Parkes, Vacuum, 24 (1974) 561.
99 D.A. Parkes, Trans. Faraday Soc., 67 (1971) 711.
100 D.A. Parkes, Trans. Faraday Soc., 68 (1972) 613.
101 D.A. Parkes, J. Chem. Soc. Faraday I, 69 (1973) 198.
102 D.A. Parkes, J. Chem. Soc. Faraday I, 68 (1972) 2121.
103 H.F. Calcote and J.L. Reuter, J. Chem. Phys., 38 (1963) 310.
104 A.N. Hayhurst and T.M. Sugden, Proc. Roy. Soc. (London) Ser. A, 293 (1966) 36.
105 H.F. Calcote, in J.L. Franklin (Ed.), Ion–Molecule Reactions, Vol. 2, Butterworths, London, 1972, Chap. 15.
106 C. Morley, Vacuum, 24 (1974) 581.
107 A.N. Hayhurst and N.R. Telford, Proc. Roy. Soc. (London) Ser. A, 322 (1971) 483.
108 Z. Herman and R. Wolfgang, in J.L. Franklin (Ed.), Ion–Molecule Reactions, Vol. 2, Butterworths, London, 1972, Chap. 12.
109 A. Ding, K. Lacmann and A. Henglein, Z. Naturforsch. A, 23 (1968) 779.
110 R.L. Champion, L.D. Doverspike and T.L. Bailey, J. Chem. Phys., 45 (1966) 4377.
111 W.R. Gentry, E.A. Gislason, B.H. Mahan and C-W. Tsao, J. Chem. Phys., 49 (1968) 3058.
112 B.R. Turner, M.A. Fineman and R.F. Stebbings, J. Chem. Phys., 42 (1965) 4088.
113 R.D. Fink and J.S. King, Jr., J. Chem. Phys., 47 (1967) 1857.
114 R.D. Fink and T.J. Broad, J. Chem. Phys., 48 (1968) 1400.
115 F.S. Klein, G.D. Lempert, E. Murad and A. Pensky, Advan. Mass Spectrom., 6 (1974) 749.
116 M. Vestal, C. Blakley, P. Ryan and J.H. Futrell, Advan. Mass Spectrom., 6 (1974) 781.
117 M. Menzinger and L. Wåhlin, Rev. Sci. Instrum., 40 (1969) 102.
118 M.A.D. Fluendy and K.P. Lawley, Chemical Applications of Molecular Beam Scattering, Chapman and Hall, London, 1973.
119 For a discussion of the results obtained with this type of detector see D.R. Herschbach, Advan. Chem. Phys., 10 (1966) 319.
120 W.B. Miller, S.A. Safron and D.R. Herschbach, Discuss. Faraday Soc., 44 (1967) 108.
121 G.O. Brink, Rev. Sci. Instrum., 37 (1966) 857.
122 R.W. Bickes and R.B. Bernstein, Rev. Sci. Instrum., 41 (1970) 759.
123 D.M. McFadden, E.A. McCullough, F. Kalos and J. Ross, J. Chem. Phys., 59 (1973) 121.
124 D.R. Jenkins and M.A. Voisey, J. Phys. E., 6 (1973) 827.

125 N.C. Blais and J.B. Cross, J. Chem. Phys., 52 (1970) 3580.
126 H.M. Powell and J.H. Heald, Jr., J. Vac. Sci. Technol., 4 (1967) 331.
127 W.J. Siekhaus, R.H. Jones and D.R. Olander, J. Appl. Phys., 41 (1970) 4392.
128 R.W. Diesen, J.C. Wahr and S.E. Adler, J. Chem. Phys., 50 (1969) 3635.
129 G.E. Busch, R.T. Mahoney, R.I. Morse and K.R. Wilson, J. Chem. Phys., 51 (1969) 449.
130 G.E. Busch and K.R. Wilson, J. Chem. Phys., 56 (1972) 3626.
131 I.T.N. Jones and K.D. Bayes, J. Chem. Phys., 57 (1972) 1003.
132 I.T.N. Jones and K.D. Bayes, J. Chem. Phys., 59 (1973) 4836.
133 L.I. Bone, Advan. Mass Spectrom., 6 (1974) 753.
134 I.T.N. Jones and K.D. Bayes, J. Amer. Chem. Soc., 94 (1972) 6869.
135 F.K. Truby, Phys. Rev., 172 (1968) 24.
136 F.K. Truby, Phys. Rev., 188 (1969) 508.
137 F.K. Truby, Phys. Rev. A, 4 (1971) 114.
138 J.A. Burt, Can. J. Phys., 50 (1972) 2410.
139 J.A. Burt, J. Chem. Phys., 57 (1972) 4649.
140 J.A. Burt, J. Geophys. Res., 77 (1972) 6280.
141 J.F.J. Todd, R.B. Turner and M.O. Norris, J. Phys. E., 6 (1973) 1110.
142 R.G. Dromey, J.D. Morrison and J.C. Traeger, Int. J. Mass Spectrom. Ion Phys., 6 (1971) 57.
143 J.D. Morrison and J.C. Traeger, Int. J. Mass Spectrom. Ion Phys., 7 (1971) 391.
144 J.D. Morrison and J.C. Traeger, Int. J. Mass Spectrom. Ion Phys., 11 (1973) 77.
145 J.D. Morrison and J.C. Traeger, Int. J. Mass Spectrom. Ion Phys., 11 (1973) 277.
146 J.D. Morrison and J.C. Traeger, Int. J. Mass Spectrom. Ion Phys., 11 (1973) 289.
147 P. Marchand, C. Paquet and P. Marmet, Phys. Rev., 180 (1969) 123.
148 W.E. Kauppila, W.R. Ott and W.L. Fite, Phys. Rev. A, 1 (1970) 1089, 1099.
149 J.R. Reeher and H.J. Svec, Advan. Mass Spectrom., 6 (1974) 509.
150 U. von Zahn and H. Tararczyk, Phys. Lett., 12 (1964) 190.
151 H. Tatarczyk and U. von Zahn, Z. Naturforsch. A, 20 (1965) 1708.
152 E. Teloy and D. Gerlich, Chem. Phys., 4 (1974) 417.
153 G. Ochs and E. Teloy, J. Chem. Phys., 61 (1974) 4930.
154 A.F. Haught and D.H. Polk, Phys. Fluids, 9 (1966) 2047.
155 R.W. Waniek and P.J. Jarmuz, Appl. Phys. Lett., 12 (1968) 52.
156 T.G.O. Berg and D.G. George, Mon. Weather Rev., 95 (1967) 884.
157 T.G.O. Berg, T.A. Gaukler and R.J. Trainor, Jr., J. Atmos. Sci., 26 (1969) 558.
158 T.G.O. Berg, Amer. J. Phys., 37 (1969) 859.
159 T.G.O. Berg and T.A. Gaukler, Amer. J. Phys., 37 (1969) 1013.
160 H.G. Dehmelt, Advan. At. Mol. Phys., 5 (1969) 109.
161 H.G. Dehmelt and F.G. Major, Phys. Rev. Lett., 8 (1962) 213.
162 F.G. Major and H.G. Dehmelt, Phys. Rev., 170 (1968) 91.
163 H.A. Schuessler, E.N. Fortson and H.G. Dehmelt, Phys. Rev., 187 (1969) 5.
164 G. Graeff, F.G. Major, R.W.H. Roeder and G. Werth, Phys. Rev. Lett., 21 (1968) 340.
165 G. Graeff, E. Klempt and G. Werth, Z. Phys., 222 (1969) 201.
166 D.A. Church and B. Mokri, Z. Phys., 244 (1971) 6.
167 H.G. Dehmelt and F.L. Walls, Phys. Rev. Lett., 21 (1968) 127.
168 D.A. Church and H.G. Dehmelt, J. Appl. Phys., 40 (1969) 3421.
169 H.G. Dehmelt and K.B. Jefferts, Phys. Rev., 125 (1962) 1318.
170 C.B. Richardson, K.B. Jefferts and H.G. Dehmelt, Phys. Rev., 165 (1968) 80.
171 K.B. Jefferts, Phys. Rev. Lett., 20 (1968) 39.
172 F.L. Walls and G.H. Dunn, J. Geophys. Res., 79 (1974) 1911.
173 F.L. Walls and G.H. Dunn, Phys. Today, 27 (8) (1974) 30.
174 J. Byrne and P.S. Farago, Proc. Phys. Soc., 86 (1965) 801.

175 J-P. Schermann and F.G. Major, NASA Rep. X-524-71-343, Goddard Space Flight Center, Greenbelt, Maryland, 1971.
176 R.F. Bonner, G. Lawson, J.F.J. Todd and R.E. March, Advan. Mass Spectrom., 6 (1974) 377.
177 R.F. Bonner, G. Lawson and J.F.J. Todd, Int. J. Mass Spectrom. Ion Phys., 10 (1972/73) 197.
178 G. Lawson, R.F. Bonner, R.E. Mather, J.F.J. Todd and R.E. March, J. Chem. Soc. Faraday I, 72 (1976) in the press.
179 M.S.B. Munson and F.H. Field, J. Amer. Chem. Soc., 88 (1966) 2621.
180 R.F. Bonner, G. Lawson and J.F.J. Todd, J. Chem. Soc. Chem. Commun., (1972) 1179.

CHAPTER XI

APPLICATIONS TO UPPER ATMOSPHERE RESEARCH

G.R. Carignan

The quadrupole mass analyzer found almost immediate applicability in upper atmospheric research. At the time of its development in the 1950's upper atmospheric research by in situ techniques was in its infancy and uncertainty in the height distribution of atmospheric constituents was a major impediment to an understanding of atmospheric physics. The implementation of mass spectrometric measurements on rockets and satellites presented great difficulties and this challenge attracted some of the leading figures in the field to space research, notably Nier, von Zahn and Brubaker.

The great need for composition measurements coupled with the popularity and consequent support of space research has resulted in the use of virtually all of the "standard" analysis techniques. The quadrupole and monopole have been in the forefront of this activity and significant contributions have been made by their measurements.

A. EARLY HISTORY OF MASS SPECTROMETRIC MEASUREMENTS IN THE UPPER ATMOSPHERE

In 1949, the Cook Research Laboratories built and installed a magnetic mass spectrometer on a captured German V-2 rocket [1]. A crude measure of the helium-to-argon ratio up to about 130 km was obtained. In February 1953, the first neutral gas mass spectrometer to obtain a spectrum in the upper atmosphere was successfully flown on an Aerobee rocket by the Naval Research Laboratory [2]. In 1954, the first ion spectra of the ionosphere were obtained [3]. An ion mass spectrometer [4] was included in the instrumentation of the third Soviet satellite, Sputnik 3. These pioneering measurements, excepting the first, were made utilizing a Bennett rf mass spectrometer [5]. These early applications of mass spectrometers in the upper atmosphere were reviewed by Johnson [6] in 1961. At that early stage of development, the results presented were mainly ratios of one constituent to another thus avoiding the great difficulty in ascribing absolute sensitivities to the instruments. The ion spectra had clearly identified the principal positive ions, O^+, NO^+, O_2^+, N_2^+ and N^+ and had reasonably well established the relative abundances. The neutral mass spectrometers had, on the other hand, produced results that were far from the truth. The stage had been set for a decade of dedicated effort before neutral gas mass spectrometry came to the fore in space exploration.

In the late 1950's, Brubaker under sponsorship of Narcisi of the U.S. Air Force, undertook development of a "Paul-type Mass Spectrometer" for space application. In March 1960, he produced what is thought to be the first mass spectrum obtained with a quadrupole filter in the United States. Some of the results of this early work were presented at an international conference in Stockholm, Sweden, in September 1960 [7]. In the course of this work, Brubaker recognized that the conditions at ion entry into a quadrupole field were important, particularly in the case of rocket and satellite applications where the instrument velocity with respect to the ambient gas is high and can be at any angle with respect to the quadrupole axis. This realization led to the concept of the "delayed d.c. ramp" mode of ionic injection which has become an important aspect of quadrupole development (see Chapter V, p. 105).

At about the same time, Schaefer and Nichols [8] were designing and constructing a quadrupole mass spectrometer for rocket application. The most important deficiency of earlier measurements of the upper atmosphere with mass spectrometers was the dearth of atomic oxygen observed. It was fairly well established from ultraviolet absorption measurements and on theoretical grounds that about half of the oxygen should exist in the atomic form in the 120 km region. Schaefer, therefore, attempted to tailor his ion source to minimize surface collisions of the ambient gas prior to ionization so that surface recombination would not destroy the atoms before measurement. The first flight of this instrument was carried out in November 1960 but produced no results for reasons not associated with the instrument itself. A few days later, Hanson, then at Lockheed, launched a rocket instrumented with a quadrupole to measure ion composition, but it also failed to produce results.

The first successful application of the quadrupole in the upper atmosphere occurred in May 1962 with a flight of Schaefer's neutral gas instrument. The preliminary results of this flight were published in 1963 [9] in the form of O/O_2 ion current ratios versus altitude. The value of this ratio measured by Schaefer at 120 km was slightly greater than one. This measurement found enthusiastic acceptance by atmospheric physicists and thereby added to the popularity of the quadrupole for upper atmospheric research.

During this period, several groups pressed the development of other mass spectrometers for space application. In 1962, Spencer et al. [10] measured nitrogen density and temperature using an omegatron, and in 1963, Nier et al. [11] flew a pair of magnetic mass spectrometers into the upper atmosphere on a single rocket. These investigators have continued their research in this area to the present and have made substantial contributions to the knowledge of the upper atmosphere. Pokhunkov [12, 13] of the USSR continued the Soviet program of composition measurements and Taylor et al. [14, 15] carried out many rocket and satellite flights using the Bennett mass spectrometer for ionosphere studies.

In 1963, von Zahn [16] described a new instrument, the monopole, which he, particularly, was to use extensively in the upper atmosphere. In 1966, it was flown together with a conventional quadrupole to measure neutral composition [17, 18]. The results were generally consistent with current thinking but the measured O_2 density was unexpectedly high and was attributed to recombination of O atoms on the instrument surfaces.

Many additional flights utilizing the monopole were made culminating in the successful application on the European satellite, ESRO IV [19, 20]. In the course of the development, a so-called "cyro ion source" [21] was designed and adapted to the monopole. The purpose of this was to cryogenically pump all gas particles having surface collisions such that the particles analyzed would be only those which had not encountered a surface. The measurements made with this combination have been successful in determining atomic oxygen abundances and in quantifying certain trace constituents which have previously been obscured by the products of surface chemistry [22]. A modification of this cryogenically cooled ion source [23] has been used effectively to pump away the shock in front of a rocket enabling quantitative measurements down to 85 km altitude [24].

At the National Aeronautics and Space Administration Goddard Space Flight Center, a program of mass spectrometer development was begun in the early 1960's. In addition to the omegatron, both magnetic and quadrupole mass spectrometers were designed for upper atmospheric research [25]. A quasi-open ion source was designed which could focus ions with transverse energies equivalent to those encountered when measuring on a satellite [26]. This attribute of the ion source permitted analysis of measurements taken when the spectrometer axis was at 90° to the satellite velocity vector, a circumstance which enhances the proportion of gas that has not suffered collisions before analysis. This ion source was employed on a magnetic spectrometer [27] on Explorer XVII, the first U.S. satellite to carry a neutral mass spectrometer. This same ion source was flown on a quadrupole mass spectrometer and used to elucidate the role of gas–surface interactions on a mass spectrometer measurement during rocket flight as well as to determine the abundances of the principal constituents [28]. The development of these upper atmosphere instruments drew heavily on the work of Hall who has contributed more than 50 treatises on upper atmosphere mass spectrometry, mostly in the form of contract reports.

The group under Spencer flew two mass spectrometers on the Explorer XXXII and one on the sixth and last Orbiting Geophysical Observatory launched in 1969. This latter instrument [29], a quadrupole, operated virtually continuously for two years and a large body of data accrued which,

References pp. 284—286

interpreted by Hedin et al. [30], greatly increased the knowledge of atmospheric variability. The data were also used by several authors (e.g. refs. 29, 31 and 32) to study dynamical processes including magnetic storms and upper atmospheric winds. More recently, a refined version of this quadrupole [33] was flown on the German AEROS satellite to provide both neutral temperature and composition measurements.

Narcisi was active in the early development of the quadrupole for upper atmosphere application. His sponsorship of Brubaker and his early application of quadrupoles in ionospheric studies helped pioneer the field of in situ measurements. The group under Narcisi concentrated on ion composition measurements, particularly in the lower part of the upper atmosphere. The high atmospheric pressure in this region dictates the use of a pump to maintain the pressure inside the mass spectrometer within its operating range. An adsorption pump was developed by Narcisi et al. [34] that met the necessary requirements. A quadrupole with such a pump was successfully flown in 1963. A description of the instrument and preliminary results were presented in 1965 [35]. This group has continued to be active in the field, having carried out many rocket flights in which abundances of both positive and negative ions were measured [36]. Their identification of metallic ions and the discovery of cluster ions constitute a major contribution to our understanding of the lower upper atmosphere. Quadrupoles have also been used by this group for neutral gas analysis on both rockets and satellites [37] with emphasis on circumventing the problems of gas surface interactions.

The quadrupole has been used successfully by Aiken and Goldberg (e.g. ref. 38) to study upper atmospheric ion composition. Another group, at the Max-Planck Institut für Kernphysik, have carried out many rocket flights using a single quadrupole instrument for measuring both neutrals and ions, alternately. This group has endeavoured to measure complex hydrated ion abundances to the lowest possible altitudes [39]. They also have flown a combination neutral and ion quadrupole analyzer [40] as part of the German AEROS satellite program.

This brief recapitulation of the early history of mass spectrometer application in the upper atmosphere does not, of necessity, include the names of all who contributed to the field. An attempt has been made to identify the major contributors, particularly those who employed the quadrupole or monopole. The interested reader is referred to other articles [36, 41] for a comprehensive treatment of the subject.

The application of mass spectrometry in the upper atmosphere has been a difficult undertaking. The convolution of the rigors of mass spectrometry with those of the use of rockets and satellites has provided a formidable challenge. Those who met this challenge have had the reward of having contributed fundamentally to the understanding of atmospheric physics.

B. THE UPPER ATMOSPHERE OF THE EARTH

The upper atmosphere, for the purpose of this discussion, will have its base at 80 km and extend upward to the point where it merges with interplanetary space. The seeming arbitrariness of the 80 km base altitude can be justified in that this is the approximate lower altitude boundary of conventional mass spectrometric measurement. Moreover, at about this altitude diffusion begins to take over from mixing. That is to say that below 80 km the principal constituents of the atmosphere are essentially in the same relative abundance as at the surface of the earth, but above, the relative abundances begin to differ. A search for knowledge of how they differ provided the basis for the extensive programs of composition measurements.

Above about 120 km, the atmosphere approaches a state of diffusive equilibrium in which the number density of each constituent decreases exponentially with altitude according to the relationship

$$n_i(z, t) = n_i(z_0, t) \frac{T(z_0, t)}{T(z, t)} \exp\left[-\int_{z_0}^{z} \frac{m_i g(z)}{kT(z, t)} dz\right] \tag{11.1}$$

where n_i is the number density of the ith constituent, z the altitude, t the time, z_0 the reference altitude, T the absolute temperature, m the molecular weight, g the acceleration of gravity and k Boltzmann's constant. Between 80 and 120 km, this relationship is modified by some residual mixing which strives to maintain a constant molecular weight. More importantly, photodissociation and ionization become major factors in establishing the state of the atmosphere. The two major consequences are (1) the increasing proportion of oxygen atoms and, (2) the onset of a sensible ionosphere. The composition of the upper atmosphere representing our current knowledge is depicted in Figs. 11.1 and 11.2 for the neutral and ionized components, respectively. Mass spectrometric measurements have contributed a major part of the knowledge that permits drawing these figures. These figures are only typical; the atmosphere is highly variable and the current thrust of atmospheric research centers on the study of this variability and the physical processes involved.

C. TECHNIQUES OF IN SITU MEASUREMENTS

(1) *Neutral gas analysis*

Access to the upper atmosphere is achieved by either sounding rockets or satellites. In both cases, the local ambient atmosphere is perturbed, principally because of the high velocity of the vehicle with respect to the ambient gas, but also because of vehicle outgassing. Moreover, as a practical matter,

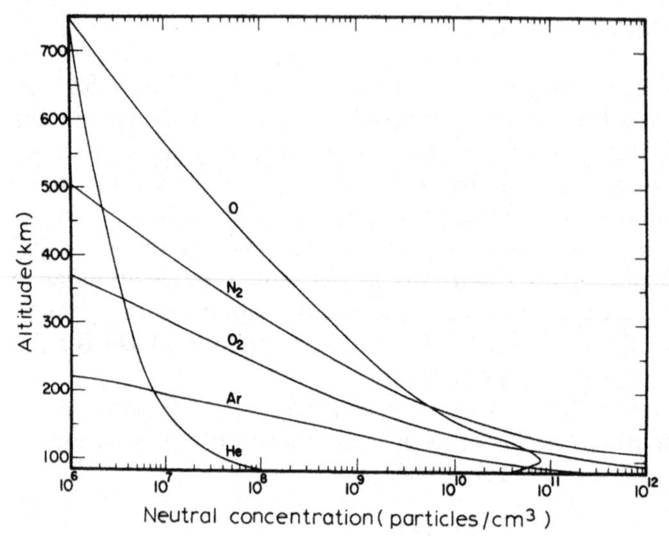

Fig. 11.1. Neutral concentrations versus altitude.

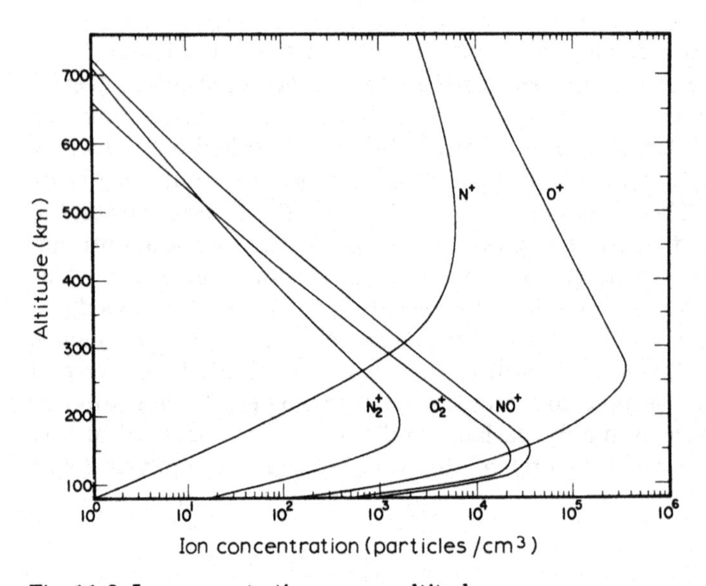

Fig. 11.2. Ion concentrations versus altitude.

in most cases the gas is analyzed after having had surface collisions where surface physics and chemistry modify the state of the gas. This is, of course, also true in ground-based applications, but the requirement to measure atomic oxygen abundance in the upper atmosphere has made the gas surface interaction problem particularly troublesome. Beyond the difficulty in measuring oxygen abundances, it has been learned that oxygen atoms combine with carbon in the metal surfaces of gauges to produce CO_2 and CO, the latter of which, at 28 amu, complicates the molecular nitrogen determination.

In the ideal case (without surface interactions) the gas density inside a gauge cavity connected to the atmosphere through a small knife-edged orifice can be related [42] to the ambient density through the relationship

$$n_s = n_a \left(\frac{T_a}{T_s}\right)^{1/2} F(S) \qquad (11.2)$$

where

$$F(S) = \exp(-S^2) + \pi^{1/2} S[1 + \mathrm{erf}(S)] \quad \text{and} \quad S = V(\cos\alpha)/c,$$

n_s is the number density in the ion source, n_a the ambient number density, T_a the ambient temperature, T_s the ion source temperature, S the speed ratio, V the vehicle velocity, α the angle between orifice normal and velocity vector and c the most probable thermal velocity of the particles. This equation predicts an enhancement in density when the gauge orifice is facing in the direction of the flight vector, a depletion when it is facing away, and the simple thermal transpiration relationship at 90° (orifice plane parallel to the velocity vector). At satellite velocities, the enhancement is greater than a factor of 10 (the value depending on the molecular weight) and the depletion is several factors of 10 below ambient.

Most of the early implementations of mass spectrometric measurements were configured such that eqn. (11.2) applied. Implicit in this method is the condition that each analyzed particle has had several collisions with the surface so that it is fully accommodated to the gauge wall temperature. In typical applications, the average number of collisions is greater than 100. Because of surface interaction modification to the sample, an ion source design which was more open had appeal and several investigators undertook the development of such sources for use on quadrupole and magnetic mass spectrometers. The equation that relates the density at the surface of a flat plate, which the open ion sources resemble, is

$$n_s = 0.5 n_a \left[1 + \mathrm{erf}(S) + \left(\frac{T_a}{T_s}\right)^{1/2} F(S)\right] \qquad (11.3)$$

assuming full accommodation and diffuse reflection. For specular reflection with zero accommodation

$$n_s = n_a[1 + \mathrm{erf}(S)] \qquad (11.4)$$

In the latter case, one half of the particles in the source have been reflected from the flat plate and in the former (full accommodation) reflected particles predominate, typically by a wide margin. Consequently, the open source designs fail to accomplish fully their principal objective, viz. the avoidance of contamination by particles which have had an opportunity to react on a surface. Moreover, the uncertainty in accommodation coefficient adds a large uncertainty to the interpretation of measured values in terms of ambient parameters.

A promising alternative technique for discriminating between direct streaming and reflected particles takes advantage of the energy associated with the relative motion between the instrument and the ambient gas. On a satellite, this relative velocity is about 8 km sec^{-1} or an equivalent energy of approximately 0.35 eV amu^{-1}. By establishing a retarding field between the ion source and the analyzer, the thermalized particles can be rejected allowing only the direct streaming component to enter the analyzer. The retarding analyzer has been used by itself [43] as an integrating mass spectrometer but with only limited success since the integration must begin at the low energies and thus a large background from the very low energy thermalized components tends to obscure the desired measurement. In combination with a mass spectrometer, however, the technique has been demonstrated to be useful [44].

In several instances, mass spectrometers flown into the upper atmosphere have performed well instrumentally but have had their results invalidated by contamination from gauge and vehicle surface outgassing. Successful investigators, almost without exception, have surmounted this difficulty by launching the spectrometer sealed in an ultra-clean evacuated condition and opening it after it has reached the upper atmosphere. This non-trivial complication has been one of the prices of successful neutral gas measurements. Also, it has been necessary to insure that vehicle outgassing is held to low levels to avoid this source of contamination.

Until recently, when retarding potential analysis of neutral mass spectra began to be employed in upper atmosphere research, investigators have found it necessary to deal with a wide variety of gas–surface interactions. Hedin et al. [45] have provided the most sophisticated analysis of this problem. Using a satellite-borne quadrupole mass spectrometer with a cavity–orifice geometry, the coefficients for adsorption, desorption and recombination for atmospheric gases were determined. The analysis takes into account the production of CO and CO_2 and shows that oxygen atoms are lost to this process in significant quantities for a period of weeks in orbit, a circumstance which greatly complicates the interpretation of measurements on rockets where the surface reactions are not only large but are changing throughout the flight.

The requirements on the analyzing field, in upper atmospheric mass spectrometry, are not stringent. The principal constituents are widely separated and, in general, operation with great resolution is not required.

(2) *Ion analysis*

In some applications, a single mass spectrometer has been used for both ion and neutral analysis by alternately measuring with the ion source off and on. Many of the difficulties encountered in neutral gas mass spectrometry are absent in ion analysis. Surface reactions in most cases are not involved since ions which strike a surface are usually neutralized. Moreover, there is no danger of contaminating the measurements with ions outgassed from dirty surfaces although photoionization of neutral contaminants has been frequently observed.

Almost without exception, ion composition is measured in terms of relative abundances often normalized to a measure of total electron density by a companion instrument and assuming charge neutrality. Ion spectra are seldom interpreted when the angle between the perpendicular to the entrance aperture plane and the velocity vector of the carrier vehicle is greater than a few degrees.

Great care must be taken to assure a proper electric field configuration at the instrument–ionosphere interface. Typically, a large conducting guard electrode is used to provide the proper field. Changes in the vehicle potential against which the instrument must operate must be compensated and some means of suppressing photoelectrons must be available for sunlit applications.

The problems to be solved in ion mass spectrometry have been more tractable than their neutral counterparts. Ion mass spectrometers, as a result, were able to make fundamental contributions to our knowledge of the upper atmosphere at an earlier date.

D. RECENT QUADRUPOLE APPLICATIONS AND DEVELOPMENTS

No single type of mass spectrometer currently dominates upper atmospheric research. The quadrupole and monopole have at least equal status with any other and at the time of this writing at least five are known to be operating on satellites in orbit.

The Atmosphere Explorer C Satellite [46] which is a major part of current experimental research in aeronomy has an instrument complement which illustrates the importance and diversity of upper atmosphere mass spectrometry. The satellite includes a Bennett ion mass spectrometer [47], a magnetic ion mass spectrometer [48], a magnetic neutral mass spectrometer [49], and two quadrupole neutral mass spectrometers, one intended for neutral temperature measurement [50] and the other for neutral composition measurement [51]. The last of these typifies the modern design of upper atmosphere quadrupole systems; its block diagram is shown in Fig. 11.3.

The instrument is coupled to the atmosphere through a knife-edged orifice into a spherical antechamber. The electron bombardment ion source, employing two redundant electron guns, samples the antechamber gas for analysis. Six-inch hyperbolic rods form the analyzing field and the resonant ions are directed to the first dynode of an off-axis multiplier. The prime measurement is pulse counting with a parallel measure of the multiplier output current using an electrometer.

References pp. 284—286

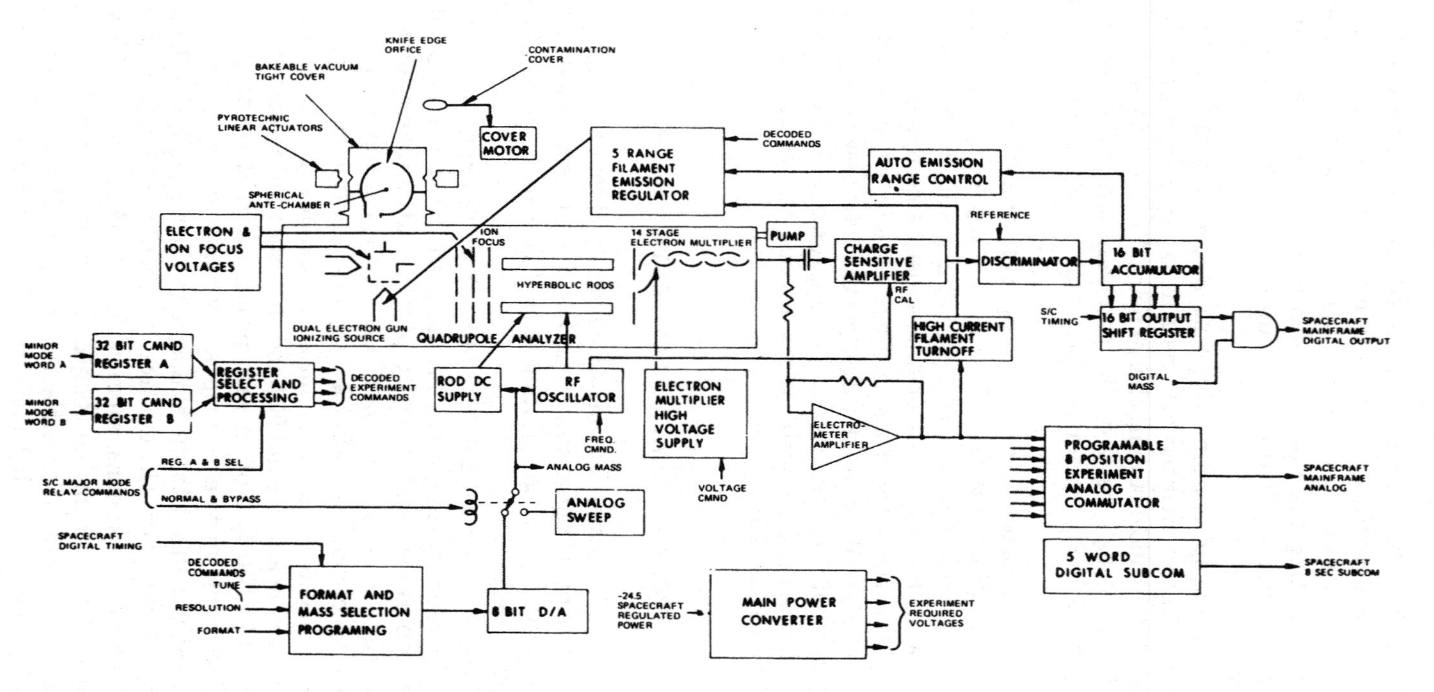

Fig. 11.3. Quadrupole system functional diagram.

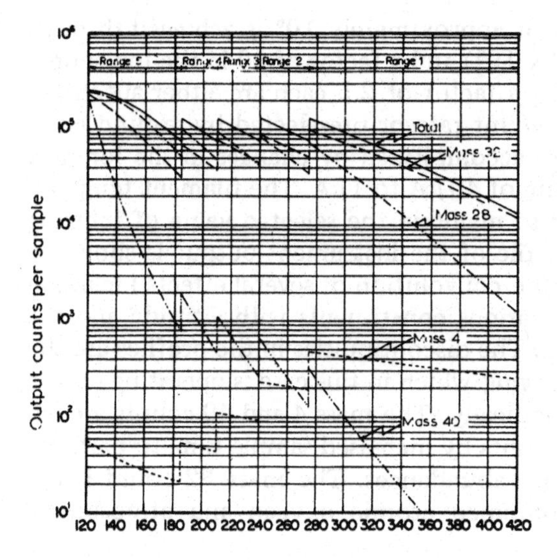

Fig. 11.4. Measured output versus altitude.

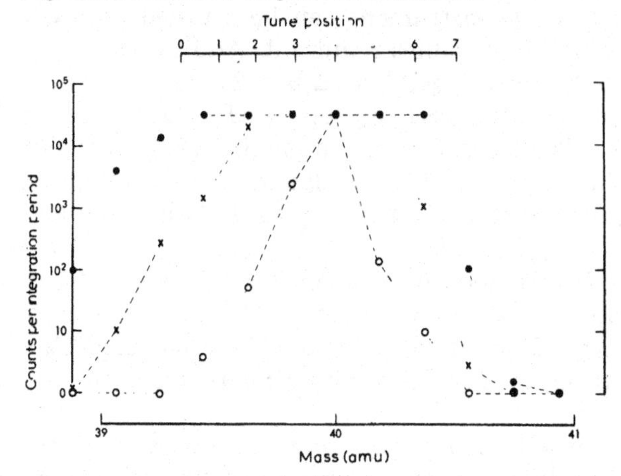

Fig. 11.5. Digital sweeps of the argon peak. ●, Resolution 7; ×, resolution 4; ○, resolution 0.

In operation, the instrument is normally tuned sequentially to the center of peaks of masses of geophysical significance, their count being integrated for a preset time, the length depending upon their abundances, and the count telemetered with a 16-bit digital word. Periodically, or on command, a full mass spectrum from 1 to 46 amu is taken in 256 digital steps of 3/16 amu each, with the value at each step telemetered. The command system provides the capability to vary the emphasis of the mass numbers being sampled, including the possibility of choosing any one of the 256 steps for continuous analysis.

The instrument dynamic range of approximately 10^6 is achieved through a 16-bit linear detection system (6×10^4) and a factor of 40 in the ion source where five sensitivities differing by a factor of 2.5 each are either automatically selected on the basis of the count rate or preselected by ground command. The sensitivity changes are accomplished by changes in the ionizing current from its nominal maximum of 40 μA to 1 μA. The filament temperature is controlled by a regulator to maintain the selected value of ionizing current. The resultant altitude profile of the instrument output is shown in Fig. 11.4. The profiles represent the convolution of several effects including the exponential variation of the various constituents with altitude and the mass dependence of that variation. The discontinuities in the profiles are the result of the automatic range changing which in this case is operating on the total count which is sampled periodically. The mass 4 and 40 counts are enhanced relative to mass 28 and 32 counts by increased sample time to partially compensate for their lower natural abundances. The mass 32 count is the sum of ions resulting from recombined atomic oxygen and atmospheric molecular oxygen.

The efficacy of the preset tuning of the instrument can be evaluated by analysis of the full spectrum and the instrument can be retuned in flight should lack of stability of any of the variables require it. Additionally, any one of eight resolutions from "pointed peak" to $\Delta M = 2$ amu at the 10% value can be selected. The instrument is normally operated at $\Delta M = 1$ amu at 10% which leads to a cross talk of less than 1 part in 10^3 at an adjacent mass. The mass 40 peak is shown in Fig. 11.5 to illustrate the tuning and resolution adjustment. The instrument is usually operated at the tune and resolution equal 4 points.

The ability to change the performance parameters through variation in electrical potentials is an attribute of the quadrupole which has been heavily exploited in the described design. This attribute is one which has made the quadrupole and monopole attractive for upper atmosphere research (see Chapter II, p. 21).

The continued application of the quadrupole and monopole in upper atmospheric research is assured. The next two Atmosphere Explorer Satellites will utilize two on each. The quadrupole is being used for astronaut breath analysis [52], (see Chapter XIII for a detailed description), and it has been chosen for flights to both Mars and Venus. In the next decade, it will have been used to analyze the atmospheres of several planets and the tails of comets. Its place in upper atmosphere research is well secured.

REFERENCES

1 M.D. O'Day, in R.L.F. Boyd and M.J. Seaton (Eds.), Rocket Exploration of the Upper Atmosphere, Pergamon, London, 1954, p. 1.

2 J.W. Townsend, Jr., E.B. Meadows and E. Pressly, in R.L.F. Boyd and M.J. Seaton (Eds.), Rocket Exploration of the Upper Atmosphere, Pergamon, London, 1954, p. 169.
3 C.Y. Johnson and E.B. Meadows, J. Geophys. Res., 60 (1955) 193.
4 V.G. Istomin, Iskusstv. Sputniki Zemli, 3 (1959) 98.
5 W.H. Bennett, J. Appl. Phys., 21 (1950) 143.
6 C.Y. Johnson, Ann. Geophys., 17 (1961) 100.
7 W.M. Brubaker, Proceeding of Fifth International Instruments Conference, Stockholm, 1961, p. 305.
8 E.J. Schaefer and M.H. Nichols, Am. Rocket Soc. J., 31 (1961) 1773.
9 E.J. Schaefer, J. Geophys. Res., 69 (1963) 1175.
10 N.W. Spencer, L.H. Brace, G.R. Carignan, D.R. Taeusch and H. Niemann, J. Geophys. Res., 70 (1965) 2665.
11 A.O. Nier, J.H. Hoffman, C.Y. Johnson and J.C. Holmes, J. Geophys. Res., 69 (1964) 979.
12 A.A. Pokhunkov, Space Res., 1 (1960) 101.
13 A.A. Pokhunkov, Planet. Space Sci., 11 (1963) 441.
14 H.A. Taylor, Jr. and H.C. Brinton, J. Geophys. Res., 66 (1961) 2587.
15 H.A. Taylor, Jr., H.C. Brinton and C.R. Smith, J. Geophys. Res., 70 (1965) 5769.
16 U. von Zahn, Rev. Sci. Instrum., 34 (1963) 1.
17 U. von Zahn and J. Gross, J. Geophys. Res., 74 (1969) 4055.
18 D. Müller and G. Hartmann, J. Geophys. Res., 74 (1969) 1287.
19 U. von Zahn, K.H. Fricke and H. Trinks, J. Geophys. Res., 78 (1973) 7560.
20 H. Trinks and U. von Zahn, Rev. Sci. Instrum., 46 (1975) 213.
21 D. Offermann and H. Trinks, Rev. Sci. Instrum., 42 (1971) 1836.
22 D. Offermann and K.U. Grossmann, J. Geophys. Res., 78 (1973) 8296.
23 D. Offermann and H. Tatarczyk, Rev. Sci. Instrum., 44 (1973) 1569.
24 T.G. Scholz and D. Offermann, J. Geophys. Res., 79 (1974) 307.
25 N.W. Spencer and C.A. Reber, Space Res., 3 (1963) 1151.
26 C.A. Reber and L.G. Hall, NASA Tech. Note TN D3211, 1966.
27 C.A. Reber and M. Nicolet, Planet. Space Sci., 13 (1965) 617.
28 H.B. Niemann, N.W. Spencer and G.A. Schmitt, J. Geophys. Res., 78 (1973) 2265.
29 D.R. Taeusch, G.R. Carignan and C.A. Reber, J. Geophys. Res., 76 (1971) 8318.
30 A.E. Hedin, H.G. Mayr, C.A. Reber, N.W. Spencer and G.R. Carignan, J. Geophys. Res., 79 (1974) 215.
31 C.A. Reber and P.B. Hays, J. Geophys. Res., 78 (1973) 2977.
32 H.G. Mayr, A.E. Hedin, C.A. Reber and G.R. Carignan, J. Geophys. Res., 79 (1974) 619.
33 N.W. Spencer, D.T. Pelz, H.B. Niemann, G.R. Carignan and J.R. Caldwell, Z. Geophys., 40 (1974) 613.
34 R.S. Narcisi, W.M. Brubaker, H.C. Poehlmann, R.P. Fedchenko and F.B. Wiens, in G.H. Bancroft (Ed.), Transactions of the 9th National Vacuum Symposium, Macmillan, New York, 1962, p. 232.
35 R.S. Narcisi and A.D. Bailey, J. Geophys. Res., 70 (1965) 3687.
36 R.S. Narcisi, in F. Verniani (Ed.), Physics of the Upper Atmosphere, Editrice Compositori, Bologna, 1971, p. 12.
37 C.R. Philbrick, R.S. Narcisi, D.W. Baker, E. Trzcinski and M.E. Gardner, Space Res., 13 (1972) 321.
38 A.C. Aiken and R.A. Goldberg, J. Geophys. Res., 78 (1973) 734.
39 A. Johannessen, D. Krankowsky, F. Arnold, W. Riedler, M. Friedrich, D. Folkestad, G. Skovli, E.V. Thrane and J. Troim, Nature (London), 235 (1972) 215.
40 D. Krankowsky, Z. Geophys., 40 (1974) 601.
41 U. von Zahn, Advan. Mass Spectrom., 4 (1968) 869.

42 F.V. Schultz, N.W. Spencer and A. Reifman, Upper Atmos. Rep. 2, Univ. of Mich. Res. Inst., Ann Arbor, Mich., 1948.
43 C.R. Philbrick and J.P. McIssac, Space Res. 12 (1971) 743.
44 A.O. Nier, W.E. Potter, D.C. Kayser and R.G. Finstad, Geophys. Res. Lett., 1 (1974) 197.
45 A.E. Hedin, B.B. Hinton and G.A. Schmitt, J. Geophys. Res., (1973) 4651.
46 N.W. Spencer, L.H. Brace and D.W. Grimes, Radio Sci., 8 (1973) 267.
47 H.C. Brinton, L.R. Scott, N.W. Pharo, III and J.T.C. Coulson, Radio Sci., 8 (1973) 323.
48 J.H. Hoffman, W.B. Hanson, C.R. Lippincott and E.E. Ferguson, Radio Sci., 8 (1973) 315.
49 A.O. Nier, W.E. Potter, D.R. Hickman and K. Mauersberger, Radio Sci., 8 (1973) 271.
50 N.W. Spencer, H.B. Niemann and G.R. Carignan, Radio Sci., 8 (1973) 284.
51 D.T. Pelz, C.A. Reber, A.E. Hedin and G.R. Carignan, Radio Sci., 8 (1973) 277.
52 W.M. Brubaker, in Proceedings of 18th Annual Conference on Mass Spectrometry and Allied Topics, University Park Press, Baltimore, 1970, p. 98.

CHAPTER XII

APPLICATIONS TO GAS CHROMATOGRAPHY

M.S. Story

A. INTRODUCTION

It is in the role as an identifier and quantifier of chromatographically separated compounds that quadrupoles today find their widest application. It is also the area of most active research on techniques of ionization, interfacing hardware and data handling. As has been pointed out in Chapter I, the quadrupole gained early popularity as a residual gas analyser and the problems to be solved in making it into a sophisticated mass spectrometer for gas chromatography effluent detection were many. Mass range and resolution, gas chromatography–mass spectrometry (GC–MS) vacuum interfacing, dynamic range, electronic and mechanical stabilities all required considerable development. On the other hand, there were characteristics inherent in the device that made the quadrupole uniquely suited to the application; high pressure tolerance, high sensitivity, low voltage ion source and the capability of being scanned very rapidly under computer control.

The solutions to these problems coupled with the quadrupole's inherent advantages have led to the dominant role that this type of mass spectrometer plays in the technique of GC–MS today. Many of the developments that solved the general operational problems of quadrupoles have been discussed in previous chapters. The first part of this chapter will deal briefly with quadrupole (mass analysing) considerations that are essential to its coupling with a gas chromatograph and then proceed to discuss developments of the ancillary equipment that make the whole spectrometer system such a powerful tool for chemical analysis.

B. GAS CHROMATOGRAPHY–QUADRUPOLE MASS SPECTROMETRY INSTRUMENTAL DEVELOPMENTS

(1) *Resolution, mass range and peak shapes of the quadrupole*

In order for the quadrupole to be of analytical use, the mass range and resolution must be high enough to cover the molecule weight range

References p. 306

encountered in chromatographic applications. For the most part, compounds of molecular weight greater than 1000 are not likely to survive gas chromatography because of the need for volatilization of the intact molecule.

Theoretically, a quadrupole can be operated at ever increasing mass range and resolution but in actual practice this is not the case. The geometry of the quadrupole structure, operating characteristics of the ion source and rf generation parameters will determine the maximum mass range and resolution attainable. For example, Fig. 6.16 showed the relationship of sensitivity and resolution ($m/\Delta m$) for a particular quadrupole structure at $m/e = 69$ and at $m/e = 502$. Past a resolution of 2 times the mass, there is considerable decrease in sensitivity for very little change in resolution and it would not be practical to use this particular quadrupole at resolutions greater than 1000 at $m/e = 500$.

As quantitation becomes more of a necessity, peak shapes, especially those considerations that effect adjacent mass intensities, become much more important. Mass range of 1000 and resolutions of 2000 are now commonplace for commercial quadrupole mass spectrometers. The developments leading to increases in mass range become even more significant as the quadrupole is applied to the detection of liquid chromatographic effluents, as liquid chromatography is often utilized for separation of larger molecular weight compounds than is gas chromatography.

(2) *Gas chromatograph interfacing*

The operation of a GC/MS system requires the coupling of a gas chromatograph that operates at 2–70 atm ml min^{-1} of carrier gas flow to a mass spectrometer whose pumping system (nominally 200 l sec^{-1}) can handle 0.5–2 atm ml min^{-1} of gas flow. This gas flow incompatability is adequately solved by a variety of molecular separators. One of these interface devices is a glass jet separator (Fig. 12.1(a)) whose typical enrichment, N (where N = sample concentration into MS/sample concentrations out of GC), characteristics are shown in Fig. 12.1(b). Yields, Y, for these types of separators are 30–40%. As has been pointed out by McFadden [1], a separator has a maximum enrichment given by

$$N_{max} = \frac{\text{yield}}{100} \times \frac{\text{Flow}_{GC}}{\text{Flow}_{MS}}$$

With typical operating parameters for this device of $\text{Flow}_{GC} = 25$ atm ml min^{-1} and $\text{Flow}_{MS} = 0.5$ atm ml min^{-1}, a maximum enrichment of 20 is predicted. Under these conditions, an electron impact ion source whose conductance to the pumping system is 5 l sec^{-1} would have about 1×10^{-3} torr as the pressure in the ionization region. This should produce little or no contribution to ionization by ion–molecule reactions.

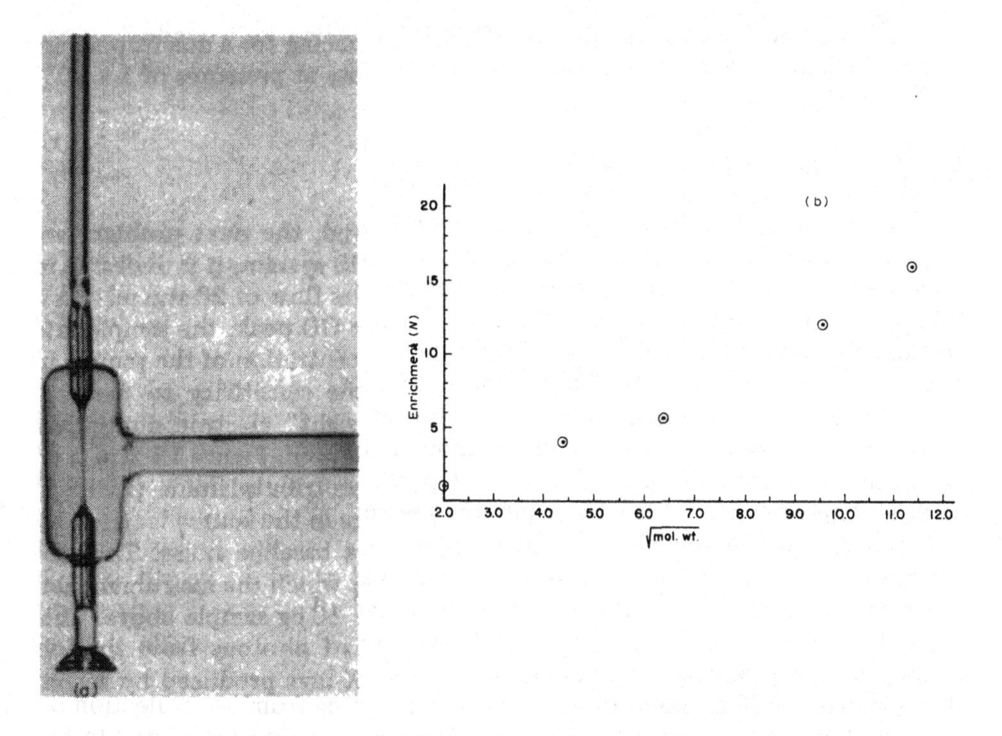

Fig. 12.1. (a) Photograph of glass molecular jet separator (Finnigan Corporation). (b) Plot of enrichment versus (molecular weight)$^{1/2}$ for a molecular jet separator, $T = 25°C$.

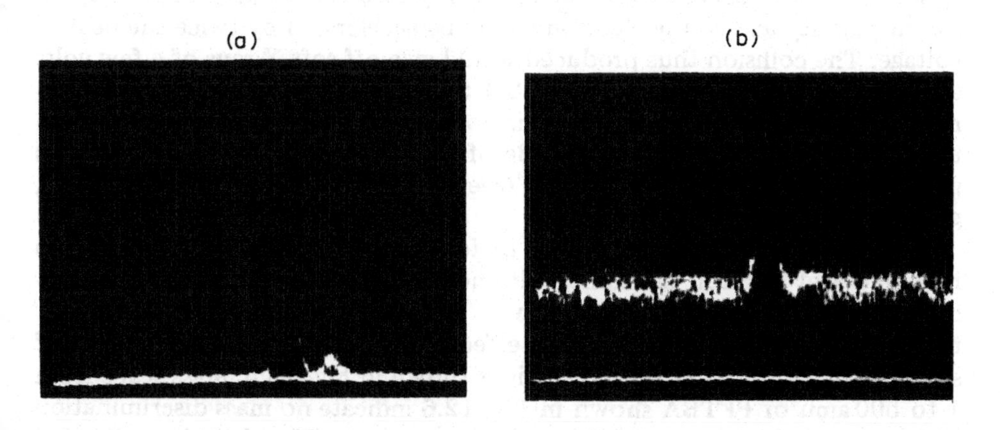

Fig. 12.2. Oscilloscope display of mass scan with "in line" multiplier [00]. (a) $P = 4 \times 10^{-6}$ torr perfluorotributylamine; sensitivity = 2×10^{-9} A/cm; sweep time = 8 msec/cm; m/e = 414. (b) As (a) except $P = 5 \times 10^{-5}$ torr due to helium being admitted to the system.

The enrichment requirements for GC–MS interfacing for a quadrupole are not as difficult because the quadrupole can operate at pressures of 1×10^{-4} to 4×10^{-4} torr helium.

(3) *Sensitivity and dynamic range*

Once the gas flow incompatabilities were solved, the next problem encountered was that of dynamic range. In a GC–MS system, it is desirable to measure as little as 10 pg of material. With a carrier gas flow of 20 atm ml min^{-1} into the separator ($Y = 0.3$, $N = 20$) and a 10 sec GC peak, the sample rate to the ion source is 3×10^{-13} g sec^{-1} and the concentration of the sample in the ion source is 336 ppb. Quadrupoles have the sensitivity to measure 1×10^{-12} g sec^{-1} but quadrupoles with "line of sight" electron multipliers suffer from background noise levels of about 1–10 ppm. Figure 12.2(a) is an oscilloscope tracing of m/e 414, 415 of perfluorotributylamine (PFTBA) with a "line of sight" multiplier. When the pressure in the source is raised by the addition of helium [Fig. 12.2(b)], there is a baseline noise. This can seriously effect the dynamic range of detection at which the instrument can perform and would not allow the analysis of the 10 pg sample above. This noise has been characterized as being made up of photons from the ion source due to filament radiation as well as soft X-rays produced by secondary electron bombardment of adjacent rods resulting from ion collection on the rods. Figures 12.3 and 12.4 show that this spurious signal is related to the initial ion production but also is dependent upon the potential of the rods. Since ions do not generate X-rays as well as they do electrons and electrons do readily generate X-rays, it is reasonable to postulate that as an ion is collected on a negative rod, an electron is ejected. The adjacent rod is opposite in polarity and so the electron would be accelerated by twice the peak rf voltage. The collision thus produced would give off soft X-rays of a few volts to 2000 V in energy. Note in Fig. 12.4 that at zero rf voltage there still remains appreciable noise current. This noise is contributed by photons from the filament striking the first dynode of the multiplier. The solution to this problem is to offset the multiplier (see also the discussion in Chapter VI, p. 139).

In order to determine what type of focusing was necessary to bring ions into an offset electron multiplier with no loss in sensitivity or mass discrimination, an experimental apparatus was constructed to allow the multiplier to be moved during operation. The effect on the background noise level of simply moving the multiplier off-axis is shown in Fig. 12.5. Mass scans from 1 to 500 amu of PFTBA shown in Fig. 12.6 indicate no mass discrimination or loss in sensitivity due to the off-axis geometry. The focusing effect of − 3000 V on the first dynode is adequate to extract all of the ions and no additional lensing is required. Dynamic range with an off-axis multiplier such as this can be as great as 10^9 to 1 as is shown in Fig. 12.7(a) and (b).

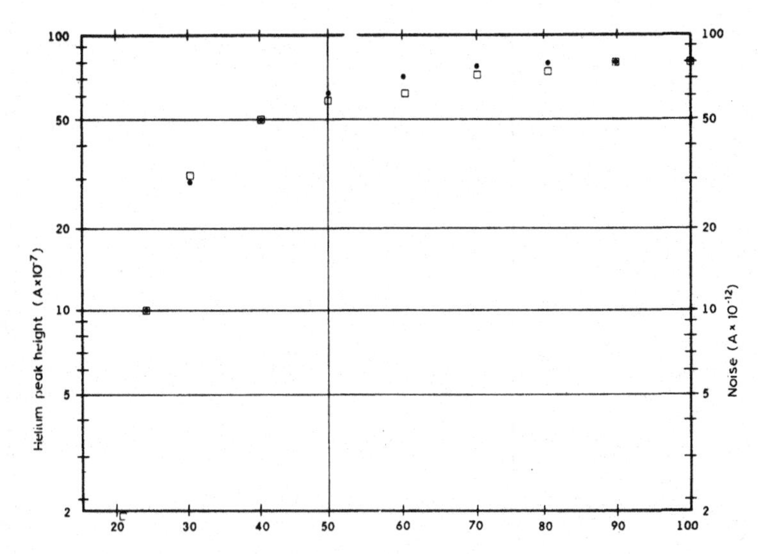

Fig. 12.3. Effect of ionization voltage on •, level of helium ion production and □, noise level.

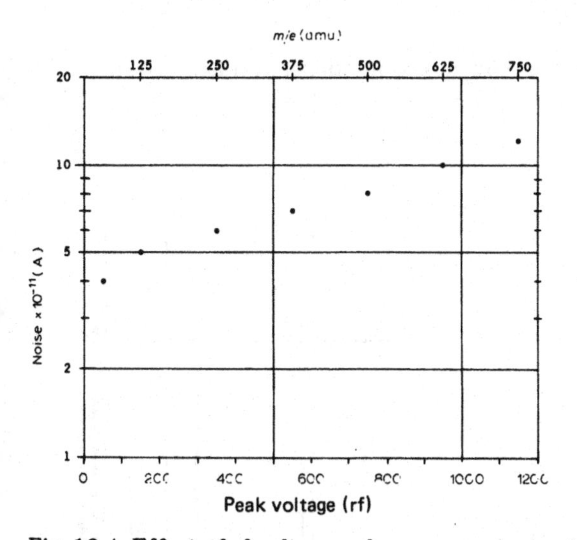

Fig. 12.4. Effect of rf voltage and mass on noise level [00].

Figure 12.7(a) is a scan of room air sampled through a variable leak valve with an offset multiplier. The ions of m/e 129^+, 131^+, 132^+, 134^+ and 136^+ are the naturally occuring isotopes of xenon present in our atmosphere at the 23, 18, 23, 9 and 7 ppb concentration, respectively. The spectrum also shows the existence of doubly charged ions from rhenium ($^{185}Re^{2+}$, $^{187}Re^{2+}$) and oxides of rhenium ($^{185}ReO^{2+}$, $^{187}ReO^{2+}$, $^{185}ReO_2^{2+}$ and $^{187}ReO_2^{2+}$) when the

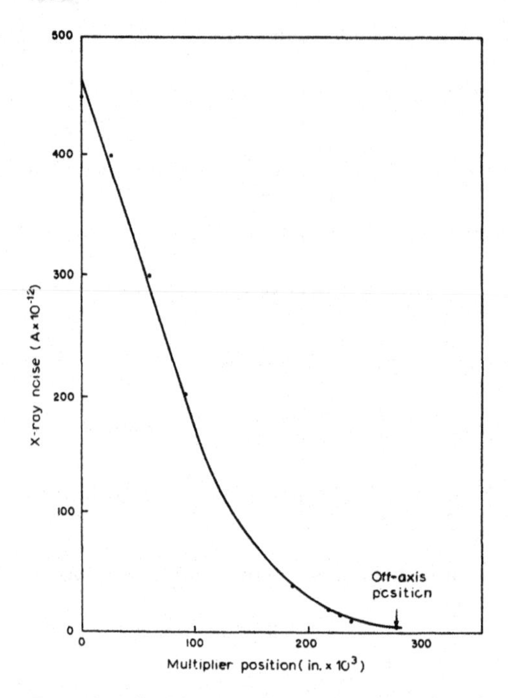

Fig. 12.5. Reduction of noise level produced by moving multiplier "off-axis" [00].

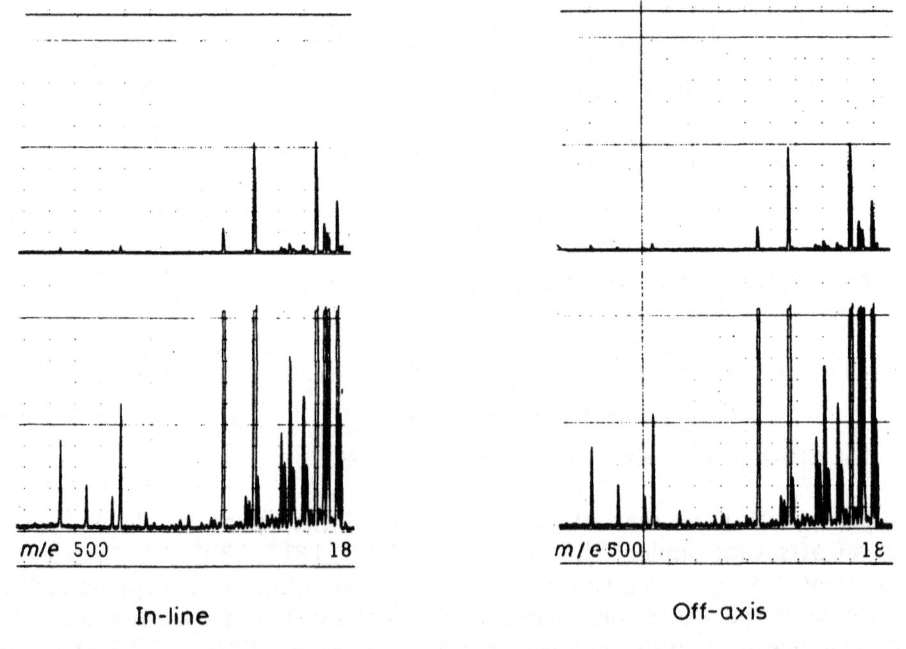

Fig. 12.6. Mass spectra perfluorotributylamine with electron multiplier in line and off-axis positions [00].

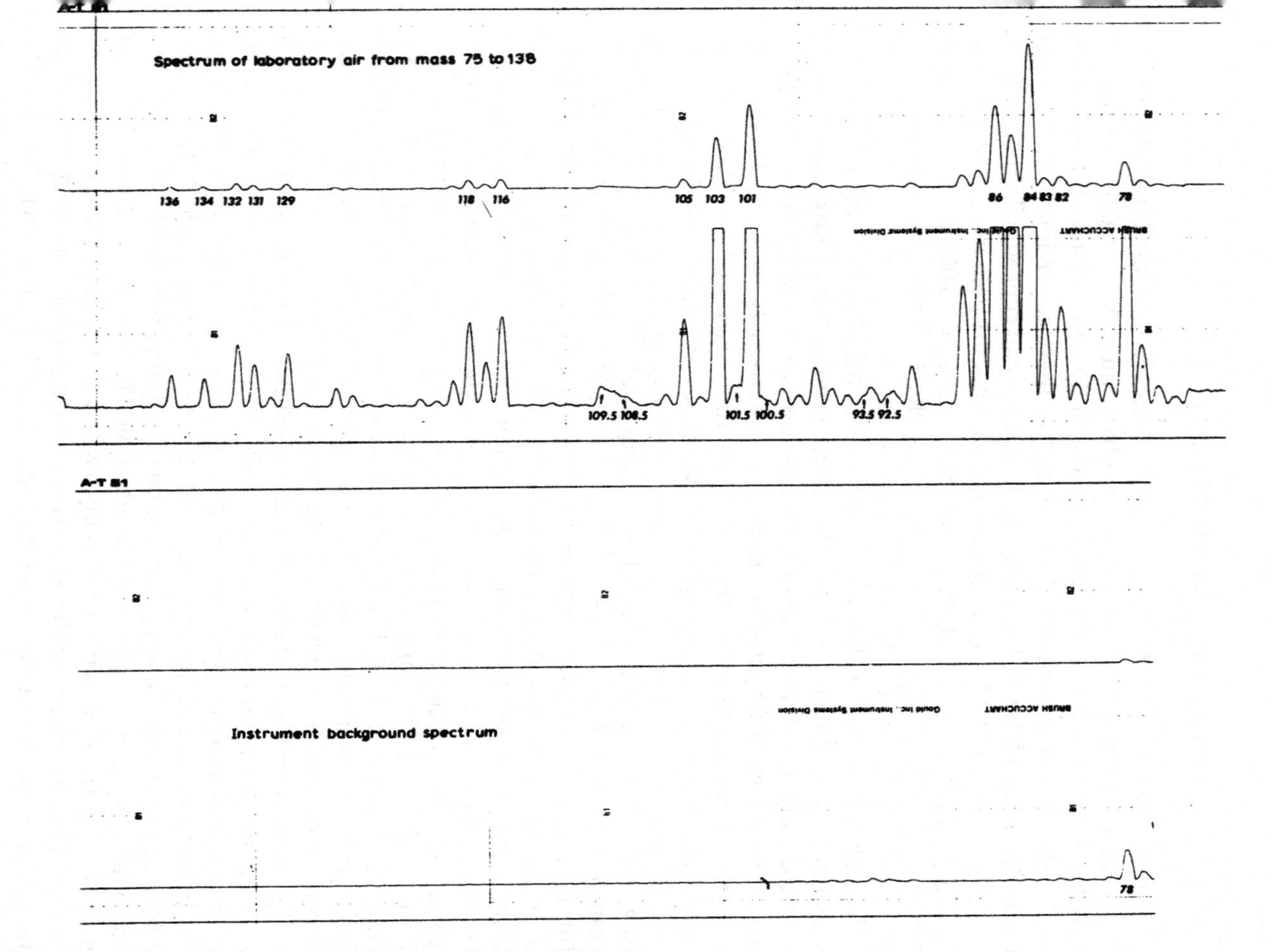

Fig. 12.7. Scan of laboratory air with off-axis multiplier [00].

filament is subjected to high oxygen partial pressures. Figure 12.7(b) is the vacuum system background scan taken after the variable leak valve was closed [2].

(4) *Electronic control requirements and capabilities*

In order to utilize fully the potential of the quadrupole for sophisticated quantitative analysis, the electronic controls must be capable of scanning rapidly, jump large mass differences very rapidly and not have the resolution (U/V ratio) or mass position, U, change after these excursions. For the techniques described later in this chapter, precisions and reproducibilities of peak heights of the order of 1% are required. For a quadrupole of $r_{rod} = 0.317$ cm, $V_{m/e\,600} = 1290$ V, $f = 1.36$ MHz and mass range of 1000 amu, a 1% peak height requirement at m/e 614 means that the U/V ratio must be regulated to better than 5 ppm for all conditions of scanning.

The ability to switch from one mass to any other mass in the range of the instrument in 1–5 msec is unique to the quadrupole and allows an analyst to monitor only those m/e values of interest in a GC–MS analysis. This technique, called mass fragmentography [3], is extremely useful in the analysis of very complex mixtures which are not fully separated by GC or require quantitation such as those found in biological or environmental studies. The technique is accomplished by setting the quadrupole at a particular m/e value (see Fig. 12.8) characteristic of the compound of interest by an external voltage source (external mass set) (Fig. 12.9) [4]. The ion current is sampled for typically 1–100 msec and the level is held in that channel (see Chapter XIII for a further discussion of multiple peak monitoring). For a second compound in the same analysis, a new channel is selected and a new mass set voltage switches the spectrometer to a new m/e value. Each channel (up to 16 in commercial equipment) can be sequenced and there is a separate output for each channel. The sample and hold circuitry keeps the output of any one channel constant until that channel is again sampled. The resulting chromatogram produced is specific only for those compounds that produce that mass fragment and hence is called mass fragmentography.

Ion monitoring of this sort enhances sensitivity because most of the time is spent sampling ion current at a m/e value of most information. There is no time spent scanning over areas of redundant or non-existent information. To utilize this technique and the capability of the quadrupole fully, the gain of the amplifier for the ion current should be programmable for each channel so that the dynamic range of the quadrupole (10^8–10^9) is not reduced to 10^4 as would be the case with a manually switched amplifier gain. It is also important that the spectrometer electronics be capable of switching and settling from one mass to another 400–500 amu from it in less than 5 msec. This technique has been used extensively to confirm the existence of trace contaminants in environmental analysis or drugs in overdose situations but it is

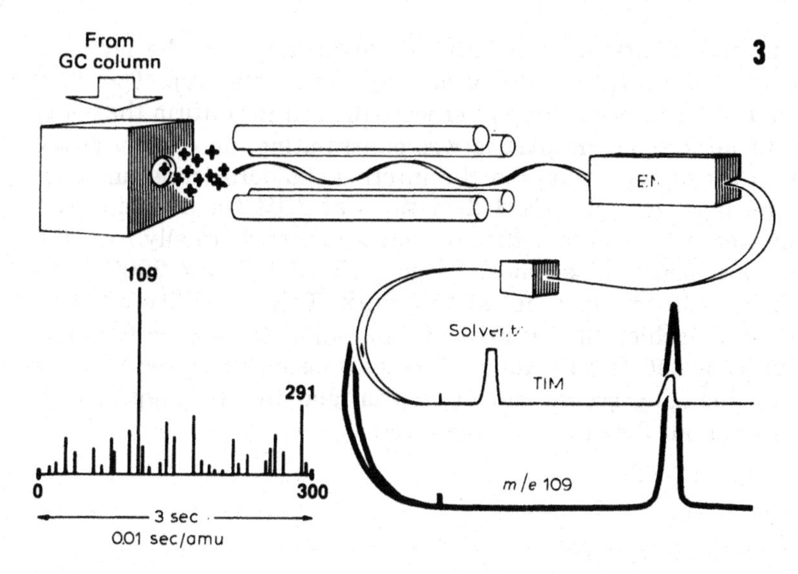

Fig. 12.8. Schematic of mass fragmentography technique (Finnigan Corporation).

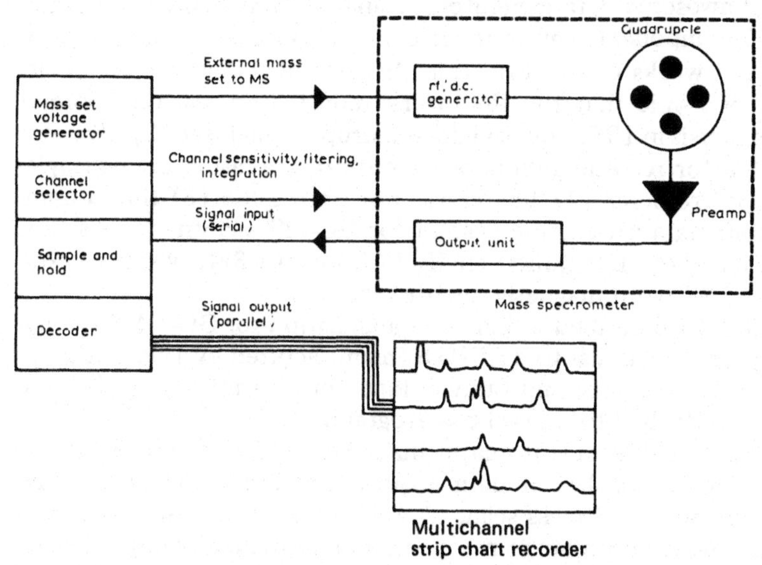

Fig. 12.9. Block diagram of a programmable multiple ion monitor device (Finnigan Corporation).

in the area of quantitation that it appears to have its greatest potential.

Strong and Atkinson [5] used the technique for measuring plasma concentrations of lidocaine and monoethylglycinexylydide (MEGX), a metabolite of lidocaine suspected of causing some of the toxicity associated with lidocaine treatment for cardiac arrhythmia. Their technique involved the use of trimecaine as an added internal standard for quantitation of the levels in plasma of lidocaine and MEGX after administration of the drug. The peak heights of two mass ions of each compound were ratioed to confirm that only the compound of interest in the plasma was contributing ion current at the *m/e* value chosen for quantitation and the internal standard trimecaine was monitored at one mass ion. Standard deviations of 3.1% for lidocaine and 7.4% for MEGX were reported over the concentration range usually found in clinical applications. Ions whose mass/charge differed by over 200% were monitored at 5 msec switching times in this work. This capability inherent to quadrupoles is not achievable by magnetic instruments because the difficulty of switching magnet currents rapidly makes it necessary to switch from mass to mass by changing acceleration potential and this technique limits their excursions to about 30% of a nominal mass.

C. AUTOMATION

The instrumental combination of a gas chromatograph with a mass spectrometer is a rather awesome data generator. A single 30 min analysis of a mixture of 50 or 60 components can produce enough spectra to require days of data reduction and weeks of identification. In view of its infancy as a mass spectrometer, it is ironic that the first work reported on real time computerized GC–MS analysis in 1967 employed a quadrupole analyser [6]. Systems had been reported for recording on tape the output of a fast scanning single-focusing mass spectrometer [7] but it was not until 1968 [8] that a computerized real time data acquisition system for GC–MS analysis for a sector instrument was reported. It required an IBM 1800 with 32K words of core storage and three 12K word magnetic disc drives.

Reynolds et al. [6] described a real time acquisition for GC–QMS system utilizing a computer very similar to a LINC 8 (manufactured by Digital Equipment Corp.) with 2K memory and built in tape unit operating a quadrupole mass spectrometer attached to a gas chromatograph.

The system was capable of automatic mass calibration including sample turn on and off, scope display of data acquisition, data acquisition, mass range capability of 500, 256 mass scans at 4 sec/scan, display any mass as a function of time (mass chromatogram), peak amplitude, and summation of all mass peaks in each scan plotted versus scan number (reconstructed gas chromatogram).

The obvious reason for the rapid development of a data system for quadrupole GC–MS was its particular suitability to being computer controlled. The spectrometer is stepped from ion peak to ion peak. No time is lost scanning between ion peaks. This maximizes sensitivity. The analyst can look at only those peaks of interest. Instead of being a passive data collector, the computer is the operator of the mass spectrometer. This allows a flexibility

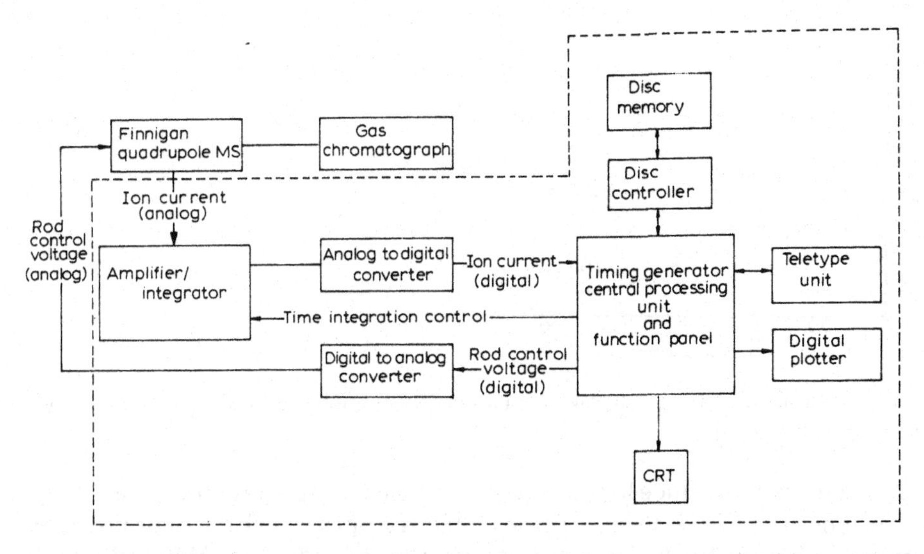

Fig. 12.10. Block diagram of an interactive GC–QMS data system (Finnigan Corporation).

that a data logging scheme cannot achieve. With the development of fast, stable electronics capable of working anywhere in a 1000 amu mass range, the computer-controlled quadrupole mass spectrometer has advanced techniques such as mass fragmentography well beyond the capability of the traditional magnetic mass spectrometer.

Current developments [9] in computerized GC–QMS allow real time acquisition and display of spectra and gas chromatogram in minicomputers with 8K core and disc storage (Fig. 12.10). Subsequent data inspection and manipulation is done interactively with immediate scope display. Plotting of hard copy data is done only *after* the operator has made decisions, subtracted background, quantitated areas of mass fragmentograms, compared isotope abundance ratios for identification of compounds [10], etc.

D. NEW TECHNIQUES

(1) *Stable isotope mass fragmentography*

The recent availability of stable isotopes has sharply increased their use in a wide variety of applications in the field of drugs and drug metabolites [11]. They have also been used extensively as internal standards for quantitation. Since the compound that is to be quantitated and the internal standard are chemically identical, the labeled standard can be added to the plasma or urine at the beginning of the sample preparation. Any losses that occur during sample work up will effect both compounds thus maintaining the quantitation. In very low level detection work, the use of larger quantities of the

References p. 306

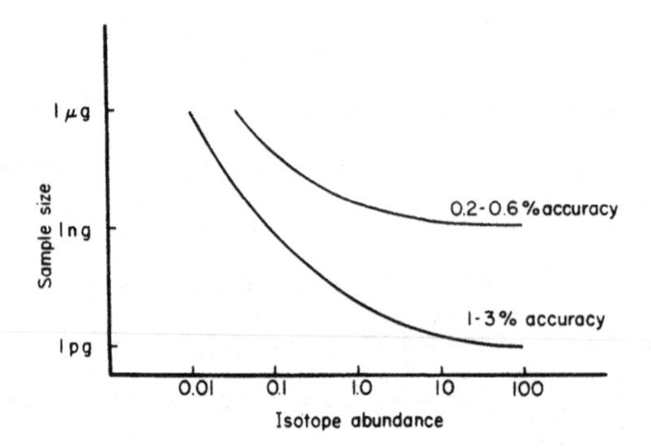

Fig. 12.11. Summary of capabilities of a GC–QMS computer system used for stable isotope quantitation [00].

labeled species will reduce loss of the non-labeled sample due to, for example, adsorption on the GC column. These stable isotopes can also be administered as tracers directly to the patient with no fear of radiation. Metabolic pathways and biological activity can thus be traced [12].

The quadrupole has added significant capabilities to these techniques. A comparison of reproducibilities, accuracies and instrumental parameters for using a QMS–computer to quantitate stable isotope ratios under static and gas chromatographic conditions has been made and compared with more traditional approaches [13].

A summary of the quantitation capabilities of a GC–QMS computer system used for stable isotope analysis is shown in Fig. 12.11 [14]. As would be predicted, the accuracy is reduced at lower levels because of ion statistics, absorptions, etc. These limits vary, of course, with the compound, availability of a fragment ion of sufficient size that is not contributed by a co-eluting compound, difficulties in chromatographing the compound, etc.

(2) *Chemical ionization*

The utilization of chemical ionization has developed from studies of ion–molecule phenomena [15] to a very useful analytical tool [16]. No ionization technique since electron impact has had as great an acceptance and use in mass spectrometry. It has especially extended the usefulness of GC–QMS. Here again, the quadrupole was the first instrument reported utilizing this technique for GC–MS work [17, 18].

At the time of writing, there are over 100 commercial GC–CIQMS instruments in use. The quadrupole's operation has unique advantages over a magnetic sector type spectrometer in adapting to this technique because of the higher analyser pressure tolerance and the low voltages employed in its source.

The technique relies upon generating chemically reactive ions in an ion source by electron impact of a reagent gas at 0.1–2 torr pressure. Because of

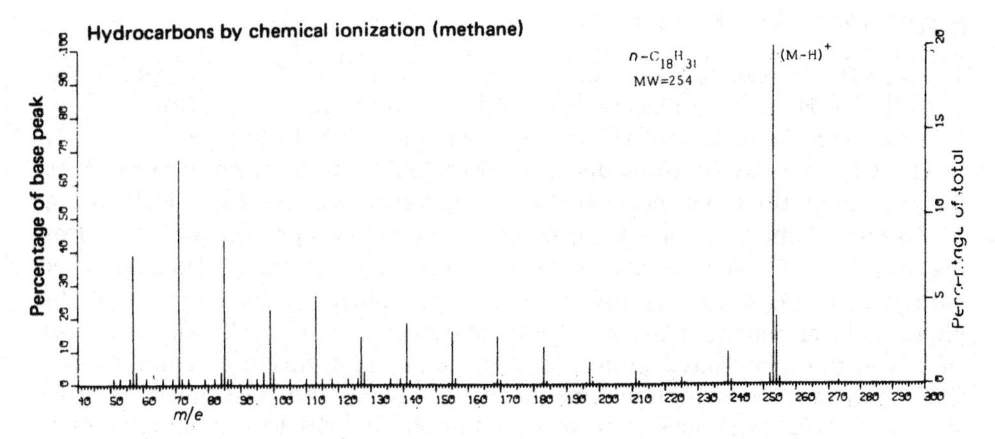

Fig. 12.12. Chemical ionization spectrum of a normal hydrocarbon.

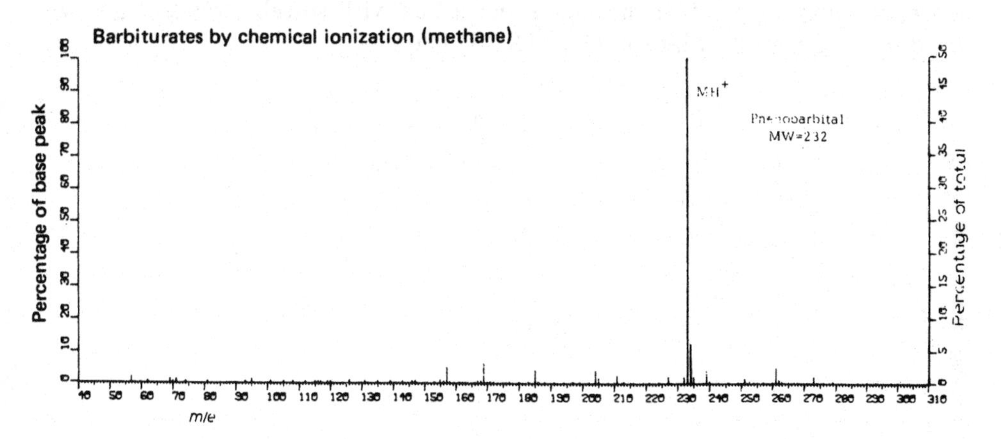

Fig. 12.13. Chemical ionization spectrum of phenobarbital.

the high pressure, these reactant ions undergo collision with the sample gas and the sample becomes ionized by proton transfer, charge exchange, or hydride transfer. For example, methane is a common reagent gas and it undergoes the following ionizations by electron impact.

$$CH_4 + e^- \rightarrow CH_4^+, CH_3^+, CH_2^+, CH^+, C^+, H_2^+, H^+ + 2\,e^-$$

These ions then react with methane as follows.

$$CH_4^+ + CH_4 \rightarrow CH_5^+ + CH_3$$

$$CH_3^+ + CH_4 \rightarrow C_2H_5^+ + H_2$$

$$CH_2^+ + CH_4 \rightarrow \begin{cases} C_2H_4^+ + H_2 \\ C_2H_3^+ + H_2 + H \end{cases}$$

$C_2H_3^+ + CH_4 \rightarrow C_3H_5^+ + H_2$

The significant reactant ions thus produced are CH_5^+ (48%), $C_2H_5^+$ (40%) and $C_3H_5^+$ (6%). Hence the plasma in the CI source produces both Brønsted and Lewis acids from methane. These ions then react with the sample.

The CI_{Me} spectra of *n*-octadecane in Fig. 12.12 results from hydride transfer and hence the ion appears at the molecular ion less one $(M - H)^+$. In Fig. 12.13 we see the CI_{Me} spectrum of phenobarbital via proton addition forming an $(M + H)^+$. The advantages of this technique are many. The simplified spectra resulting from the low energy of the reactions such as that of the phenobarbital enhances the sensitivity of detection as virtually all the ionization is in the protonated molecular ion. In complex mixture analysis where there may not be complete separation, it reduces the possibility of interference due to fragments from other compounds [19] thus making identification easier. This is especially important when used for mass fragmentography on biological systems. Many compounds of interest do not show any molecular ion in electron impact spectra but with suitable reagent gas will produce a quasimolecular ion (Fig. 12.14) [20].

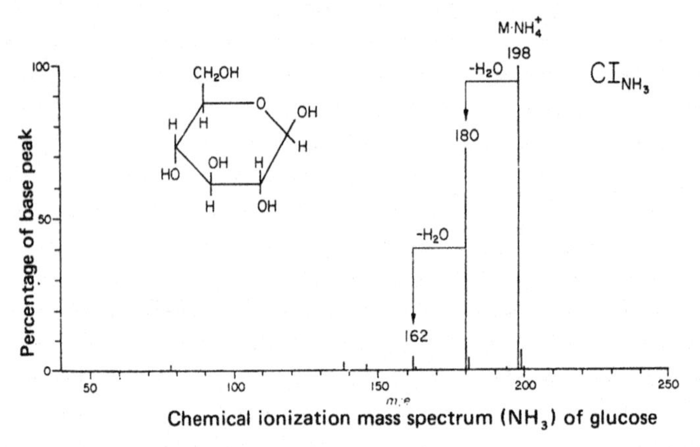

Fig. 12.14. Chemical ionization spectrum of glucose with ammonia as the reagent gas [00].

This technique also allows changes in the gas flow considerations. In the discussion of GC gas flow, separators, and ion source characteristics it was pointed out that source pressure was kept low enough to prevent ion–molecule reactions. If it was desired to eliminate the separator, a larger pump could be put on an EI source; however, the conductance out of the source is small and the pressure in the ionization region would become so high that ion–molecule reactions would predominate. This is undesirable for an electron impact source but is necessary for a chemical ionization source. If a diffusion pump of sufficient pumping speed to handle 20 atm ml min^{-1} is allowed to differentially pump the ion source and the conductance of source is adjusted

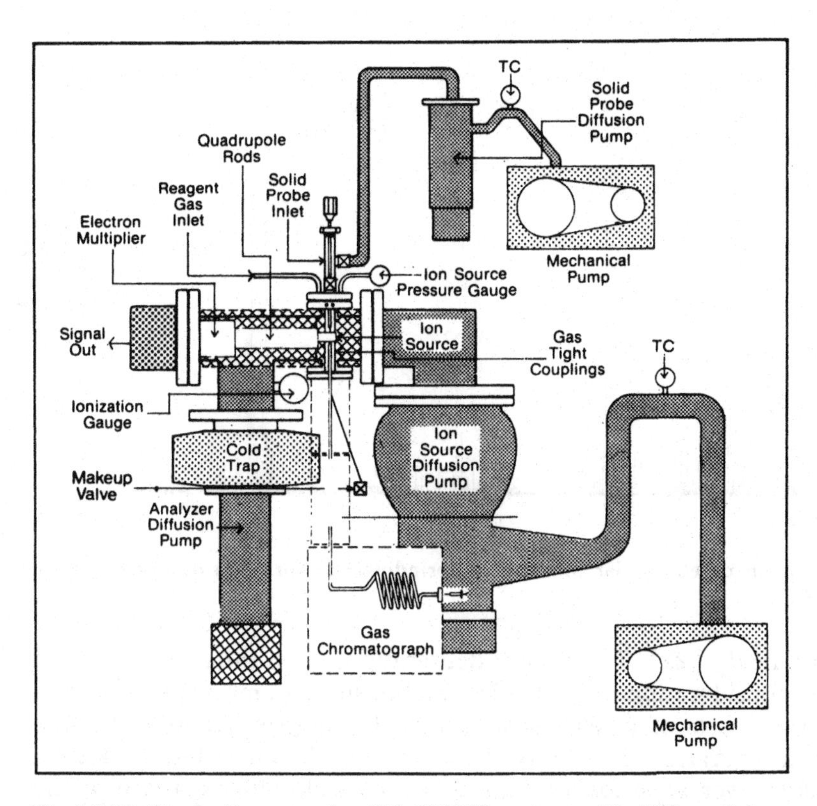

Fig. 12.15. Block diagram of a GC–CIQMS system with differential pumping (Finnigan Corporation).

to maintain a 0.5–1.0 torr pressure inside while allowing a 10^{-4} torr or less pressure at the filament, the source can handle gas flows necessary for packed column GC work. At the same time, the pressure in the analyser can be as high as 10^{-4} torr and maintain sensitivity and resolution. In addition, if a gas such as methane or isobutane is used as a carrier gas for the GC, it can also be the reagent gas for chemical ionization. Such a configuration is shown in Fig. 12.15. The transfer line between the GC and the MS when operating with the source at 1 torr is at sufficiently high pressure to permit the use of vacuum pump and valve (unheated) to evacuate any unwanted portion of the GC effluent such as the solvent. This will greatly reduce the rate at which this type of source contaminates when large numbers of samples are being analysed such as in clinical studies. The chemical ionization source has contributed even greater versatility to the technique of mass fragmentography [21]. Because of the predominant protonated molecular ion appearance in the spectra, additional advantages can be obtained from the use of stable isotopes with chemical ionization. This is especially true for detection and structural definition of a drug and drug metabolites in biological fluids.

For example, an antiarrythmic drug aprindine is labeled with deuterium

References p. 306

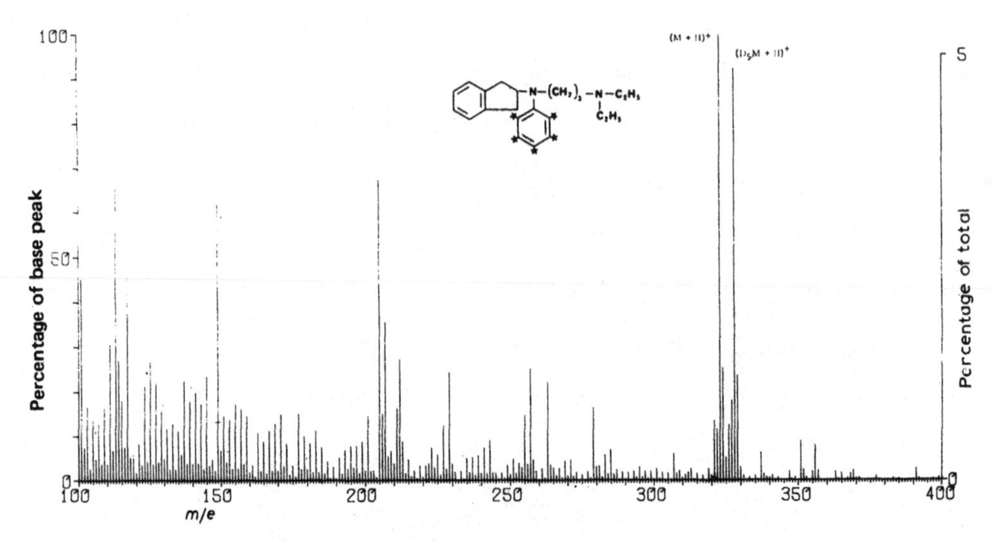

Fig. 12.16. Spectrum of equimolar mixture of aprindine and aprindine-d_5 corrected for background [00].

to give aprendine-d_5 [22]. The sites of deuteration, ★, on the phenyl ring are selected after consideration of probable metabolism. The mass spectrum of a mixture of equal parts of labeled and unlabeled drug (Fig. 12.16) shows the doublet with a separation of 5 amu. The mixture is then ingested. Biological fluids are taken over a period of time and analysed. The existence of the doublet in the CI mass spectrum greatly simplifies the locating of these eluting compounds even though the general analysis is very messy. In addition, the location of the sites of metabolic substituents which may be added during metabolism is made easier. Two of the metabolites which were recovered had added 16 to the molecular weight but had eluted at different times during the GC run. The difference between the two compounds was immediately apparent because in one the doublet was still 5 apart but in the other it was 4 apart. Thus ring hydroxylation had taken place on the unlabeled bicyclic undanyl portion of the molecule and in the other case on the labeled phenyl ring portion of the molecule. This displaced one of the deuteriums and reduced the doublet interval to 4.

An interesting application of quantitation by mass fragmentography CIQMS is described by Knight and Matin [23] for compounds that are unstable under gas chromatographic conditions. By use of solid probe sampling, tolbulamide (ether extract from plasma) and a deuterium-labeled internal standard are vaporized directly into the chemical ionization source. The mass ions of interest are monitored and areas of the mass fragmentogram measured and compared with an analytical curve. Even though plasmas are extremely dirty mixtures and gas–liquid chromatography was not possible, the compound specificity of the CIQMS mass fragmentographic method prevented

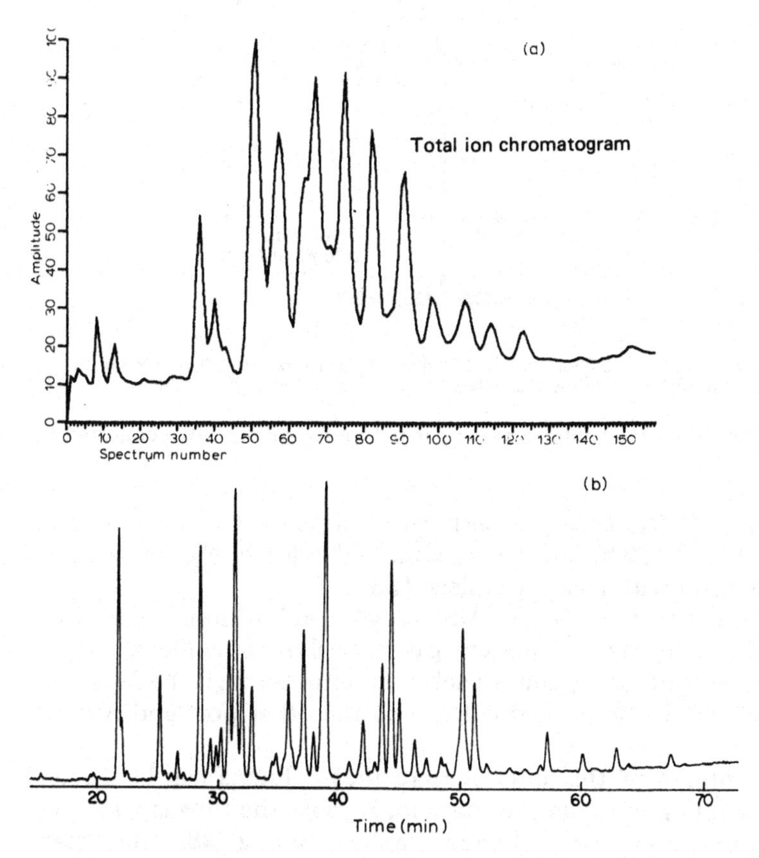

Fig. 12.17. Chromatogram of mixture of polychlorobiphenyls. (a) Aroclor 1254 by packed column GC–MS. (b) Aroclor 1254 by capillary column GC–MS, PZ-176, 1°C/min (200°C).

interference from unwanted materials and allowed good quantitation. The quantitative and chemical information afforded by these types of analyses have brought the mass spectrometer into consideration for routine use in the clinical laboratory for the first time (see Chapter XIII).

(3) *Capillary columns*

Capillary column GC–QMS is currently undergoing rapid development. The interest is generally brought about by techniques in production and use of high efficiency glass columns [24–33], improved injectors [34] and the unique GC–CIQMS interfacing [35]. Because of the low flow rates (1–2 atm ml min^{-1}) the effluent can be brought into combination EI–CI sources and the source pressure will be low enough to allow an EI spectra to be run. Figure 12.17 shows one of the obvious advantages of using a capillary column.

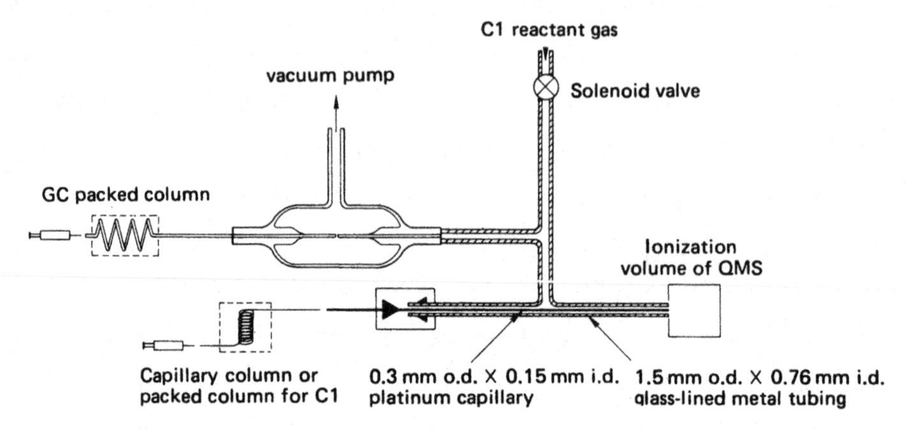

Fig. 12.18. Diagram for a co-axial interface for packed and capillary GC connected to EI–CIQMS [00].

Figure 12.17(a) is a GC–QMS analysis performed on Aroclor 1254, a mixture of polybiphenylethers (PCB's), done with a packed column. Figure 12.17(b) is the same mixture run with a glass capillary [36].

With an interface between the GC and MS as suggested by Blum and Richter [37] in Fig. 12.18, one can add reagent gas to analyse a capillary column effluent by chemical ionization, run a packed column through the separator for EI spectra or switch the packed column to the other port and run the effluent directly in.

Additional advantages of this interface are that columns can be disconnected from the coupling to the ion source and, because the flow is restricted by the small inside diameter of the platinum capillary tubing [38], the system is not vented to atmospheric pressure. This also allows the use of a reagent gas that is reactive with the liquid phase of the capillary column as the reactant gas and yet not have the two come in contact with each other. Blum and Richter have experienced serious background signal and baseline instabilities in the TIC gas chromatograms due to the interaction of H_2O and D_2O (reagent gases) with silicon liquid phases in the interface region.

Figure 12.19 shows the TIC gas chromatograms using this interface showing equal performance with respect to gas chromatogram resolution and retention time for EI, CI_{CH_4} and CI_{D_2O} analyses of a monoterpene mixture.

E. CONCLUSION

As each technical problem has been solved or new device developed, new areas of application have opened for quadrupole mass spectrometry. Now well established as the instrument of choice for GC–MS, the device appears to have a dominant role in new developments currently being investigated.

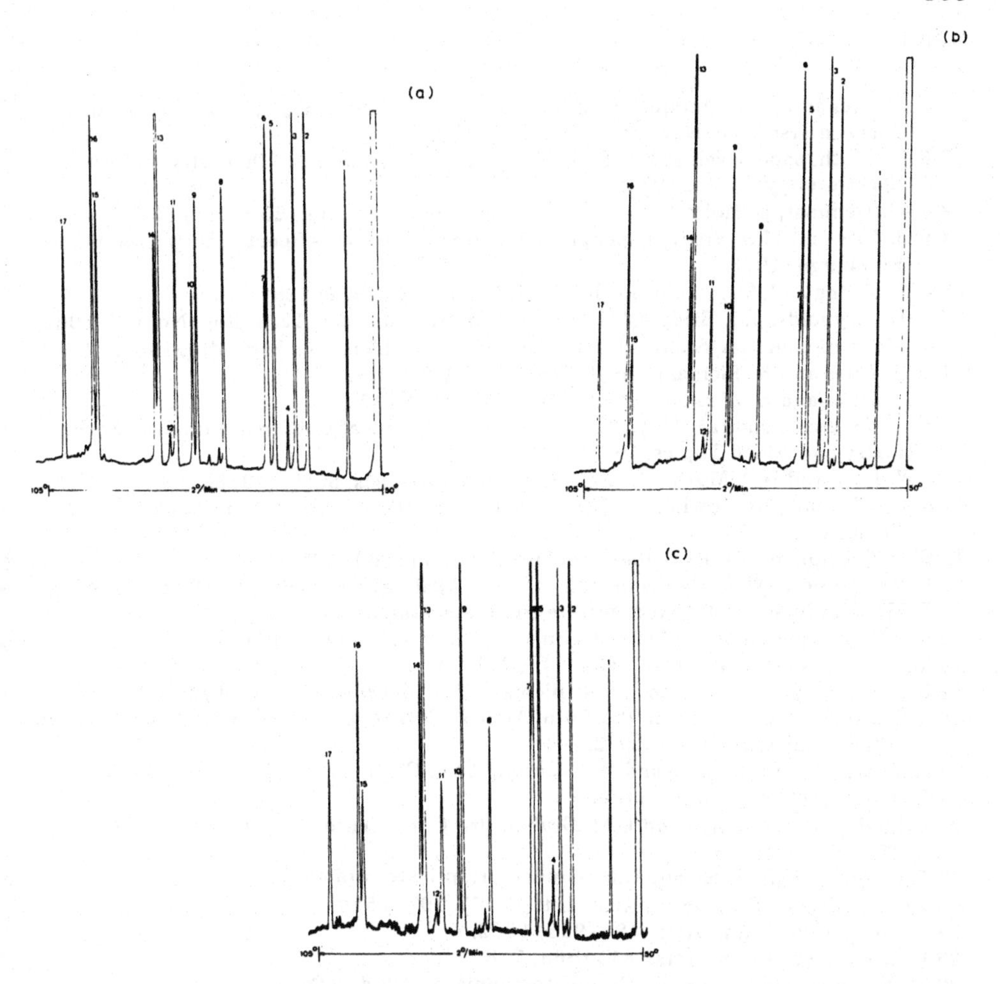

Fig. 12.19. Total ion current chromatograms of a monoterpene analysis utilizing (a) EI, (b) CI (methane), and (c) CI (deuterium oxide) [00].

Of these emerging methodologies, i.e. new EI–CI source designs, atmospheric ionization, rf plasma ionization [39], field desorption and LC–MS, all have been done with a quadrupole and some uniquely so. It is no longer necessary for the advocates of the quadrupole to prove its adequacy.

There will certainly be improvements in performances and new techniques for quadrupole systems, but the greatest potential of this device is in its special configuration for specific analyses. The complexities of our scientific endeavors and our environment have lead to the design of machines to fit a specific problem or analysis rather than fitting the analysis to the instrument.

REFERENCES

1 W.H. McFadden, Techniques of Combined Gas Chromatography–Mass Spectrometry, Wiley-Interscience, New York, 1973, p. 168.
2 R.D. Smith, paper given at Pittsburgh Conference on Analytical Chemistry and Applied Spectroscopy, March 1974.
3 C.-G. Hammar, B. Holmstedt and R. Ryhage, Anal. Biochem., 25 (1968) 532.
4 W.J. Fies and M.S. Story, paper given at International Mass Spectroscopy Conference, Edinburgh, 1973.
5 J.M. Strong and A.J. Atkinson, Jr., Anal. Chem., 44 (1972) 2287.
6 W.E. Reynolds, J.C. Bridges, T.B. Coburn and R.B. Tucker, Tech. Rep. No. IRL-1062, Dept. of Genetics, Stanford University School of Medicine, Palo Alto, 1967.
7 R.A. Hites and K. Biemann, Anal. Chem., 39 (1967) 965.
8 R.A. Hites and K. Biemann, Anal. Chem., 40 (1968) 1217.
9 V. DaGragnano and H.P. Hotz, paper given at 21st Annual Conference on Mass Spectrometry and Allied Topics, 1973.
10 J.W. Rote and W.J. Morris, J. Ass. Offic. Anal. Chemists, 56(1) (1973) 188.
11 M.G. Horning, J. Nowlin, K. Lertratanangkoon, R.N. Stillwell and R.M. Hill, Clin. Chem. 19 (1973) 845.
12 R.M. Caprioli and D. Rittenberg, Biochemistry, 8 (1969) 3315.
13 R.M. Caprioli, W.J. Fies and M.S. Story, Anal. Chem., 46 (4) (1974) 453A.
14 R.M. Caprioli and E.J. Heron, private communication.
15 M.S.B. Munson and F.H. Field, J. Amer. Chem. Soc., 88 (1966) 1621.
16 M.S.B. Munson, Anal. Chem., 43 (11) (1971) 29A.
17 G.P. Arsenault, J.J. Dolhun and K. Biemann, Chem. Commun., (1970) 944.
18 M.S. Story, paper given at the 20th Annual Conference on Mass Spectrometry and Allied Topics, June 4–9, 1972.
19 H.M. Fales, G.W.A. Milne and T. Axenrod, Anal. Chem., 42 (12) (1970) 1432.
20 R. Foltz, private communication.
21 R.L. Wolen, E.A. Ziege and C.M. Gruber, Jr., Clin. Pharmacol. Therap., 17(1) (1975) 15.
22 R.L. Wolen, E.A. Ziege and C.M. Gruber, Jr., private communication.
23 J.B. Knight and S.B. Matin, Anal. Lett., 7 (7) (1974) 529.
24 K. Grob, Helv. Chim. Acta, 48 (1965) 1362.
25 K. Grob, Helv. Chim. Acta., 51 (1968) 718.
26 M. Novotny and U. Tesarik, Chromatographia, 1 (1968) 332.
27 M. Novotny and A. Zlatkis, J. Chromatogr. Sci., 8 (1970) 346.
28 R.G. Mathews, R.D. Schwarts, M. Novotny and A. Zlatkis, Anal. Chem., 43 (1971) 1161.
29 M. Novotny and A. Zlatkis, Chromatogr. Rev., 14 (1971) 1.
30 J. Merle d'Aubign, C. Landault and G. Guiochon, Chromatographia, 4 (1971) 309.
31 G.A.F.M. Rutten and J.A. Luyten, J. Chromatogr., 74 (1972) 177.
32 G. Alexander and G.A.F.M. Rutten, Chromatographia, 6 (1973) 231.
33 A.L. German, C.D. Pfaffenberger, J.-P. Thenot, M.G. Horning and E.C. Horning, Anal Chem., 45 (1973) 930.
34 K. Grob and G. Grob, Chromatographia, 5 (1972) 3.
35 W. Blum and W.J. Richter, Tetrahedron Lett., 11 (1973) 835.
36 E.C. Horning, private communication.
37 W. Blum and W.J. Richter, private communication.
38 N. Neuner-Jehle, F. Etzweiler and G. Zarske, Chromatographia, 6 (1973) 211.
39 B. Hoegger and P. Bommer, Int. J. Mass Spectrom. Ion Phys., 13 (1974) 35.

CHAPTER XIII

MEDICAL AND ENVIRONMENTAL APPLICATIONS

G. Lawson

Mass spectroscopy has advanced over the past twenty years from its original specialized application in physics research to common everyday use by chemists and biochemists and is now being tentatively applied to medical research. The development of the quadrupole approach has been even more recent, perhaps over the last five or ten years, and so it is not surprising that the environmental and particularly the medical applications are still in the initial stages.

There are two excellent accounts by Milne [1] and by Waller [2] of the application of mass spectrometry (as a technique) to medical and biochemical research, but the majority of the data reported was obtained with large double-focusing magnetic instruments. Unlike the medical field, where in general copius amounts of sample are available, environmental research demands ever-increasing detection capabilities and to provide these, physicists are currently engaged in designing improved ion sources and detectors.

A. MEDICAL APPLICATIONS

The delay in the application of mass spectrometric techniques to clinical and medical research can be attributed to the size and cost of earlier instruments. Their complexity meant that reliable quantitative data could only be obtained by persons highly skilled in physical sciences. Perhaps the major problem was ineffective system design since in most medical applications the mass spectrometer should be used as a component in a routine to provide physiological data. As a result, the mass spectrometer characteristics, such as the input and output modes, should be compatible with the routine requirements. Only over the past few years have attempts been made to design a system using this approach. Mosharrafa [3] has published a set of "guide-lines" defining the optimum system parameters, of which the ability to transport the mass spectrometer to other test equipment, patients' bedsides and operating theatres is highly desirable. The instrument must therefore be self-contained, requiring no water cooling. Simplicity of operation is of paramount importance if routine use by medical personnel is envisaged. Ideally, all that would be required of the operator is a knowledge of the mass number of the constituent to be analysed; all other parameters should be preset.

One of the most stringent requirements of any analytical instrument is that it should have a rapid response time (usually defined as the time required for the output to record 90% of a step change in sample constituent at the inlet). In a mass spectrometer, the rise time is primarily a function of the physical properties of the inlet system and, to a lesser extent, is dependent on the electronic circuitry of the analyser and detector and, of course, on the type of sample being analysed. Response times of less than 0.1 sec may be required for breath-by-breath analysis of respiratory gases.

Instrumental reliability and reproducibility are also of extreme importance and, therefore, provision must be made to permit calibration "in the field". Reproducibility over lengthy periods involves great attention being paid to instrumental stability during the design stage, particularly for the ion source, where stable beam configurations must obtain. This must be combined with easy replacement of the filament assembly without loss of ion source alignment.

In physiological research, the analyser should be capable of determining partial pressures to an accuracy of ± 0.5%. As indicated in Fig. 13.1, the mass range requirements vary widely, from between m/e 12–50 for respiratory analysis to values in excess of m/e 2000 for biochemical analysis.

The following sections describe those areas of medical applications of mass spectrometry in which quadrupoles are becoming increasingly employed.

(1) *Respiratory gas analysis*

The lung is responsible for the introduction into the body of its main energy source, oxygen, and the removal of its main metabolite, carbon dioxide. Any investigation into the function or malfunction of the lungs must have, as its starting point, an accurate knowledge of inspired and expired breath composition and of how this composition changes during the course of respiration.

Several techniques for the continuous monitoring of one component of respiratory gases at a time, were available by 1950, notably thermal conductivity, infrared absorption, paramagnetic susceptibility and spectral emission [4, 5]. About this time, it was recognized that mass spectrometry was also a promising analytical method [6, 7].

Fowler [8] and West [9] have discussed the characteristics a mass spectrometer must possess for it to be of value in respiratory studies and these are summarized below.

(i) A mass range (in amu) of at least 25–50 and preferably 4–50, which would enable all gases of physiological interest, N_2, O_2 and CO_2, together with gases of experimental value, He, A, and N_2O, to be monitored. A minimum acceptable resolution of 20 on a 10% valley definition would also be necessary.

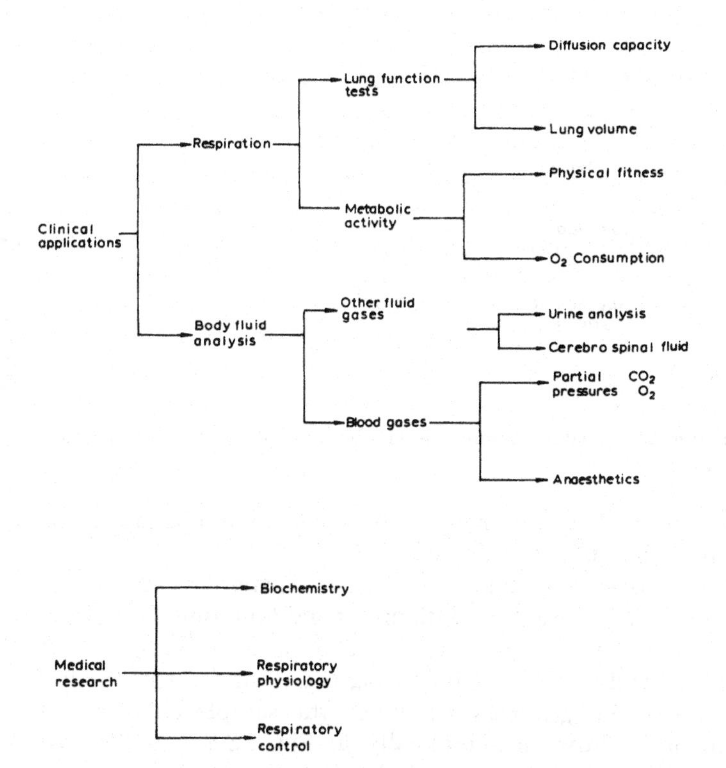

Fig. 13.1. Possible applications of mass spectrometry in clinical and medical research.

(ii) The ability to record continuously the partial pressures of at least three gas components.

(iii) A sensitivity such that full scale deflection is produced by a gas component of less than 5 mol.% with a signal-to-noise ratio such that the accuracy of measurements is not impaired. Combined with this, the instrument should give a linear response to the partial pressure of the gas component at the sampling point with an accuracy of $\pm$ 2%.

(iv) The sampling system should have a response time of 100 msec or less without affecting the gas composition either by fractionation or condensation [10] along its length which should be such as to facilitate easy connection between patient and apparatus. The inlet should draw off a gas flow which is less than 1% of the subjects' turnover (50 ml min^{-1} or less) and maintain a constant sample chamber pressure in the face of changes in gas composition, viscosity, temperature and water vapour content.

(v) A vacuum system such that partial pressures in the ionizing region are proportional to those in the sample chamber, irrespective of changes in viscosity, and capable of following partial pressure changes in the sample within the time available (i.e. between breaths). Unfortunately, to date no commercial instrument satisfies all these requirements, but despite the complexity

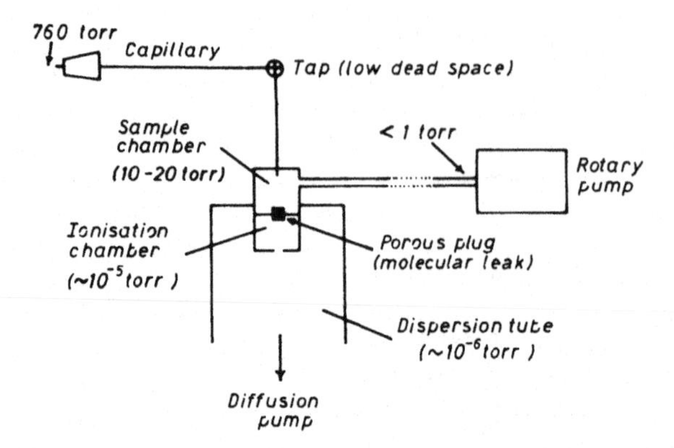

Fig. 13.2. Schematic diagram of a heated stainless steel capillary inlet system for respiratory gas analysis applications.

and stringent requirements of its electronics, the quadrupole appears to be the optimum choice for many uses in the medical field.

The most frequently chosen inlet system is the heated stainless steel capillary tube [11], typically 0.25 mm internal diameter and 2 m long. One particular advantage of this system is the pressure drop along its length from atmospheric to approximately 10 torr eliminating the need for any further pressure reduction before the gas passes through the sample chamber. A satisfactory arrangement is shown schematically in Fig. 13.2 and detailed analyses of the performance and design considerations have been given by Fowler and Hugh-Jones [7].

A magnetic instrument with multiple collectors is the only way so far developed to give truly continuous peak monitoring by mass spectrometry, but such an instrument generally lacks any versatility. The quadrupole has a very fast cycle time which can be reduced further to give the impression of continuously monitoring either four or eight peaks.

In many instruments, the ramp voltage controlling the rf and d.c. potentials (see Chapter VI) may be replaced by an external analogue voltage, usually in the range 0–10 V, thus enabling external control of the m/e value transmitted by the quadrupole. Two applications of this technique are illustrated below. The relationship between a conventional quadrupole mass spectrum and the linear ramp are shown in Fig. 13.3(a), whilst Fig. 13.3(b) represents the effect on the spectrum produced when the control voltage has the form indicated. By replacing the linear ramp with a short series of discontinuous ramps, a much abbreviated spectrum is produced with only the peaks of interest being monitored. Using this technique, redundant information is eliminated and the scan speed over the regions of interest is consequently increased.

A variant on this method is the use of a stepped voltage function [12] under which circumstances the ion current appears in histogram form, Fig.

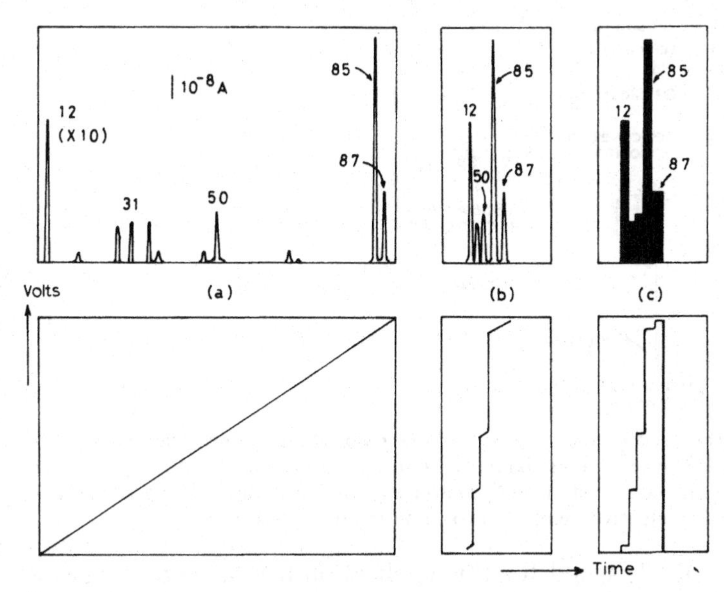

Fig. 13.3. Modes of action of the programmable peak selector for a quadrupole mass filter. The upper traces depict the various types of ion current output and the lower traces show the control voltage waveforms needed to generate the outputs shown.

13.3(c). The units which provide this method of operation [13] generally incorporate a series of channels containing sample-and-hold circuits such that at each voltage step the output data are gated into the appropriate channel. This effectively gives continuous monitoring of a number of ion intensities.

One commercially available respiratory gas analyser designed around a quadrupole mass filter [14] (Figure 13.4) comprises a completely self-contained vacuum system requiring neither water cooling nor liquid nitrogen and has complete bake-out facility and fail-safe protection. The inlet system is a heated stainless steel capillary tube. The analyser has a resolution of ca. 120 and a scan speed of 1 msec when used in conjunction with the multipeak monitor as shown. The response time for this apparatus is quoted as < 100 msec. The Bendix Pulmonary Function Analyser is a similar instrument but is unique in the sense that the mass filter is coupled to a flow meter, thereby enabling the measurement of many lung functions particularly the oxygen consumption.

In one completely digitally controlled modular quadrupole system [15] for both respiratory and blood gas analysis, the design concept has been such that each discrete control parameter may be replaced by changing a single module, considerably reducing instrument down time.

Brubaker and Marshall [16, 17] have described a miniaturized quadrupole breath analyser small enough to allow the entire vacuum system including the mass filter and the ion pump to fit under an astronaut's chin, inside his

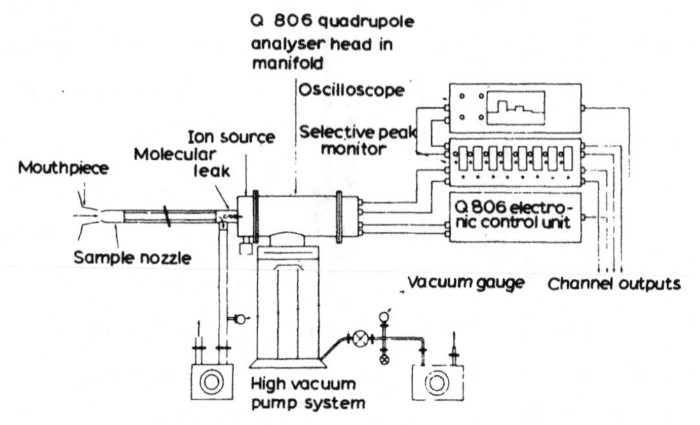

Fig. 13.4. A block diagram of a commercially available respiratory gas analyser comprising (1) vacuum system with switch panel, bake-out control, combined vacuum gauge and fail-safe facility, (2) heated inlet system with heater control, (3) quadrupole analyser and control unit, (4) 8-channel selective peak monitor, and (5) oscilloscope.

helmet. The quadrupole designed for this application has 5.08 cm long rods with a hyperbolic cross-section to provide a high-performance instrument. Peak monitoring was restricted to CO_2, N_2, O_2 and H_2O and the entire scan period, including time for automatic zeroing of the output amplifiers from the Faraday-cup collector, requires about 7.5 msec, a short interval compared with the 50 msec response time constant of the inlet system. In order to obtain stable outputs for each breath component, the peaks must have broad tops and the supply voltages should be extremely stable. The use of a crystal-controlled oscillator, precision rectification and feedback control produced a quadrupole with a transmission efficiency which remained constant for a year without adjustment. The only problem encountered with this instrument was the oxidation of hydrogen to water by the hot filament. This was alleviated by coating the filament with lanthanum hexaboride to reduce its operating temperature and replacing the closed ion source with an open design, thereby increasing the sensitivity of the apparatus to oxygen.

The majority of the experimental work on respiratory gas analysis has involved the use of fixed multi-collector magnetic instruments but there is now evidence of a move towards the quadrupole.

(2) *Blood gas analysis*

Determination of the concentration of gases in the blood at various locations in the circulatory system is of great importance as an aid in evaluations of diseases of the heart and lung. Present methods for determining these concentrations are generally time-consuming and require the handling of blood samples drawn under controlled conditions. Provided with a suitable sampling system, mass spectrometry makes such determinations an in vivo possibility. Another application of such a technique is the continuous monitoring of the partial pressures of volatile anaesthetics in the blood of a patient undergoing surgery. The versatility of the quadrupole is a further asset in that it can measure the concentrations of any gas within its mass range. It could therefore be used to determine indirectly the blood flow within certain regions of the circulatory system by the injection of small amounts of non-respiratory gases, for example N_2O, Ar or He.

In 1966, Woldring et al. [18] described the application of mass spectrometry to the continuous in vivo analysis of blood gas tensions. These investigations have been continued and expanded by Brantigan et al. [19] and Wald et al. [20, 21]. As with respiratory gas analysis, the quadrupole is only now beginning to have any impact on this particular field of investigation [3, 22]. The type of analyser required and its performance specifications are met by those instruments produced for respiratory analysis, but it is worthwhile describing the types of inlet systems that are used.

Blood gas analysis poses probably the most stringent requirements of any on a mass spectrometer inlet system; not only must it permit the diffusion of dissolved gases through a membrane whilst preventing the passage of any liquid but it must also be harmless to the patient. Several techniques have been employed, all using diffusion membranes or porous plugs in stainless steel or teflon tubing. Woldring et al. [18] described a technique involving gas sampling through a latex membrane applied to the tip of a standard polyethylene catheter, whereas Brantigan et al. have used silicon rubber plugs [19] and teflon membranes [19a]. Wald et al. have also reported the use of silicon rubber membranes [22]. A very simple catheter design has been reported by Key [23] in which a composite capillary tube, 1.5 mm o.d., of nylon lined with polyurethane is used. A porous bronze plug is fitted into the end of the tube which is then dipped into a silicon rubber dispersion, removed and heated at 150°C to give a very fine polymerized membrane. The basic construction of these probes is shown schematically in Fig. 13.5. In all these instances the major constructional details of the inlet systems were basically those shown in Fig. 13.2.

Similar techniques to those already mentioned could be used to measure the partial pressures of gases dissolved in other body fluids such as urine or cerebrospinal fluid. Information related to the gas pathways, exchange mechanisms and the roles played by these gases in the human body might be of clinical value in tests to evaluate organ functions. In this rapidly expanding area of application of mass spectrometric techniques, the quadrupole seems destined to play a major part [15, 23].

(3) *Drug detection and analysis*

The proliferation over the past few years of the drugs available for sedative, tranquilising and analgesic purposes combined with narcotics abuse, has

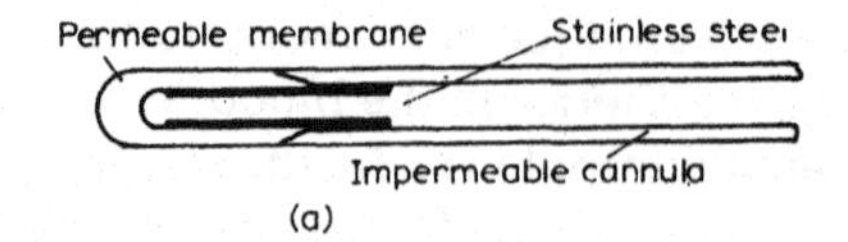

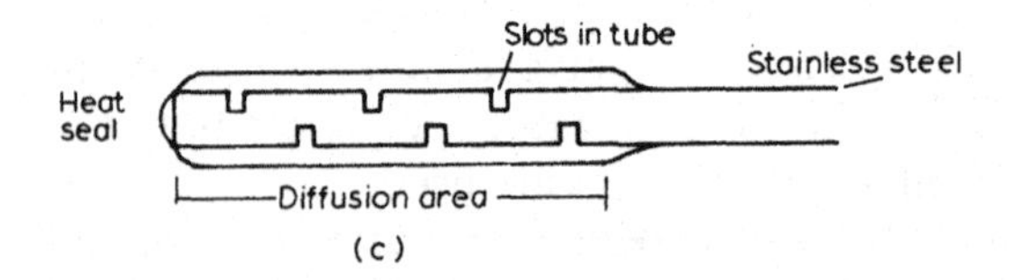

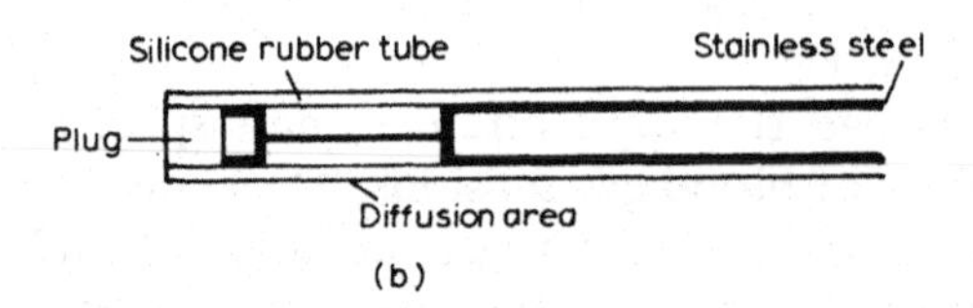

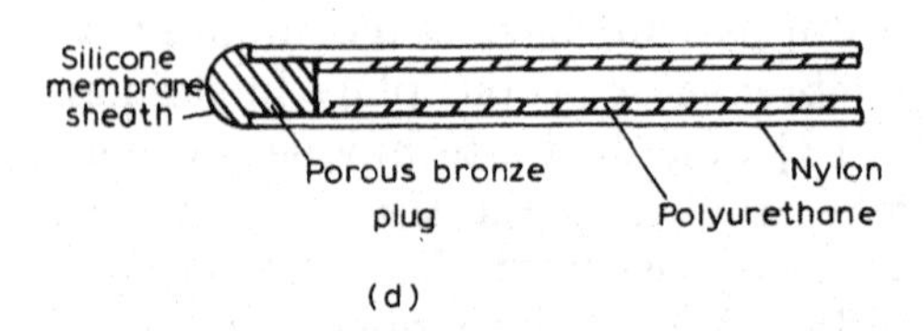

Fig. 13.5. Schematic diagram of the semipermeable region of the probes used in blood gas analysis. (a), (b), and (d) Silicon rubber membranes; (c) teflon membrane.

produced a large number of overdose cases, raising problems for both hospitals and forensic laboratories. Rapid and accurate identification of the drug concerned is essential in treating casualties and it is at this stage that both the gas chromatograph–mass spectrometer and the chemical ionization mass spectrometer (see Chapter XII) are becoming widely used. Since many of the common drugs have parent and/or fragment ions at the same mass unit or just one or two units apart, precise mass analysis is an absolute necessity.

A commercially available instrument [24] produced specifically to tackle this problem is designed around a quadrupole with a mass range of 0–800 amu in a single sweep. The minimum scan rate of the instrument is 3 sec over the entire mass range with a mass location stability of ± 0.15 amu/24 h and there is a mass marking facility. Sample introduction is by gaseous inlet, solid probe or by gas chromatographic sampling. The quadrupole utilises a chemical ionization source employing isobutane as the reagent gas operating at 1 torr pressure. Isobutane has proved [25] a simple reliable and efficient reagent for each drug and its metabolite and no serious overlap has been found for any of these compounds or other natural products found in urine. The entire

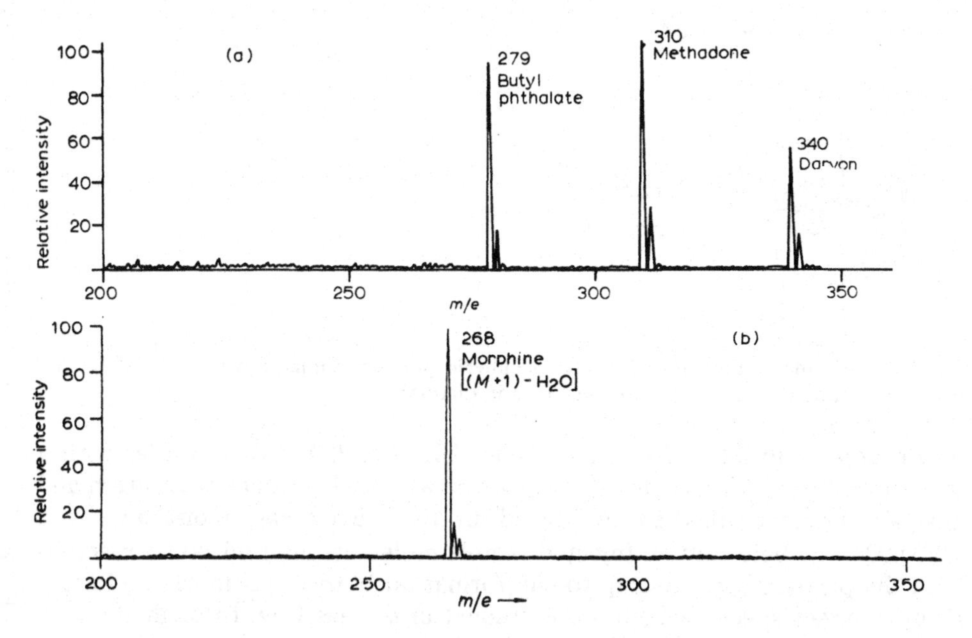

Fig. 13.6. (a) Chemical ionization (isobutane) mass spectrum of stomach extract containing 1 ppm Methadone and 2 ppm Darvon. (b) Mass spectrum of a remote breath analysis from a patient after a morphine injection. Chemical ionization using isobutane.

system is designed to minimize down time; all key voltages are independently monitored and fail-safe provisions are incorporated. The application of this system to the analysis of stomach content is illustrated in Fig. 13.6(a), where 1 ppm of Methadone and 2 ppm of Darvon are easily identified in a simple chloroform extract of a raw urine sample. Since the peaks in the spectrum represent 2×10^{-4} of the total sampling time, the sensitivity of the instrument is of the order of 0.01 ppb for analysis of extracts. A further demonstration of this sensitivity is the detection of drugs in exhaled breath. A patient, who had received an injection of 5 grains of morphine, exhaled into a bubbler containing a solvent for several minutes after injection. The extract from the solvent was analysed and the mass spectrum, Fig. 13.6(b), representing one tenth of one exhalation, clearly indicates the presence of morphine.

Horning et al. [26] have designed a novel mass spectrometer system with an atmospheric pressure ion source for the detection of picogram quantities of compounds of biological interest. The apparatus is shown in schematic form in Fig. 13.7. The source body is constructed of stainless steel incorporating a glass lining to prevent decomposition on the metal surfaces since the entire source is usually operated at 200°C. The carrier/reactant gas employed was nitrogen passed over a molecular sieve to remove water and oxygen. The ions were extracted from the source via a 0.0025 cm diameter pinhole aperture in a nickel disc of the same thickness directly into the

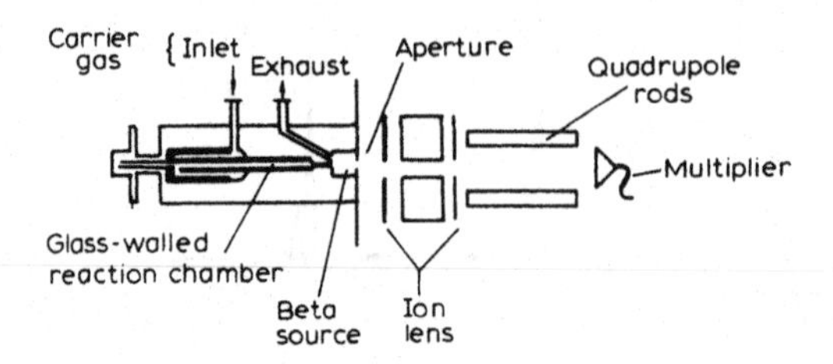

Fig. 13.7. Schematic diagram of the atmospheric pressure ionization source (API). The reaction volume is about 1 cm long and 1 cm in diameter.

quadrupole analyser, a Finnegan model 1015. A PDP 8 E computer with a laboratory-built interface and display scope was used for data acquisition and analysis. Primary ions are produced in the carrier gas, moist air, by a 12.5 mCi ^{63}Ni beta source. Injected sample molecules are ionized by reaction with the primary ions leading to the formation both of positively and negatively charged species which are entrained in the gas flow through the pinhole aperture into the mass analyser. The high-vacuum region of the apparatus was maintained at approximately 10^{-5} torr although the actual pressure in the ion focusing lens and quadrupole region was estimated to be an order of magnitude higher. A conventional electron impact ion source was also incorporated to permit calibration spectra to be run. Complete freedom of ion lens and quadrupole bias voltages was provided to permit optimisation of conditions for either positive or negative ion monitoring.

The ion–molecule reactions leading to the formation of sample ions are fairly complex and are summarized below.

(a) Positive ions (data from Good et al. [27])

Primary ions

$$N_2 + e^- \rightarrow N_2^+ + 2\,e^-$$

$$H_2O + e^- \rightarrow H_2O^+ + 2\,e^-$$

$$O_2 + e^- \rightarrow O_2^+ + 2\,e^-$$

Ion–molecule reactions

$$N_2^+ + 2\,N_2 \rightarrow N_4^+ + N_2$$

$$N_4^+ + H_2O \rightarrow H_2O^+ + 2\,N_2$$

$$H_2O^+ + H_2O \rightarrow H_3O^+ + OH$$

$$H_3O^+ + n\,H_2O \xrightarrow{N_2} H^+(H_2O)_{n+1}$$

Sample ionization

$Sample + N_4^+ \rightarrow (Sample)^+ + 2\ N_2$
$Sample + H_3O^+ \rightarrow H^+(Sample) + H_2O$
$Sample + H^+(H_2O)_n \rightarrow H^+(Sample) + n\ H_2O$

(b) Negative ion formation (after Karasek et al. [28])

(i) In air

Primary ions

$e^-_{fast} + M \rightarrow 2\ e^-_{slow} + M^+ \quad (M = N_2, O_2, H_2O, etc.)$
$e^-_{slow} + O_2 \rightarrow O_2^-$

Ion–molecule reactions

$O_2^- + H_2O \rightarrow (H_2O)\,O_2^-$
$(H_2O)\,O_2^- + n\ H_2O \rightarrow (H_2O)_{n+1}O_2^-$ (n generally = 2)

Sample ionization

$Sample + (H_2O)_nO_2^- \rightarrow (Sample)\,O_2^- + n\ H_2O$

(ii) In pure dry nitrogen

$e^-_{fast} + N_2 \rightarrow 2\ e^-_{slow} + N_2^+$

Sample ionization

$Sample + e^-_{slow} \rightarrow (Sample)^-$

Experimental work with reference compounds indicated that two circumstances were particularly favourable for the detection of specific types of compound without interference from solvents or from many naturally occurring substances. In the positive ion mode of operation, basic drugs were detectable through MH^+ ion formation. In the negative ion mode, compounds with high electron affinities, barbiturates and pesticides, for example, were easily detectable. The use of negative ion detection has, as an added advantage, the reduction of the solvent peak intensities, as can be seen from Fig. 13.8 which compares the two modes of operation for a range of drugs. So far very little interference is seen when both large (500 ng) and small (2 ng) amounts of sample are ionized at the same time, but high concentrations of sample ions and molecules may lead to cluster formation if these are stable under the ion source conditions. A much improved sensitivity has recently been reported [29] with a minimum detectable sample concentration of 150 fg in ~ 2 μl of benzene. This technique of employing a high or atmospheric pressure ion source is particularly exciting and, as will be seen later in this chapter, has many possible applications.

A great deal of analysis of both drugs and other biologically significant

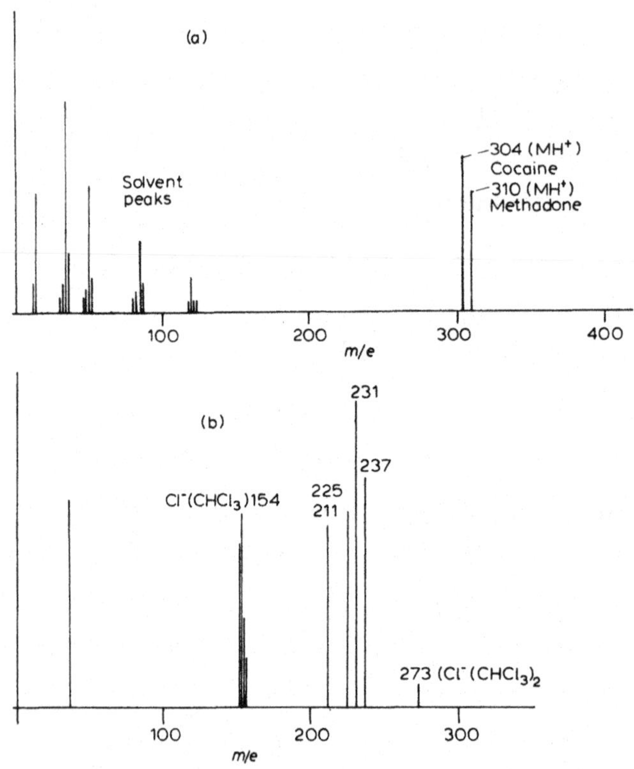

Fig. 13.8. Examples of the detection of drugs. (a) Positive ion mass spectrum of chloroform solution containing 15 ng of cocaine and 12 ng of methadone; (b) negative ion mass spectrum of a chloroform solution of four barbiturates: butabarbital, m/e = 211, 8 ng; pentobarbital, m/e = 225, 8 ng; phenobarbital, m/e = 231, 10 ng; secobarbital, m/e = 237, 10 ng.

materials has been carried out using the powerful analytical capabilities of the gas chromatograph–quadrupole combination. For further details on the technique, the reader should consult Chapter XII.

(4) "*Fingerprinting*" *of bacteria*

Very little work has been published on the "fingerprinting" of complex biological material by pyrolysis mass spectrometry since Zemany [30] reported the characterization of albumin and pepsin in 1952. These proteins were pyrolized on a hot filament in a closed borosilicate glass flask which was subsequently coupled to a mass spectrometer. No data concerning the reproducibility of this procedure were given and the apparent lack of activity in this field since then is most probably explained by the advance of cheaper instrumental techniques for preparing pyrolysis fingerprints, notably pyrolysis gas–liquid chromatography (Py–GLC). Despite significant progress, Py–GLC techniques are limited by an obvious lack of reproducibility and an inherent lack of inter-laboratory standardization [31]. Nevertheless, some of the more sophisticated pyrolysis techniques, such as that employing a filament developed by Levy [32] and the Curie-point technique reported by Simon and Giacobbo [33], are, in principle, suitable for standardization since nearly all the parameters influencing the temperature–time profile can be adequately defined.

Meuzelaar and Kistemaker [34] have reported results obtained by the direct coupling of a Curie-point pyrolysis system to a quadrupole mass filter with data output direct to a multichannel signal averager. The apparatus consists of a vacuum housing pumped by two $70 \, l \, sec^{-1}$ oil diffusion pumps with mounting flanges for a small quadrupole and a re-entrant high frequency heating coil. Provision is also made for the sample, in a quartz tube, to be inserted through two three-way metal ball valves and two viton sliding seals. Repetitive scans of the pyrolisis products obtained at 510°C, from samples of freeze-dried bacteria, gave spectra in which the peak intensity ratios between spectra were of the order of 1.09 ± 0.04.

It thus appears that pyrolysis-mass spectrometry is a promising method for fingerprinting biological material, but it is regrettable that so much structural information [35] is lost during the ionization process. The use of chemical and field ionization techniques appear to be the obvious experimental approach to this subject.

B. ENVIRONMENTAL MONITORING

Much of the research in the field of environmental monitoring has been directed towards the development of a "mechanical nose" capable of responding not only to odours but also to extremely hazardous materials and compounds of very low vapour pressure: Such a technique must obviously include all the advantages of its biological counterpart, particularly in terms of size, detection limits and unambiguous data output. There are analytical systems which can detect two or three selected gases at extremely low concentrations but in general, no single system is available which can determine the detailed composition of the atmosphere in a rapid and efficient manner.

The minimum levels of detection by quadrupole mass spectrometry have been drastically reduced but only at the expense of increasing apparatus size and complexity. The applications which have, so far, been reported are summarized in Table 13.1 which shows that the range of techniques employed is almost as extensive as the types of analysis performed.

The types of apparatus which have been devised for the research listed in the table will be discussed in the order in which they are summarized, with the exception of upper atmosphere research (see Chapter XI) and those fields utilizing gas chromatography–quadrupole techniques (see Chapter XII).

References pp. 331—333

TABLE 13.1

Summary of the applications of quadrupole type instruments in environmental monitoring

Application	Type of analysis	Technique
Pollutants in air	PCBs	PC–quadrupole*
	Freons	Thermal analysis–monopole
Upper atmosphere	Composition	Direct monitoring quadrupole–monopole
	Reactions, Clusters	Flowing afterglow–quadrupole, PC–quadrupole
Pollutants in water	Pesticides	GC–quadrupole
	Heavy metals	Argon discharge–quadrupole
Soil analysis	Pesticides	GC–quadrupole
Hazardous materials	Explosives	PC–quadrupole
	Cryogenic fuels	Direct monitoring quadrupole
Technological	Car exhausts	Direct monitoring quadrupole
	Combustion products	Molecular beam sampling–quadrupole
	Space-craft interior	Direct monitoring–quadrupole

*PC denotes plasma chromatography.

(1) *Detection of pollutants in air*

Air pollution has been defined as the presence in the air of substances generated by man which interfere with his comfort, safety and/or health. This problem was recognized as early as 1315, when King Edward I of England tried to prohibit the burning of sea coal because of the noxious fumes. He was not successful and such problems have persisted over the centuries because cheap forms of raw energy are always in demand. Before measures to control pollution can be taken, the exact nature of the contaminant must be determined. For gaseous pollution, one of the obvious possible monitoring systems is the mass spectrometer.

Pernicka and McGowan [36] have critically reviewed the currently available mass spectrometer instrumentation with a view to the construction of a portable monitoring system. Both the quadrupole and the monopole possess the advantage of being fairly small and lightweight, but were severely criticized for their low resolution and their sensitivity to particulate contamination [37]. One quoted advantage of the quadrupole was the possibility of periodic operation in the total pressure mode (zero resolution) to enable calibration checks of total and partial pressures to be made. In order to obtain the performance required, these authors chose a double-focusing magnetic instrument.

The plasma chromatography–mass spectrometer (PC–MS) combination

[38], the forerunner of the atmospheric pressure ionization instrument discussed earlier, introduced a completely new concept in mass spectrometry: that of divorcing the ion source from the analyser and operating the former at atmospheric pressure. The reason behind this move is fairly easy to understand since reducing the pressure of the sample to be analysed by seven orders of magnitude similarly reduces the number of molecules per cm^3 of the pollutant to a vanishingly small number. The average ionization efficiency of a mass spectrometer is only 10^{-2} to 10^{-3} and normally the majority of the ionization, producing O_2^+, N_2^+, H_2O^+ etc., would be effectively wasted. It is therefore advantageous if these charges can be transferred to non-ionized sample molecules. In the plasma chromatograph, the ionized species undergo many collisions in the ion–molecule reactor region (Fig. 13.9) leading to the production of more sample ions either by charge transfer or ion–molecule reactions, in accordance with the reactions previously described (Section A.(3), p. 316).

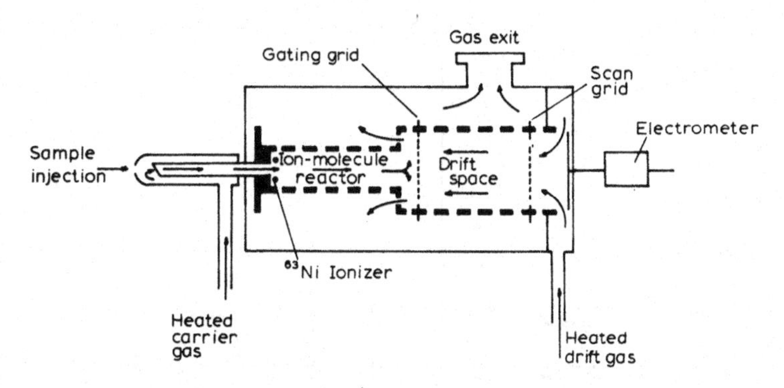

Fig. 13.9. Schematic diagram of the plasma chromatograph.

When the gating grid is opened, the ions leave the source and drift down the tube under the influence of the potential gradient and separate according to mobility in a manner analogous to the time-of-flight instrument or a gas chromatography column from which the name of the apparatus is derived. The counter-flowing drift gas employed with this apparatus is usually dry nitrogen and this effectively quenches any further reaction once the ions have left the reaction region.

It should be readily appreciated that by simply changing the sign of the potential gradient along the drift tube, the monitoring of either positive or negative ions becomes possible. The combination of the plasma chromatograph and mass spectrometer is quite simply achieved by replacing the collector of the PC with a pinhole orifice into a vacuum chamber to allow the ions to effuse directly into the mass filter as shown schematically in Fig. 13.10. Typical positive and negative ion mass spectra, recorded by the Finnegan 1015 quadrupole, are shown in Fig. 13.11 for a mixture of 10^{-7} mole parts

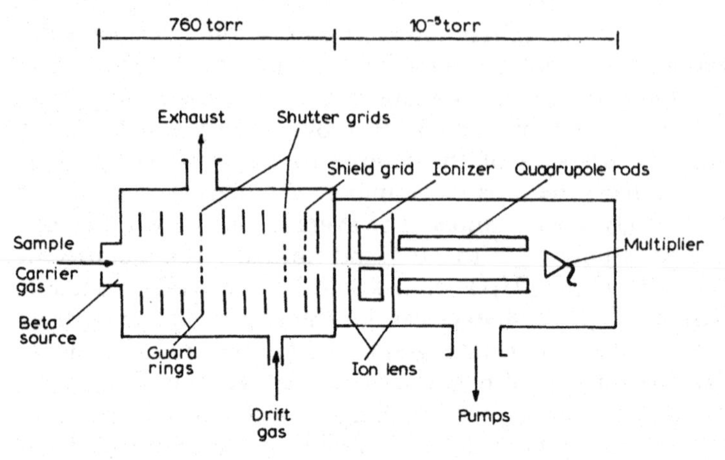

Fig. 13.10. Schematic representation of the plasma chromatograph–mass spectrometer combination.

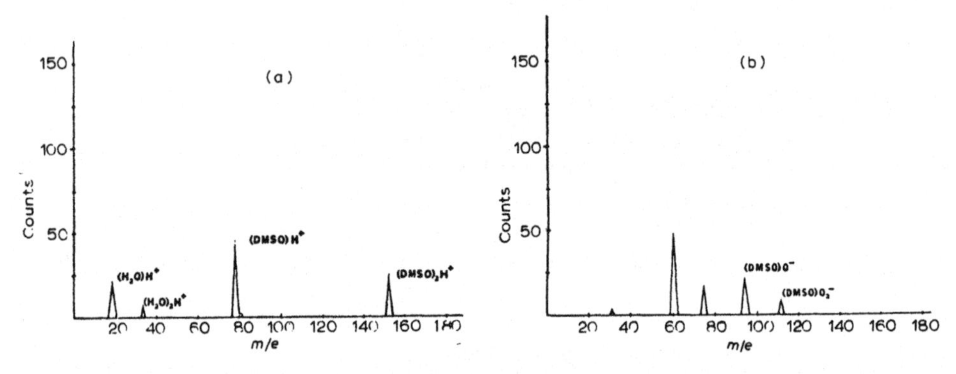

Fig. 13.11. (a) Positive and (b) negative ion mass spectrograms of dry air containing $< 10^{-7}$ mole parts of dimethyl sulphoxide (DMSO).

of dimethylsulphoxide (DMSO) in air, in which the protonated species is the largest peak in the positive ion spectrum whereas (DMSO) O_2^- is of fairly low intensity.

Several applications of PC–MS in the environmental field have been reported recently [39], particularly the detection of polychlorinated biphenyls and TNT. Polychlorinated biphenyls (PCB) have assumed a prominent position in environmental studies because of their high toxicity and the mounting evidence of their gross contamination of the global ecosystem. Because of the many compounds and isomers possible in this class, their analysis is usually a difficult and lengthy procedure largely involving mass spectrometric and gas chromatographic methods [40, 41].

The detection of trace amounts of TNT in air is a currently relevant and difficult analytical problem that has been explored by PC–MS, with favourable results [42, 43]. The vapours evolving from TNT in equilibrium at room temperature place only 10^{-9} g in each 100 ml of air directly over the sample. Experiments with the PC–MS to date, show a detection capability of 10^{-9} g with a signal-to-noise ratio of 1000:1 (possibly 10^{-11} g, signal-to-noise 10:1) in the negative ion mode with the negatively charged molecular ion being the most abundant. In these experiments, the carrier gas was air which produces plenty of $(H_2O)\,nO_2^-$ ions, the reactant ion in this case. The capabilities of instruments such as the plasma chromatograph operating at atmospheric pressure are still in their infancy and much further research is required particularly into types of carrier/reactant gas.

Enhancing the sensitivity of a mass spectrometer to a trace constituent usually involves some form of pre-concentration: a gas chromatography column, a selective membrane, a plasma chromatograph or, as in the work reported by Schubert [44], a freeze-out [45] (thermal analysis) technique. Stainless steel cylinders containing the air to be sampled were connected by a leak valve to a vacuum system with 110 l sec^{-1} ion pump and a General Electric Monopole with a mass range of 2–300 amu. Mass spectra of the sample gas were recorded in three steps, the first being directly from the cylinder at atmospheric pressure. After this, the leak valve was closed, the cylinder cooled to 80K and after 15 min the cylinder was pumped down to about 3×10^{-2} torr in order to remove most of the oxygen and nitrogen content. The closed cylinder was then heated to room temperature, the leak valve opended and the spectrum recorded (step II). The cylinder was reclosed, recooled to 80K and mass spectra were recorded, with the leak valve fully open, at every 5K rise as the cylinder was heated to room temperature (step III). This series of spectra enabled the identification of the individual mass spectra superimposed upon one another in steps I and II to be made.

The results presented include the analysis of three synthetically contaminated prepurified nitrogen samples using Freon 113 at 16 ppb and C_6H_{14} at 130 ppb and 9 ppb, and a sample of air taken from a central telephone office. These data, Table 13.2, clearly show a detection capability of <1 ppm. Although this series of experiments is time-consuming when performed manually, a suitable computer-controlled instrument complete with data analysis could be designed.

(2) *Water analysis*

Although the principle of mass spectrometry is attractive to the analyst, its practical application is "simple" only in the case of gaseous samples. The analysis of aqueous solutions in particular, has until recently only been performed indirectly by evaporation and spark source techniques. As usual, the problems and the answers lie in the mode of ionization and the introduction of the sample into a high-vacuum environment.

TABLE 13.2

Analysis air samples

Sample	Measured impurity	ppm
N_2 + 0.12 ppm C_6H_{14}	C_6H_{14}	0.2
N_2 + 0.009 ppm C_6H_{14}	None detectable	
N_2 + 0.016 ppm Freon 113	Freon 113	0.016
Telephone central office	Freon 12	0.23
	Freon 113	0.065
	$C_2Cl_4H_2$	<0.01
	C_2H_6	2.0
	C_6H_6	1.6

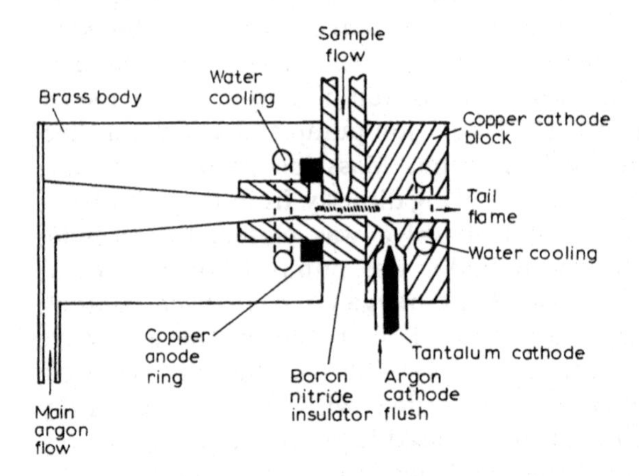

Fig. 13.12. Capillary arc discharge source.

One approach has been to use a wall-stabilised d.c. plasma [46] maintained in an argon carrier gas, in which a core temperature of about 4000K may be achieved from a well-smoothed 300 V d.c. supply at 11 A. The capillary arc unit used for this work has been described by Gray [47, 48] and is reproduced in Fig. 13.12. The discharge occurs in the bore of the main insulator and is approximately 1 cm long and less than 3 mm in diameter. Three separately metered argon flows are fed to the arc; a small flow, $0.1\,l\,min^{-1}$, to cool the tantalum cathode, a main flow $1.0\,l\,min^{-1}$, along the discharge channel and the sample flow, $1.5\,l\,min^{-1}$, which is introduced tangentially into the centre of the arc. The main body of the source, which also forms the anode, and the block supporting the cathode, through which the tail flame emerges, are made from water-cooled copper.

For an element with an ionization potential less than 10 V, under the sample introduction conditions used, the degree of ionization would be 100% for solutions of $1\,\mu g\,cm^{-3}$ or less, reducing to about 15% at $100\,\mu g\,cm^{-3}$

[49]. At lower ionization potentials, the degree of ionization increases; at 5 V, for example, it is complete for solutions up to about 1% concentration. The aqueous solutions are introduced into the sample stream by direct nebulization using either a pneumatic or ultrasonic nebulizer. The gas stream containing the sample droplets is passed through a glass chamber heated to about 150°C and then through a water-cooled condenser to remove as much excess water from the sample as possible. Vaporization followed by ionization occurs in the arc giving the plasma, from which optimum sampling conditions are obtained with an orifice diameter of 75 μm admitting the ions into the first vacuum chamber at a pressure of less than 10^{-3} torr. Immediately after passing through the orifice, the ions are focused by a cylindrical electrode to pass through a 2 mm aperture into the second chamber maintained below 10^{-5} torr. Further cylindrical electrodes ensure that the ion beam is co-axial with the quadrupole mass filter system which is mounted in this chamber. The analyser has a mass range of 300 amu with similar resolution (10% valley) incorporating an off-set "channeltron" multiplier operating in the saturated mode.

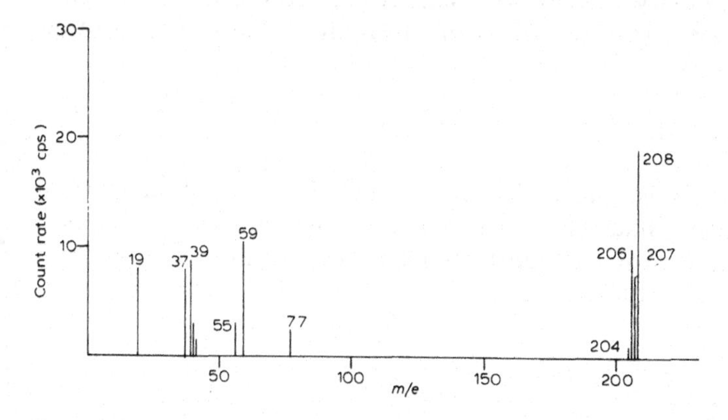

Fig. 13.13. Typical plot obtained directly from a solution containing 0.5 μg ml^{-1} each of cobalt and lead.

A typical plot obtained directly from a solution containing 0.5 μg ml^{-1} each of cobalt and lead is reproduced in Fig. 13.13 which shows the spectrum to be remarkably simple, doubly charged ions being completely absent whilst *m/e* 37 and 55 are $(H_2O)_2H^+$ and $(H_2O)_3H^+$ resulting from ion-molecule reactions in the first vacuum stage. The detection limits for a number of elements determined for this apparatus and expressed as values equivalent to twice the standard deviation of the background are given in Table 13.3 where they are compared with atomic absorption data corrected to 1971 [50].

As an ion source for trace analysis in aqueous solutions, the atmospheric

TABLE 13.3

Comparative elemental detection limits ($\mu g\ ml^{-1}$) for aqueous solutions (2σ values)

Element	Atomic absorbtion	Gray
Ag	0.0004	0.0001
As	0.1	0.0005
Cd	0.01	0.001
Co	0.005	0.00004
Cu	0.005	0.0008
Hg	0.5	0.14
Mg	0.0003	0.001
Mn	0.002	0.0009
Pb	0.01	0.00005
Se	0.1	0.02
Zn	0.002	0.008

pressure plasma arc sampling system shows considerable promise. Very high sensitivity is potentially available and sample introduction is direct and convenient without preliminary preparation. This source combined with a programmed quadrupole mass analyser offers an alternative to atomic absorption spectroscopy for multielement analysis.

(3) *Technological applications*

This section is really a miscellany of monitoring applications of an environmental nature, including exhaust gas measurements from combustion engines, molecular beam sampling of gas constituents inside engine cylinders, and space-craft contamination studies.

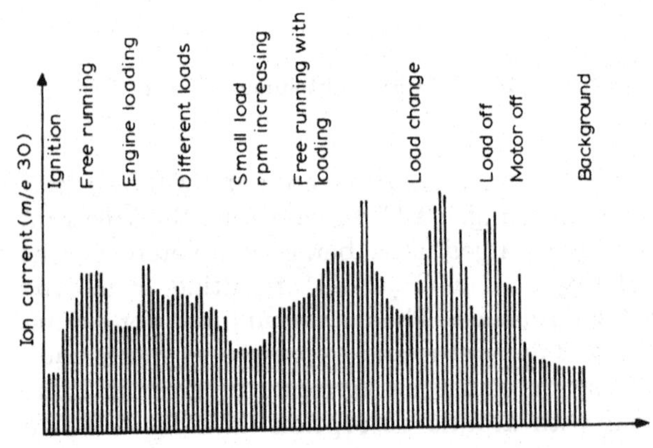

Fig. 13.14. Nitric oxide concentration monitored as a function of the operating conditions of an internal combustion engine.

Current concern regarding the exhaust emissions from internal combustion engines has led to fairly extensive monitoring experiments with particular emphasis placed on NO_x measurements. One of these [51] involved the direct analysis of the exhaust gases by admitting them through a two-stage inlet system, to prevent discrimination, into a vacuum system containing a Balzers QMG 101A quadrupole. Provision was made to vary the loading on the engine by means of a brake and thus enable analysis of the NO emissions as a function of engine speed (Fig. 13.14). More recent investigations have shown that the NO_2 concentration is virtually zero in exhaust gases, only forming when the gases cool down in the atmosphere, and thus NO_2 measurements in exhaust systems do not appear to have the significance that they have in atmospheric measurements. Accordingly, no loss of information results from a restriction to the measurement of NO only [52].

A more direct monitoring of the actual variation in NO concentration inside the cylinder of a combustion engine was reported by Young et al. [53], who sampled the gases from a model aircraft engine, through a pinhole drilled in the cylinder wall, directly into an EAI Q250 quadrupole. Whilst the engine is running, a small percentage of the molecular beam entering the quadrupole is ionized by electron bombardment and mass analysed. The output for 512 scan cycles from the quadrupole is stored in a signal averager and the beam is then blocked by a beam flag. A further 512 background cycles are recorded and subtracted from the data previously stored to give the required output data format.

It was found that the maximum NO peak intensity occurred about 55° (crank angle) after ignition, during the expansion part of the cycle. Similar results were obtained for CO_2, H_2O and CO although some difficulty was experienced because of energetic ions overtaking the more slowly moving ones.

The monitoring of the background gases [54] in space vehicles has assumed greater importance as the duration of manned space flights has increased. Similar concern regarding the outgassing of components in sophisticated satellite equipment leading to experiment failure has also been expressed [55]. An extensive investigation of vacuum weight loss and contamination tests of materials used in space missions has been reported by Poehlmann [56]. The data recorded included continuous in situ sample weight loss, mass spectrometric fingerprints of outgassing species and continuous mass versus time measurements of products condensing on a temperature-controlled quartz crystal microbalance using the apparatus shown in Fig. 13.15. Experiments were performed over the temperature range 30–90°C showing that, of the polymers examined, the fluorocarbons, polycarbonates and polyimides, generating H_2O, CO_2, H_2 and hydrocarbons up to m/e 287, generally produced less contamination than the silicone-based materials which evolved H_2, CH_3 and silicones to m/e 489 at 60°C.

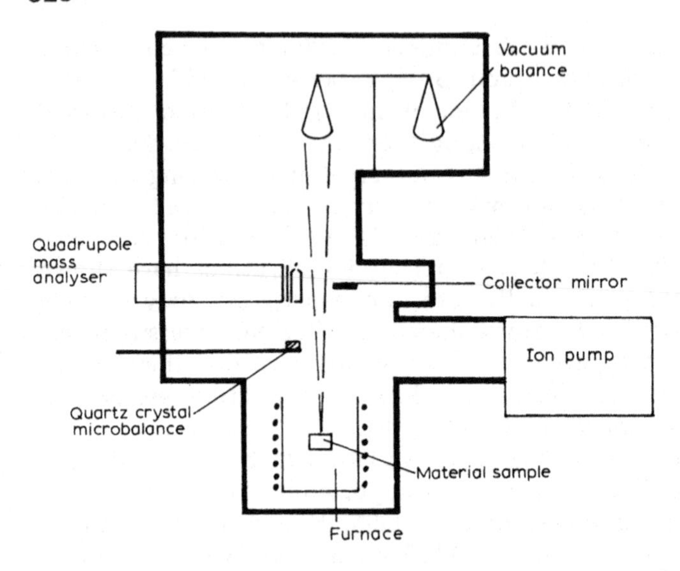

Fig. 13.15. Schematic diagram of the apparatus reported by Poehlmann.

TABLE 13.4

Instrumental sensitivity*

Apparatus	Sensitivity
Quadrupole–Faraday cup	1 in 10^6
Quadrupole–electron multiplier	1 in 10^8
GC–quadrupole–electron multiplier	1 in 10^9
GC–CI–quadrupole–electron multiplier	1 in 10^{10}
API–quadrupole–electron multiplier	1 in 10^{10}
PC–quadrupole–electron multiplier	1 in 10^{11}
Human nose	1 in 10^{13}

* Sensitivities are compound-dependent and thus these figures are only approximate.

C. CONCLUSIONS

In this chapter, an attempt has been made to review many of the applications of quadrupole type instruments in medical and environmental research. The development of the required instrumentation has engendered ever-increasing sensitivity (Table 13.4) such that the idea of a mechanical nose looks even more feasible.

Despite the sparse applications of quadrupoles in medical research to date there is little doubt that the number of applications will increase particularly in intensive care units and anaesthetic studies.

The current emphasis on environmental monitoring presents an opening for the further development of high sensitivity instruments, especially if low cost medium mass range/resolution devices can be produced. The detection of fires and hazardous or toxic environments are other areas demanding similar attention, as does the "sniffing" of drugs and explosives. In this instance, the aim would be to have a highly compact robust and portable instrument capable of operation by non-scientific personnel or by remote control and telemetry networks. The direct control of the operation and data output of the quadrupole by small computers is a field which is only just developing and there is clearly considerable potential for application in remote monitoring or automated system control.

NOTES ADDED IN PROOF.

A. MEDICAL APPLICATIONS

One of the major obstacles impeding the much wider use of mass spectrometers in both respiratory and blood gas analysis is the much greater capital equipment expenditure required in comparison with the cost of infrared or polarographic electrode techniques. Both of these systems may be operated in a continuous mode but are generally designed for one specific application. Separate pH electrodes must be used for each species monitored whilst the infrared instruments, in a manner analogous to the fixed collector magnetic mass spectrometer, are set to monitor specific gases.

The justification of the increase in cost of the mass spectrometer-based analyser in terms of the wide monitoring capability, within its specified mass range, appears to ignore the fact that such facilities may not be required. For example, it is of little comfort to a clinician who is only interested in monitoring respiratory species up to m/e 44 to know that the instrument is capable of analysis to m/e 200 when trying to obtain the necessary funds. It may well be that subsequently the instrumental capabilities will be more extensively used, but it is dubious whether funding is ever based on such a premise.

This problem appears to have been appreciated by Edwards Instruments Limited, Eastbourne, U.K., who have recently launched a new quadrupole-based Medical Analyser, R.M.S. 3. The prototype instrument is designed around a miniature quadrupole residual gas analyser and, as such, has a somewhat reduced performance. The mass range is quoted as 2—200 amu with a resolution of approximately 70. The use of fragment ion monitoring excludes most of the problems expected from an instrument with such low resolution. The remaining general performance specifications are comparable with other medical mass spectrometers currently available. It can be mounted either on a bench top or used in trolley form and requires only an electrical supply for the fully protected vacuum system. Initial models of this instrument have two monitoring modes

(a) complete scans across the entire mass range; and

(b) fixed monitoring of oxygen and one other species dependent on the operators' choice.

References pp. 331—333

Development work is in hand to produce a more sophisticated peak switching unit to increase the performance specifications still further. The major advantage of this system, however, is the price. The complete unit costs less than one third of the price of the currently available quadrupole instruments without a necessarily significant deleterious reduction in performance.

B. BLOOD GAS ANALYSIS

Much interest has lately been shown in *non-invasive* analysis of blood gases, particularly for continuous monitoring of oxygen tensions in sick babies. Miniature polarographic electrodes to measure oxygen tension have been developed, but no suitable carbon dioxide electrode has been produced. In the clinical situation, it is more advantageous to monitor the partial pressures of both oxygen and carbon dioxide in the blood stream. Electrochemically this situation would require two separate probes and measuring systems which is impractical for routine use.

Parker and Delpy [57] have measured these concentrations by detecting the gases diffusing through the skin using both magnetic and, latterly, quadrupole-based analysers. The probe reported consisted of a small cylindrical cup, the open end of which was sealed with a gas-permeable membrane supported on a porous substrate. A heating coil and thermistor kept the temperature of the probe at around 43°C. A standard capillary inlet system connects the inner chamber of the probe to the mass spectrometer. The data obtained for the partial pressures of oxygen and carbon dioxide in the blood, when monitored with the probe placed on the forearm, showed very good agreement with that obtained from more conventional measurements.

This technique has two distinct advantages over the electrochemical method: only one probe would need to be attached to the skin, and the mass spectrometer can analyse any species which diffuses through the skin, particularly anaesthetic gases or even emitted vapours.

C. ANAESTHETIC STUDIES

Several recent studies have called attention to the presence of trace concentrations of anaesthetic gases in the operating theatre atmosphere and indicated a potential hazard to exposed personnel [58—60]. Two techniques are commonly used for anaesthetic administration, a non-rebreathing system in which spent gases are vented outside the theatre and a semi-closed circle filter system. Exposure to halothane, particularly, may occur owing to the porosity of the plastic tubing employed in these systems.

Cohen et al. [61] have employed a quadrupole mass filter to analyse air samples taken at nine different locations in an operating theatre. These samples were taken through a length of polyethylene tubing and were repeated at half hourly intervals over a period of several days. A comprehensive survey of halothane concentrations in the operating theatre and in end-tidal samples obtained from anaesthetists and nurses was reported. A comparison was also made between a non-rebreathing and a semi-closed circle filter system.

The results obtained showed an order of magnitude decrease in halothane concentration for the same sampling position when the semi-closed anaesthetic system was employed. This concentration varied between 13 ppm (non-rebreathing) and 0.47 ppm for the semi-closed technique. The mean end-tidal concentrations of halothane in nurses and anaesthetists before and during the working day were found to be as shown.

	Initial concn. (ppm)	Mean concn. (ppm)
Nurses	0.01 ± 0.0006	0.21 ± 0.04
Anaesthetists	0.08 ± 0.03	0.46 ± 0.07

A second approach to the problem of reducing the possible exposure to halothane has been reported by Bushman et al. [62]. This technique comprises a virtually completely closed rebreathing system in which a quadrupole mass filter controls the quantity of anaesthetic and oxygen in the loop. The analyser monitors the oxygen level and when this decreases below a chosen level, "demand valves" are opened to admit more oxygen to the loop. Carbon dioxide is removed by conventional filters. Fail-safe facilities are incorporated into the system which has been successfully used on a number of occasions.

REFERENCES

1 G.W.A. Milne (Ed.), Mass Spectrometry: Techniques and Applications, Wiley-Interscience, New York, 1971.
2 G.R. Waller, Biochemical Applications of Mass Spectrometry, Wiley-Interscience, New York, 1972.
3 M. Mosharrafa, Res. Develop., 21 (1970) 24.
4 J.C. Lilly, Methods Med. Res., 2 (1950) 131.
5 J.C. Lilly, Amer. J. Physiol., 161 (1950) 342.
6 K.T. Fowler, A mass spectrometer for the rapid and continuous analysis of gas mixtures in respiratory research, Ph.D. Thesis, University of London, 1958.
7 K.T. Fowler and P. Hugh-Jones, Brit. Med. J., 1 (1957) 1205.
8 K.T. Fowler, Phys. Med. Biol., 14 (1969) 185.
9 J.B. West, Bull. Physio-Pathol. Resp., 3 (1967) 381.
10 P. Scheid, H. Slama and J. Piper, J. Appl. Physiol., 30 (1971) 258.
11 R.G. Bickel, C.F. Diener and H.L. Brammel, Clin. Aviat. Aerosp. Med., 41 (1970) 203.

12 J.R. Arthur and J.J. Leporte, Rev. Sci. Instrum., 37 (1966) 794.
13 J.J. Norton and R.D. Smith, in D. Price (Ed.), Dynamic Mass Spectrometry, Vol. 2. Heydon, London, 1971, p. 212.
14 Twentieth Century Electronics, King Henry's Drive, New Addington, Croydon, Surrey, U.K.
15 S. Marlowe, Colloquium on Mass Spectrometry. Digest No. 1973/24, Institution of Electrical Engineers, London, 1973, p. 5.
16 W.M. Brubaker, Final Report, NASA Contract NA59-8371, 1969.
17 W.M. Brubaker and J.H. Marshall, 20th ASMS Annual Conference on Mass Spectrometry and Allied Topics, June 1972, p. 226.
18 S. Woldring, G. Owens and D. Woolford, Science, 153 (1966) 885.
19 J.W. Brantigan, V.L. Gott, M.L. Vestal, G.J. Fergusson and W.H. Johnston, J. Appl. Physiol., 28 (1970) 375.
19a J.W. Brantigan, V.L. Gott and M.N. Martz, J. Appl. Physiol., 32 (1972) 276.
20 A. Wald, W.K. Hass, F.P. Siew and D.H. Wood, Med. Biol. Eng., 8 (1970) 111.
21 A. Wald, W.K. Hass, J. Ransohoff and D.H. Wood, Trans. Amer. Neurol. Ass., 98 (1970) 255.
22 A. Wald, W.K. Hass and J. Ransohoff, J. Ass. Advan. Med. Instrum., 5 (1971) 325.
23 A. Key, Colloquium on Mass Spectrometry, Digest No. 1973/24, Institution of Electrical Engineers, London, 1973, p. 4.
24 Scientific Research Instruments, Baltimore, Md., U.S.A.
25 G.W.A. Milne, H.M. Fales and T. Axenrod, Anal. Chem., 43 (1971) 1815.
26 E.C. Horning, M.G. Horning, D.I. Carroll, I. Dzidic and R.N. Stillwell, Anal. Chem., 45 (1973) 936.
27 A. Good, D.A. Durden and P. Kebarle, J. Chem. Phys., 52 (1970) 212.
28 F.W. Karasek, W.D. Kilpatrick and M.J. Cohen, Anal. Chem., 43 (1971) 1441.
29 D.I. Carroll, I. Dzidic, R.N. Stillwell, M.G. Horning and E.C. Horning, Anal. Chem., 46 (1974) 706.
30 P.D. Zemany, Anal. Chem., 24 (1952) 1709.
31 N.B. Coupe, C.E.R. Jones and S.G. Perry, J. Chromatogr., 47 (1970) 291.
32 R.L. Levy, Thesis, Israel Inst. Technol., Haifa, 1963.
33 W. Simon and H. Giacobbo, Chem. Eng. Technol. 37 (1965) 709.
34 H.L.C. Meuzelaar and P.G. Kistemaker, Anal. Chem., 45 (1973) 587.
35 W. Simon, P. Kriemler, J.A. Völlmin and H. Steiner, J. Gas Chromatogr., 5 (1967) 53.
36 J.C. Pernicka and G.F. McGowan, Technology Utilization Ideas for the 70's and Beyond, A.A.S. Science and Technology Series, Vol. 26, American Astronautical Society, Tarzana, Calif., 1971, p. 161.
37 L.F. Herzog, Int. J. Mass Spectrom. Ion Phys., 4 (1970) 337.
38 M.J. Cohen and F.W. Karasek, J. Chromatog. Sci., 8 (1970) 330.
39 F.W. Karasek, Anal. Chem., 46 (1974) 710A.
40 L.M. Reynolds, Bull. Environ. Contam. Toxicol., 4 (1969) 128.
41 F.W. Karasek, Anal. Chem., 43 (1971) 1982.
42 F.W. Karasek and D.W. Denney, J. Chromatogr., 93 (1974) 141.
43 F.W. Karasek, Res. Develop., 25 (1974) 32.
44 R. Schubert, Anal. Chem., 44 (1972) 2084.
45 M. Shepherd, S.M. Rock, R. Howard and J. Stormes, Anal. Chem., 23 (1951) 1431.
46 J.L. Jones, R.L. Dahlquist and R.E. Hoyt, Appl. Spectrosc., 25 (1971) 628.
47 A.L. Gray, in D. Price and J.F.J. Todd (Eds.), Dynamic Mass Spectrometry, Vol. 4, Heyden, London, 1976.
48 A.L. Gray, Analyst, 100 (1975) 289.
49 P.W.J.M. Boumans, Theory of Spectrochemical Excitation, Hilger, London, 1966, pp. 156–232.
50 V.A. Fassel, XVI Colloquium Spectroscopium International, Plenary Lecture, Hilger,

London, 1971.
51 H. Eppler, Balzers Tech. Rep. VC2, July, 1972.
52 R.M. Campan and J.C. Neerman, Continuous Mass Spectrometric Determination of Nitric Oxide in Automotive Exhausts, Ford Motor Co., U.S.A.
53 W.S. Young, Y.G. Wang, W.E. Rodgers and E.L. Knuth, Technology Utilization Ideas for the 70's and Beyond, A.A.S. Science and Technology Series, Vol. 26, American Astronautical Society, Tarzana, Calif. 1971, p. 161.
54 E.W. Budde, Int. Aerosp. Abstr. No. A71-20067, (1971) 1023.
55 M.P. Thekaekara, Space Simulation, NASA Spec. Publ. SP-128, 1972, p. 780.
56 H.C. Poehlmann, Inst. Phys. Phys. Soc. Conf. Ser., 6 Part 2 (1968) 809.
57 D. Delpy and D. Parker, Lancet, (1975) 1016.
58 B. Hallen, H. Ehrner-Samuel and M. Thomason, Acta Anaesth. Scand., 14 (1970) 17.
59 H.W. Linde and D.L. Bruce, Anaesthesiology, 30 (1969) 363.
60 A.I. Vaisman, Eksp. Khir. Anest., 3 (1967) 44.
61 E.N. Cohen, C.E. Whitcher and J.R. Trudell, Anaesthesiology, 35 (1971) 348.
62 To be published.

APPENDICES

APPENDIX A

Parameters characterizing the acceptance ellipses for the mass filter with no fringing fields. Initial phases from 0 to $\pi/2$ are given. Other phases are obtained by changing the sign of A. In each case, $q = 0.706$.

	y direction			*x* direction		
Initial phase	B	Γ	A	B	Γ	A
(i) $a = 0.23342$						
0.0	8.17	0.12	0.0	27.93	0.036	0.0
0.1π	9.62	2.29	4.58	23.56	6.63	− 12.45
0.2π	14.08	6.70	9.66	14.35	16.03	− 15.14
0.3π	21.45	8.10	13.14	6.11	18.00	− 10.44
0.4π	29.21	3.55	10.13	1.44	15.40	− 4.60
0.5π	32.55	0.031	0.0	0.072	13.98	0.0
(ii) $a = 0.236098$						
0.0	16.47	0.061	0.0	52.91	0.019	0.0
0.1π	19.08	4.02	8.70	47.06	11.98	− 23.72
0.2π	27.71	12.88	18.87	29.10	30.97	− 30.01
0.3π	42.25	16.29	26.21	12.58	35.65	− 21.16
0.4π	57.97	7.63	21.01	3.02	30.71	− 9.57
0.5π	65.31	0.015	0.0	0.036	27.64	0.0
(iii) $a = 0.236651$						
0.0	27.03	0.037	0.0	86.53	0.012	0.0
0.1π	31.31	6.54	14.27	74.14	18.88	− 37.40
0.2π	45.46	21.07	30.93	45.84	48.83	− 47.30
0.3π	69.30	26.66	42.97	19.80	56.22	− 33.35
0.4π	95.07	12.49	34.44	4.72	48.43	− 15.08
0.5π	107.10	0.0093	0.0	0.023	43.60	0.0

Approximate acceptance ellipse parameters for the mass filter for transmission at 50% of the initial phases in the presence of fringing fields of various lengths. In each case $q = 0.706$.

Length of fringing field (rf cycles)	B	Γ	A	ε
$a = 0.23342$; y direction				
0.5	3.72	0.29	0.30	0.0057
1.0	16.75	0.10	0.84	0.028
2.0	24.82	0.58	3.67	0.019
3.0	40.00	1.11	6.59	0.0090
4.0	43.16	1.26	7.34	0.0039
6.0	51.79	1.92	9.91	0.00034
$a = 0.23342$; x direction				
0.5	2.00	1.09	1.09	0.0033
1.0	4.19	0.33	0.60	0.0043
2.0	4.61	0.34	0.76	0.0063
3.0	7.27	0.22	0.77	0.0087
4.0	6.58	0.21	0.64	0.0089
6.0	6.58	0.21	0.62	0.0089
$a = 0.2361$ y direction				
2.0	32.14	0.79	4.94	0.0064
$a = 0.2361$; x direction				
2.0	5.10	0.35	0.88	0.0020
$a = 0.23665$; y direction				
2.0	42.50	1.02	6.50	0.0028
$a = 0.23665$; x direction				
2.0	5.92	0.39	1.14	0.00089

Paired values of a (upper) and q (lower) satisfying the conditions $\beta(a, q) = 1/d$; $\beta(-a, -q) = p/d$, for $2 < d < 13$, $p \leqslant d$[a]

Values of d	Values of p: 1	2	3	4	5	6	7	8	9	10	11	12
3	0.000000	0.117076	0.183624									
	0.451105	0.656266	0.752057									
4	0.000000	0.078407	0.166227	0.206116								
	0.344959	0.522848	0.673016	0.732687								
5	0.000000	0.053806	0.126186	0.190656	0.216980							
	0.278436	0.429064	0.575816	0.683276	0.723310							
6	0.000000	0.038706	0.095019	0.156067	0.204406	0.223000						
	0.233150	0.362294	0.495503	0.610395	0.689585	0.718107						
7	0.000000	0.029026	0.073030	0.125161	0.175636	0.212867	0.226671					
	0.200422	0.312945	0.432374	0.542783	0.633358	0.693652	0.714933					
8	0.000000	0.022510	0.057492	0.100984	0.147256	0.189023	0.218429	0.229069				
	0.175699	0.275178	0.382516	0.485382	0.576545	0.649155	0.696401	0.712857				
9	0.000000	0.017941	0.046272	0.082550	0.123236	0.163704	0.198534	0.222274	0.230721			
	0.156378	0.245420	0.342500	0.437523	0.525286	0.601176	0.660405	0.698336	0.711428			
10	0.000000	0.014622	0.037966	0.068448	0.103772	0.140909	0.176176	0.205510	0.225041	0.231906		
	0.140869	0.221402	0.309820	0.397543	0.480593	0.555717	0.619571	0.668668	0.699745	0.710402		
11	0.000000	0.012139	0.031673	0.057528	0.088154	0.121500	0.155027	0.185808	0.210768	0.227097	0.232785	
	0.128150	0.201625	0.282696	0.363872	0.441966	0.514574	0.579307	0.633616	0.674900	0.700802	0.709641	
12	0.000000	0.010235	0.026802	0.048948	0.075590	0.105293	0.136268	0.166407	0.193377	0.214824	0.228667	0.233455
	0.117532	0.185067	0.259856	0.335236	0.408554	0.477963	0.541700	0.597892	0.644554	0.679707	0.701613	0.709061

[a] These are operating points for double focusing.

APPENDIX D

Coefficients from C_0 to C_6 used in the analytical solution to the equation of motion in the mass filter.

q_y	q_x	β_y	β_x	C_0	C_{-2}	C_2	C_{-4}	C_4	C_{-6}	C_6
0.706		0.00		1.0	− 0.16763	− 0.16763	0.00729	0.00729	− 0.00014	− 0.00014
		0.02			− 0.17089	− 0.16449	0.00750	0.00708	− 0.00015	− 0.00014
		0.04			− 0.17427	− 0.16145	0.00773	0.00688	− 0.00015	− 0.00013
		0.08			− 0.18143	− 0.15569	0.00821	0.00651	− 0.00016	− 0.00012
	− 0.706		0.92	1.0	0.79504	0.08515	0.06065	0.002454	0.00167	0.00004
			0.96		0.89098	0.08288	0.06990	0.00240	0.00196	0.00004
			0.98		0.94383	0.08176	0.07508	0.00235	0.00212	0.00003
			1.00		1.00000	0.08065	0.08065	0.00230	0.00230	0.00003

APPENDIX E

An example of a simple computer program for calculating ion trajectories by numerical (Runge–Kutta) integration of the Mathieu equation. The program is written in Fortran IV G.

W(1), W(2)	are the initial positions in the transverse directions.
DW(1), DW(2)	are the initial velocities
Z1	is the initial phase
H	is the integration step size in terms of ξ
A, Q	are a, q
NT	is the total number of integration steps
M4	the time in cycles; the positions and velocities are printed when the counter M1 = M4
M3	equals zero for a single calculation; equals 1 for repeated use of the program
M2	if M3 ≠ 0, the calculation is repeated for the initial phase differing by $\pi/20$ until M2 equals 20

```
    DIMENSION V(2), DV(2), W(2), DW(2)
    GRAD(V,Z)=(A+2.0*Q*COS(2.0*Z)*V
300 READ (1,210)M2,M3,M4,NT,Z1,H
210 FORMAT(312,I4,2F9.6)
    READ(1,200)W(1),DW(1),W(2),DW(2),A,Q
200 FORMAT(4E11.4,2F9.6)
    Z=-Z1

    WRITE(3,110)H
110 FORMAT('OTHE INTEGRATION STEP IS',4X,F9.6)
    WRITE(3,120)A,Q
120 FORMAT('OVALUES OF A AND Q ARE',2F9.6)
 50 V(1)=W(1)
    V(2)=W(2)
    DV(1)=DW(1)
    DV(2)=DW(2)
    WRITE(3,130)V(1),V(2),DV(1),DV(2),Z
130 FORMAT('OINITIAL VALUES FOR POSN, VELOCIY, PHASE',
      4E12.4,4X,F9.6)
    M1=0
    DO 20 N=1,NT
    DO 10 I=1,2
    A1=DV(I)*H
```

```
      B1=GRAD(V(I),Z)*H
      A2=(DV(I)+0.5*B1)*H
      B2=GRAD(V(I)+0.5*A1,Z+0.5*H)*H
      A3=(DV(I)+0.5*B2)*H
      B3=GRAD(V(I)+0.5*A2,Z+0.5*H)*H
      A4=(DV(I)+B3)*H
      B4=GRAD(V(I)+A3,Z+H)*H
      V(I)=V(I)+(A1+2.0*A2+2.0*A3+A4)/6.0
      DV(I)=DV(I)+(B1+2.0*B2+2.0*B3+B4)/6.0
      A=-A
      Q=-Q
      EN=N*H/3.14159
 70 M1=M1+1
 60 IF(M1.LT.M4)GO TO 10
    M1=0
    WRITE(3,100)EN,V(1),V(2),DV(1), DV(2)
100 FORMAT(F10.6,4E14.4)
 10 CONTINUE
 20 Z=Z+H
    IF(M3.EQ.0)GO TO 400
    IF(M2.EQ.20)GO TO 500
    Z=(M2*3.14159/20.0)
    M2=M2+1
    GO TO 50
500 IF(M3.EQ.1)GO TO 300
400 STOP
    END
```

APPENDIX F

This list of quadrupole patents is not necessarily complete. Generally only the U.S. patents are quoted. A few others are noted at the end.

U.S. Patents

3,784,814 (1974)	T. Sakai, M. Sakimura and Y. Ino. Quadrupole mass spectrometer having resolution variation capability ($U = \alpha_M V - K$)
3,767,914 (1973)	R.K. Mueller and J.P. Carrico Continuous injection mass spectrometer
3,740,553 (1973)	N.R. Whetten Method and apparatus for measuring size

	distribution of particles using a two-dimensional a.c. electric field.
3,725,700 (1973)	W.R. Turner Multipole mass filter with artifact reducing electrode structure
3,699,330 (1972)	P.F. McGinnis Mass filter electrode (coated glass electrodes)
3,648,046 (1972)	D.R. Denison and C.F. Morrison Quadrupole gas analyser comprising four flat plate electrodes
3,641,340 (1972)	W.J. Van Der Grinten and G. Jernakoff Multichannel readout mass spectrometer
3,629,573 (1971)	J.P. Carrico and P.F. McGinnis Monopole/quadrupole mass spectrometer
3,621,242 (1971)	L.D. Ferguson and J.P. Carrico Dynamic time-of-flight mass spectrometer
3,617,736 (1971)	E.F. Barnett, W.S.W. Tandler and D.L. Handler Quadrupole mass filter with electrode structure for fringing field compensation
3,614,420 (1971)	R.H. Dillenbeck Monopole mass spectrometer (mounting means)
3,560,734 (1971)	E.F. Barnett, W.S.W. Tandler and W.R. Turner Quadrupole mass filter with fringing field penetrating structure
3,555,273 (1971)	J.T. Arnold Mass filter apparatus having an electric field the equipotentials of which are three-dimensionally hyperbolic
3,555,271 (1971)	W.M. Brubaker and C.F. Robinson Radiofrequency mass analyser of the non-uniform electric field type
3,553,451 (1971)	P.M. Uthe Quadrupole in which the pole electrodes comprise metallic rods whose mounting surfaces coincide with those of the mounting means.

3,527,939 (1970)	P.H. Dawson and N.R. Whetten Three-dimensional quadrupole mass spectrometer and gauge
3,501,631 (1970)	J.T. Arnold Charged particle trapping means employing a voltage divider and a plurality of simple conductors
3,501,630 (1970)	W.M. Brubaker Mass filter with removable auxiliary electrode
3,473,020 (1969)	W.M. Brubaker Mass analyser having series aligned curvilinear and rectilinear analyser sections
3,473,019 (1969)	W.M. Brubaker Mass analyser with extension means to decrease the distance between electrode surfaces
3,473,018 (1969)	W.M. Brubaker Mass analyser using two spaced tubular and coaxial electrodes
3,457,404 (1969)	P.M. Uthe Quadrupole mass analyser (ion injection aperture)
3,418,464 (1968)	W.M. Brubaker and F.B. Wiens Duopole mass filter
3,413,463 (1968)	W.M. Brubaker Resolution control in multipole mass filter
3,410,998 (1968)	R.L. Watters Electrical control circuit for a scanning monopole mass analyser
3,410,997 (1968)	W.M. Brubaker Multipole mass filter (with ion deflection means)
3,371,205 (1968)	C.E. Berry Multipole mass filter with a pulsed ionizing electron beam
3,371,204 (1968)	W.M. Brubaker Mass filter with one or more rod electrodes separated into a plurality of insulated segments

3,350,559 (1967)	J.R. Young, J.B. Hudson and R.L. Watters Monopole mass spectrometer having one ceramic electrode coated with metal
3,334,225 (1967)	R.V. Langmuir Quadrupole mass filter with means to generate a noise spectrum exclusive of the resonant frequency of the desired ions
3,328,146 (1967)	W. Haenlein Method of producing an analyser electrode for a quadrupole mass spectrometer
3,321,623 (1967)	W.M. Brubaker and C.F. Robinson Multipole mass filter having means for applying a voltage gradient between opposite electrodes
3,284,629 (1966)	U. von Zahn Mass filter having an ion source structure with pre-selected relative potentials applied thereto (retardation at ion entry)
3,284,628 (1966)	K-G. Gunther (as 3,284,629)
3,280,326 (1966)	K-G. Gunther Mass filter with sheet electrodes on each side of the analyser rod
3,280,325 (1966)	C. Brunnée, L. Delgmann and K. Kronenberger Mass filter with particular circuit means connected to the electrodes for establishing the ion deflecting field
3,235,724 (1966)	W.M. Brubaker Quadrupole mass filters with introductory ion accelerating field proportional to the quadrupole electric field
3,221,164 (1965)	K-G. Gunther Ion source for mass filter wherein the electron and ion beam axes are the same
3,197,633 (1965)	U. von Zahn Method and apparatus for separating ions of different specific charges (monopole)

3,129,327 (1964)	W.M. Brubaker Auxiliary electrodes for quadrupole mass filters
3,105,899 (1963)	K-G. Gunther, H. Freller and G. Titze Electric mass filter (mounting)
3,075,076 (1963)	K-G. Gunther Gas analysing method and apparatus
2,950,389 (1960)	W. Paul, H-P. Reinhard and H. Frohlich Method of separating ions of different charges
2,939,952 (1960)	W. Paul and H. Steinwedel Quadrupole mass spectrometer

Other patents

USSR 285,322 (1970)	L.E. Zaslovskii et al. Quadrupole mass spectrometer
USSR 379,279 (1973)	B.S. Kristov et al. Quadrupole mass spectrometer with end zone shielding
Canada 946,527 (1972)	J.B. Hiller, J.A. Richards and R.M. Huey Means for effecting improvements to the mass filter (rectangular excitation)
Canada 973,282 (1975)	P.H. Dawson A high resolution focusing dipole and mass spectrometer
Great Britain 1367—638 (1974)	G.W. Ball Ion filter used in a mass spectrometer
Great Britain 1379—514, 1379—515	G.W. Ball Mass spectrometers for gas analysis using a quadrupole ion filter

INDEX

a, definition of, 13, 14, 15
Acceptance ellipses, at the ion source, 106
—, mass filter, 25—30, 73, 86—92, 95—106, 335, 336
—, monopole, 44, 109
Afterglows, 249, 251
Air pollution, 320—323
Alternating gradient fields, 3
Amplitude, maximum, 24—29, 41, 42, 65, 73, 87
—, with delayed d.c. ramp, 97
—, with fringing fields, 97
Anaesthetic studies, 330
Angular distribution of reaction products, 253
Angular momentum, in ion trap, 48
Aperture, effective, 21, 24, 26
Atmospheric pressure, in particle trap, 201
Atmospheric pressure ionization, 305, 315—317, 321, 325
Atomic frequency standard, 264
Atomic oxygen, 274
Automation, 296—297, 329
Auxiliary oscillatory circuit, 49, 66, 74—76

Background, ion trap, 52
—, mass filter, 34, 290—292
—, monopole, 157
Bacteria, 318
Beam guides, 5, 62, 257, 259
Beam transport, 5, 62, 257, 259
Biological samples, 294, 315—317
Blood gas analysis, 295, 312—313, 329, 330
Body fluid analysis, 313
Bolometric technique, 263
Boxcar detector, 53, 189, 191
Breath analysis, 284, 308—312, 329
Bunching of ions, 32

Capacitance of mass filter, 150
Capillary arc source, 324
Capillary column chromatography, 303—304
Catalytic studies, 248—249

Characteristic exponent, 16, 68
Chemical ionization, in GC—MS, 298—304, 314
—, in ion trap, 57, 203, 266
Clinical applications, 303, 307—318, 329
Cluster formation, 251, 276, 325
Collisions of ions, 77, 206, 207, 219
Combustion engines, 326—327
Computation, *see also* Matrix methods
—, of maximum amplitudes, 24, 27, 41, 65
—, of non-linear resonances, 113—117
—, of performance, 80
—, —, ion trap, 54—56, 91, 92, 114
—, —, mass filter, 31—36, 90—92, 97, 113—117
—, —, monopole, 42—44
—, of stability boundaries, 65
—, of trajectories, *see* Numerical integration
Computer control, 296—297, 329
Contamination of electrodes, 46, 96, 125, 128, 135, 170
Crossed beams, 253
Cryogenic ion source, 275
Cylindrical housing, 81, 118
Cylindrical co-ordinates, 80

Data acquisition, 296—297, 316, 329
Delayed d.c. ramp, 5, 33, 96, 97, 105—107, 144, 274, 342
Dipole, focussing, 5, 60, 344
Distortions, *see* Field distortions
Drug analysis, 294, 297, 301, 313—318
Duopole, 45, 342
Dynamic analysers, 2
Dynamic focussing, 2
Dynamic range, *see* Trace analysis

Ejection of ions, 52—53, 188, 198, 202
Electrodes, *see also* Rods
—, contamination, 46, 96, 125, 128, 135, 170
—, end-caps, 12
—, misalignment, *see* Field distortions
—, mis-shaping, 95
—, ring, 12

Electrometers, 136
Electron attachment, 251
Electron beam in ion trap, 52—56, 198
Electron impact studies, 256
Electron-induced desorption, 131, 244
Electron—ion recombination, 195, 264
Electron multiplier, disadvantages, 139
—, in mass filter, 136—140, 290
—, offset, 35, 138, 290—293, 325
Ellipses, *see* Acceptance ellipses
Emittance, definition, 28, 87
—, ion source, 30, 99, 106
—, mass filter, 34
Energy analyser, 5, 62, 239
Enrichment, in GC—MS, 288
Environmental analysis, 294, 303, 319—328, 330
Equation of motion, 12—15, 47; *see also* Mathieu
Evaporation of droplets, 201, 261
Exit, from mass filter, 34
Explorer satellite, 281, 284
Explosive detection, 322

Faraday cup, 136
Field distortions, 79, 95, 109—119, 122, 124, 125—130
—, fourth order, 111, 112, 114, 116
—, monopole, 118—119
—, resolution and, 114—116, 125
—, sixth order, 111, 112
—, third order, 111, 112, 114, 115, 116
Field ionization, 131, 305
Finite difference equation, 81
Flames, 252—253
Flash desorption, 244, 248
Focussing, dipole, 5
—, energy dependent, 62
—, exact double, 6, 34, 57—59, 71, 229, 237, 260, 337
—, partial, 37—39, 43
Frequency, characteristic, 18, 70
—, ion trap, 57
—, mass filter, 13, 21, 121—122
—, monopole, 41, 178
—, stability, 146
Fringing fields, 79, 95—109, 141, 341
—, entrance, 6, 33, 35, 100, 141, 143
—, exit, 34, 95, 103
—, matrix for, 99
—, monopole, 44, 95, 107—109
—, three-dimensional, 83, 97, 101—102
—, trapping ions in, 102—103

Gas chromatography, 4, 287—306, 314, 328
—, interfacing, 287, 288—290

Harmonic in waveform, 66, 68, 85, 95, 113, 164
Hexapole distortion, 111
High pressures, ion trap, 199, 220
—, mass filter, 77, 140
—, monopole, 176
Hill equation, 14, 66—69, 85
Hyperbolas in phase-space, 87
Hyperbolic electrodes, 10, 23, 129

Induced voltages, 51
Initial conditions, 24—29, 86—90
Initial phase, 15, 18, 24, 37, 41, 42, 54, 86, 97
Initial velocities, 28, 56, 88
Inlet systems, 308, 309, 313, 314
Integration, *see* Numerical integration
Ion collection, difficulties, 34
Ion counting, 138, 177, 242, 281
Ion energy, after analysis, 34, 36, 126, 132
—, ion trap, 204, 219—222
—, mass filter, 24, 124, 135
—, monopole, 38, 44, 158, 159, 171, 172
Ion entrance, *see* Fringing fields
—, mass filter, 102
—, monopole, 107, 158, 172
Ion exit, mass filter, 34, 103
—, monopole, 158, 172
Ion imaging, 34, 57, 62, 237—238
Ionization gauge, 243
Ionization period, 52
Ion lifetime, 196, 207
Ion—molecule reactions, 52, 190, 196, 203, 204, 206, 210, 220, 249, 253, 258—260, 265—267, 298, 316, 321
Ionosphere, 3, 273, 276, 280
Ion source, alignment, 131—135, 144
—, capillary arc, 324
—, cold cathode, 134
—, coupling to mass filter, 99, 103—107, 131
—, cryo, 275
—, electron impact, 131—134
—, field desorption, 131, 305
Ion storage, *see* Ion trap
Ion trajectories, *see also* Numerical integration
—, for exact focussing, 58—60

—, in quadrupole field, 15, 17, 18, 86
—, ion trap, 48, 213
—, mass filter, 21—22, 97
—, monopole, 39—40, 108—109, 153
Ion trap, A,B,Γ values, 87—88
—, applied frequency, 57
—, as ion source, 5, 57, 190, 203—204, 209, 258, 265—267
—, continuous mode, 198
—, cylindrical, 56
—, electrodes, 11, 81
—, equations of motion, 14, 47
—, external ion detection, 5, 52—56
—, ion density, 196, 215, 217
—, ion energies, *see* Ion energy
—, loss processes, 198, 204—210
—, mass selective detection, 49—52, 185—187
—, mass selective storage, 52—56, 181, 188—190, 342
—, mesh, 57, 184
—, mixed mode, 198
—, novel features, 56
—, pulsed mode, 198
—, race-track, 181
—, resonant detection, 5, 49—52, 185—187
—, six-electrode, 181, 190, 203
—, space charge, *see* Space charge
—, static, 193
—, storage ring, 193
—, synchronization of pulse-out, 53, 189, 198, 202
—, three-electrode, 181—190
—, —, construction, 182—184
—, —, electron beam, 52, 183, 198
—, —, rf power supply, 184
—, velocity distributions in, 204, 219—222
iso-β lines, 18, 26, 66, 70—71
iso-μ lines, 70—71
Isotope separation, 74
Isotopes in GC—MS, 297, 302

Jet separator, 289

Laplace's equation, 9, 81
Limits to sensitivity, 328
Linearity of scan, 21, 24
Liquid chromatography, 288, 305

Magnetic fields, axial, ion trap, 194
—, —, mass filter, 62, 76
—, monopole with, 43, 174
—, offset detector with, 140, 246
Mass discrimination, electron multiplier, 139, 290
—, mass filter, 100, 105, 143—144
—, monopole, 173
Mass filter, A,B,Γ values, 29, 87—88, 99, 335, 336
—, background, 34, 138, 290—293
—, electrodes, 10; *see also* Rods
—, equations of motion, 13; *see also* Mathieu
—, field, 10
—, high pressure limit, 140
—, ion detection, 136—140, 290—294
—, limited length, 34, 89, 91, 98
—, mass range, 121—124, 287
—, peak shapes, *see* Peak
—, resonances, *see* Resonance
—, sensitivity and mass, 100, 105, 143—144
—, sensitivity and resolution, 141—142
—, space charge, *see* Space charge
—, power supply, *see* Power supplies
Mass fragmentography, 294, 297—298, 301—303, 310
Mass range, mass filter, 121—124, 287
—, monopole, 155, 165
—, —, extended, 41, 155, 165, 167—168
Mathieu equation, 14, 15, 37, 46, 47, 65, 69—75, 339
—, complete solution, 72—74
—, inhomogeneous, 49, 75
Mathieu function of integral order, 16
Matrix methods, 26, 31, 36, 54, 65, 83—92, 99—105
Medical applications, 303, 307—319, 329
Meissner equation, 35, 69, 83
Metastable ions, 203, 257—258, 265
Microparticle storage, 5, 199—203, 261—262, 340
Modulation, molecular beam, 244
—, transmitted signal, 27, 31, 32, 106, 108, 109
Molecular beams, 244, 253—254
Molecular separators, 288, 289
Monopole, 5, 6, 37—46, 153—180, 343
—, advantages, 45, 179
—, apertures, 43, 172
—, disadvantages, 45, 179
—, electrodes, 11, 159, 178
—, equations of motion, 14; *see also* Mathieu
—, exit aperture, 43, 158, 172
—, extended mass range, 41, 155, 165, 167—168
—, field distortions, 118—119
—, focussing, 5, 57—60, 173
—, four-fold, 5, 45
—, fringing fields, 107—109

—, mass discrimination, 43, 173
—, mass dispersion, 39, 164—165
—, mass scale, 41, 164—166
—, mechanical construction, 159—161
—, peak shapes, 42, 108, 169—171
—, performance, 37—44
—, quadruple, 5, 45
—, relative sensitivities, 43, 46, 173
—, resolution, 43, 165—169, 171
—, scanning, 38, 40, 44, 161, 165—169
—, upper atmosphere and, 275
Multiple peak monitoring, 21, 294, 310

Negative ions, 139, 176, 251, 252, 255, 264, 265, 317, 321—322
Noise, *see also* Background
—, amplifier, 136
—, electron multiplier, 137, 292
Numerical integration, 65, 80—83
Numerical matrix methods, 83—92

Octopole, 253, 258—260
Octopole distortion, 111
Operating line, mass filter, 20, 21, 23
—, monopole, 38
Oscillations in ion density, 217
Oscillator, rf, 147, 164
Oscillatory time-of-flight analysers, 61—62, 225—236
Outgassing, 134, 327

Partial pressure limit, 137, 243
Particle trapping, 5, 199—203, 261—262, 340
Patents, 340—344
Path-stability spectrometers, 2
Peaks, asymmetry, 33, 34, 55, 56, 98, 108, 170
—, bell-shaped, 32
—, flat-topped, 27
Peak shapes, ion trap, 55—56, 115, 116
—, mass filter, 27, 32—34, 74, 126, 127, 133, 287
—, monopole, 42—43, 108
—, split, 110, 114—117, 126, 127, 133, 197
—, with fringing field, 42, 98
Peak tailing, 33, 35, 42, 98, 169
Peak width, 122, 158, 169—170
Periodicity, 16—19
Phase, initial, 15, 18, 24, 37, 41, 42, 54, 86, 97
Phase-space dynamics, 65, 79, 83—92
—, fringing fields and, 98—105
—, ion energy calculation, 222
—, ion trap, 54—56, 83
—, mass filter, 25—31
—, monopole, 44, 109
—, SIMS design and, 90, 99, 247
Photodetachment, 255—256
Photofragment spectroscopy, 255
Photoionization, 255—256, 277
Plasma chromatography, 320—323
Plasma confinement, 264
Plasma oscillations, 197, 217
Plasma, sampling of, 249—252
Polychlorinated biphenyls, 303, 320, 322
Potential distribution, 10, 46
Power, mass filter, 129, 150
—, monopole, 168, 178
Power supplies, ion trap, 184
—, mass filter, 129, 145—151
—, monopole, 161—164, 342
—, rectangular excitation, 36, 144
Pre-cursor peaks, 6, 69, 118, 126, 127
Programmable peak selection, 21, 294, 310
Pseudo-potential theory, 77, 210—214
Pyrolysis mass spectrometry, 318—319

q, definition of, 13—15
Quistor, *see* Ion trap, three-electrode

Race-track quadrupole, 181
Rectangular excitation, 5, 35—36, 83, 144, 344
Recursion relations for C_{2n}, 66, 70, 338
Residual gas analysis, 52, 137, 187, 196, 197, 241
Resolution, electronic adjustment, 21
—, exact focussing monopole, 57—59
—, high, devices for, 6, 36, 57—60, 243
—, limitations to, 22, 122—129, 141—142
—, mass filter, 21—23, 122—129, 141—142, 287
—, monopole, 38, 42, 158, 167, 169
—, time-of-flight, 233—235
Resonance, induced linear, 74—76, 185—188
—, non-linear, 110—118, 197; *see also* Field distortions
Resonant detection, 5; *see also* Ion trap
Respiratory analysis, *see* Breath
Response time, 136—138
Retardation analyser, 280
Retardation of ion entry, 100, 159, 343
rf power, *see* Power supplies
rf spectroscopy, 262—263
Rods, *see also* Electrodes

—, accuracy, *see* Field distortions
—, alignment, 111, 122, 124—127, 129, 161
—, circular concave, 130
—, diameter, optimum, 117, 122, 129
—, round, 11, 23, 112, 117—118, 125, 129—130
Runge—Kutta integration, 80, 85, 339

Satellite peaks, 6, 69, 118, 126
Scanning, frequency, 21, 41
—, ion trap, 51, 52, 187, 189
—, non-linear, 24, 165, 340
—, monopole, 38, 40, 161, 165—169
—, voltage, 21, 41, 151
Secondary ion mass spectrometry, 90, 99, 131, 244, 246
Sequential ionization, 267
Sinusoidal excitation, 13, 69
Solenoid spectrometer, 5, 63
Space-charge, ion trap, 76, 196, 198, 214—216
—, mass filter, 76, 77, 216
—, pseudo-potential calculation, 77, 214—216
—, stability diagram with, 77, 217
Spacecraft contamination, 326
Space research, 3, 273—284, 311, 327
Stability, boundary, 16, 68—72
—, second zone, 17, 18, 88
Stability diagram, composite, 19—20, 46—47
—, Mathieu, 13, 16, 18, 20, 23, 46—47
—, Meissner, 35—36
—, monopole, 38
—, six electrode trap, 193
Static fields, 5, 62—63
—, cylindrically symmetric, 62
—, ion trap, 62, 193, 264
—, rotationally symmetric, 5, 62
—, twisted quadrupole, 5, 62
Storage ring trap, 193
Surface studies, 243—249, 280
Synchronization of electron beam, 53, 189, 198
Synchronization of pulse-out, 53, 189, 198, 202

Temperature of stored ions, 263
Thermal analysis, 323
Thermogravimetric techniques, 248
Three-dimensional quadrupole, *see* Ion trap
Time-of-flight spectrometers, 61—62, 225—240, 341
—, arrival time, 61, 226, 230—231
—, bunching factor, 62, 232
—, experimental results, 233
—, ion displacement, 227
—, mass dispersion, 231
—, resolution, 231—233
TNT, 322
Trace analysis, 137, 246, 294, 323, 325, 328, 331
Transmission, in mass filter, 22—28, 98—105
—, in monopole, 37, 42, 107—109
Trapping efficiency, 54—56, 206
Trapping in fringing fields, 102—103

Upper atmosphere, 3, 273—286
—, analysis, ionic, 273, 280—281
—, —, neutral, 277—280, 281

Vacuum techniques, 3, 241—243
Vaporisation, 248, 325
Velocity analysis, 5, 62, 236, 237, 247, 253, 260
Viscous drag, 76, 202
Voltage stability, 145—146

Water analysis, 323—326
Wronkian determinant, 49, 73

Zero blast, 35

Printed in the United States
By Bookmasters